Springer

Milano
Berlin
Heidelberg
New York
Barcelona
Hong Kong
London
Paris
Singapore
Tokyo

P. Podio-Guidugli
M. Brocato
(Eds)

Rational Continua, Classical and New

A collection of papers dedicated to Gianfranco Capriz on the occasion of his 75th birthday

Springer

P. Podio-Guidugli
Università di Roma Tor Vergata
Roma, Italy

M. Brocato
IEI – CNR
Area della Ricerca di Pisa
Pisa, Italy

Springer-Verlag Berlin Heidelberg New York
a member of BertelmannSpringer Science+Business Media GmbH
ISBN-13: 978-88-470-2233-1 e-ISBN-13: 978-88-470-2231-7
DOI: 10.1007/978-88-470-2231-7

Softcover reprint of the hardcover 1st edition
http://www.springer.de

Library of Congress Cataloging-in-Publication Data
Rational continua, classical and new / P. Podio Guidugli, M. Brocato (eds.).
p. cm.
Includes bibliographical references and index.

1. Thermodynamics. 2. Continuum mechanics. I. Podio-Guidugli, Paolo. II. Brocato, M., 1962-

QC311.2 .R38 2002
536'.7--dc21

2002023679

Cover design: Simona Colombo, Milan
Typesetting: Bürosoft/Text- und DTP-Service, Berlin

SPIN: 10837718

Foreword

Gianfranco Capriz was born in Gemona del Friuli on October 16, 1925. After graduating *summa cum laude* in mathematics at the Scuola Normale Superiore in Pisa (1948) and successfully attending a one-year doctoral course there (1949), he was appointed by Mauro Picone as a researcher at the Istituto Nazionale per le Applicazioni del Calcolo in Rome (1951–56). At the Institute, while working at his first research papers, he also served as a programmer in the staff operating the first general purpose computer ever installed in Italy.

In Rome he met Barbara, who was shortly to become his wife, and became acquainted with Ennio De Giorgi, Gaetano Fichera, Tristano Manacorda, Carlo Pucci, Michele Sce, and Edoardo Vesentini, with all of whom he was to maintain friendly and scientific relationships thereafter. In the same period he started his research activity in rational mechanics under the supervision of Antonio Signorini.

From Rome he moved to Stafford (UK) to work for the English Electric Company (1956–62) as a research mathematician and a programmer of DEUCE, the engineered version of the pilot machine ACE, originally designed by Alan Turing. This period of his life ended when Capriz was asked by Sandro Faedo to return to his country to contribute to the creation in Pisa of the largest concentration ever in Italy of research and development activities in computer science and information technology. As early as 1954, at the suggestion of Enrico Fermi, the construction of the first Italian scientific computer had been decided, and the task assigned to the Centro Studi Calcolatrice Elettronica (CSCE), based in Pisa. In 1961, the product of this effort, the Calcolatrice Elettronica Pisana (CEP), was inaugurated; one year later, CSCE became part of the Italian National Research Council. For two decades, from 1963 to 1983, Capriz was to serve as the Director of CSCE (later to be transformed into the Istituto di Elaborazione della Informazione) and then of CNUCE (Centro Nazionale Universitario di Calcolo Elettronico).

In those busy years, Capriz, who had been given the chair of rational mechanics at the University of Pisa in 1966, had also a central role in the creation of a school in continuum physics, which was one of the outcomes of another inspired initiative of Faedo, namely, the revival at the highest levels of mathematical activities in Pisa, with the appointments of A. Andreotti, J. Barsotti, E. Bombieri, S. Campanato, G. Prodi, G. Stampacchia, and Vesentini at the University, and of De Giorgi at the Scuola Normale.

Capriz never ceased to do research, not even while he was the President of TECSIEL (1983–92), a company of the IRI group, where computer networks were studied and, in particular, the OSI standards first effected and installed (OSIRIDE network, 1984). In addition, he repeatedly served as visiting professor abroad (at the Johns Hopkins University, the University of Minnesota and the Carnegie Mellon University in the US; at the University of Manitoba, in Canada; and as Erskine Professor at the University of Canterbury, in New Zealand). He was Vice-President of UMI, the Unione Matematica Italiana (1976–82), President of ISIMM, the International

Society for the Interactions of Mechanics and Mathematics (1997–99), and President of AIMETA, the Associazione Italiana di Meccanica Teorica ed Applicata (1999–2001). He is presently a corresponding member of the Accademia dei Lincei and a *professor emeritus* at the University of Pisa.

When he first met Clifford A. Truesdell in the middle sixties, Capriz had already worked on such diverse subjects as computational mechanics, lubrication, creep, vibrations and stability of rotating shafts, stability and numerical computations in hydrodynamics, viscoelasticity, and the manufacture of ceramics. After meeting Truesdell, his scientific interests were more and more directed toward the analysis of fundamental and innovative problems in continuum mechanics, especially, materials with memory, problems with live loads, non-linear vibrations of strings, mixtures, and a host of problems involving the continuum descriptions of microstructures: continua with voids; liquids with bubbles; granular materials; continua with vectorial, affine, or spherical structure; bodies with continuous distribution of dislocations; Cosserat continua; and liquid crystals. The book on *Continua with Microstructure* edited by C. Truesdell for Springer in the series Tracts in Natural Philosophy, summarizes about fifteen years of his scientific achievements in the field of the title, and contains innumerable suggestions for further research.

Both Capriz' broad scientific production and the variety of themes he dealt with during his career bear witness to the agility and sharpness of his mind, ready to capture weaknesses and pitfalls, as well as his ability to spot promising possibilities, sometimes deeply hidden in continuous models, no matter whether classical or just proposed, and to convert them into new challenging research tasks: whence the title of this tribute volume.

A mathematician and an engineer, a philosopher and a manager, a leader and a friend: all this Gianfranco Capriz is to those who have the good fortune, honor and pleasure to work with him.

Preface

A selected number of prominent researchers, all in close personal and scientific contact with Gianfranco Capriz, have been invited to contribute to this volume, on a subject of their choice. They are, in alphabetical order: *P. Biscari, G. Cimatti, S.C. Cowin, C. Davini, R.L. Fosdick, P. Giovine, J.T. Jenkins, R.J. Knops, I. Müller, D.R. Owen, M. Šilhavý, G. Vergara Caffarelli, E.G. Virga, K. Wilmanski,* and *H. Zorski.* It is because of the outstanding quality of their effort, and that of their coauthors, that this book not only meets to the full its purpose as homage but also offers–so we believe–a rather unique and variegated collection of papers in modern continuum mechanics.

Many contributions are in research areas in which Gianfranco Capriz has been active, but not all. Among the latter papers, those by Biscari and Zorski, the first and last in the list, exemplify well how useful concepts from continuum mechanics can be in modeling and analyzing bioaggregates. The papers by Cimatti and Davini are also rather remote by theme from Gianfranco's own research; yet, they will certainly appeal to his taste for mathematical analysis, when applied to concrete problems of continuum physics and structural mechanics. Gianfranco has displayed such taste all along in his scientific life, first of all in dealing with questions from the theory of elasticity, linear or non-linear. It is not by mere coincidence that papers in elasticity comprise a relatively large subgroup in this book, a group including the works by Cowin, Fosdick (coauthored with Dunn and Zhang), Knops, Šilhavý, and Vergara Caffarelli (with Carillo and Podio-Guidugli). In particular, the paper by Cowin is devoted to find which response symmetries of a linearly elastic material are compatible with the presence of a geometrically organized distributed microstructure; for this reason, this paper may serve as a bridge from elasticity to one of Gianfranco's favorite subjects, continua with microstructure. Three papers in the book deal with this subject, those by Giovine, Jenkins and LaRagione, and Virga; a fourth paper, by Owen, discusses the nonstandard type of microstructure due to nonsmooth submacroscopic *disarrangements*. Finally, the paper by Wilmanski, which is about porous media, another class of microstructured continua, focuses on an issue especially dear to Gianfranco's heart as a rational mechanist, namely, the role of inertial interactions in the governing equations of a thermomechanical theory; although the occasion is a study of heat conduction within the framework of extended thermodynamics, the theme of Müller's contribution with Barbera is the same.

Rome–Paris, September 2002

Paolo Podio-Guidugli
Maurizio Brocato

Preface

List of Contributors

- **Elvira Barbera**, FB 6, Thermodynamik, Technische Universität Berlin, 10623 Berlin, Germany
- **Paolo Biscari**, Dipartimento di Matematica, Politecnico di Milano, Piazza Leonardo da Vinci 32, 20133 Milano, Italy and Istituto Nazionale di Fisica della Materia, Via Ferrata 1, 27100 Pavia, Italy
- **Sandra Carillo**, Dipartimento di Metodi e Modelli Matematici per le Scienze Applicate, Università di Roma "La Sapienza", Via Scarpa 16, 00161 Roma, Italy
- **Giovanni Cimatti**, Department of Mathematics, Via Buonarroti, 2, 56100 Pisa, Italy
- **Stephen C. Cowin**, The Center for Biomedical Engineering and The Department of Mechanical Engineering, The School of Engineering of The City College and The Graduate School of The City University of New York, New York, NY 10031, USA
- **Cesare Davini**, Dipartimento di Ingegneria Civile, Università degli Studi di Udine, via delle Scienze 208, 33100 Udine, Italy
- **J. Ernest Dunn**, Scientific Consulting, 167 W. Dawn Drive, Tempe, AZ 85284, USA
- **Roger Fosdick**, Department of Aerospace Engineering and Mechanics, University of Minnesota, Minneapolis, MN 55455, USA
- **Pasquale Giovine**, Dipartimento di Meccanica e Materiali, Università di Reggio Calabria, Via Graziella, Località Feo di Vito, 89060 Reggio Calabria, Italy
- **James T. Jenkins**, Department of Theoretical and Applied Mechanics, Cornell University, Ithaca, NY 14853, USA
- **Robin J. Knops**, Department of Mathematics, Heriot-Watt University, Edinburgh EH14 4AS, Scotland
- **Luigi La Ragione**, Dipartimento di Ingegneria Civile e Ambientale, Politecnico di Bari, 70125 Bari, Italy
- **Ingo Müller**, FB 6, Thermodynamik, Technische Universität Berlin, 10623 Berlin, Germany
- **David R. Owen**, Department of Mathematical Sciences, Carnegie Mellon University, Pittsburgh, PA 15213, USA
- **Paolo Podio-Guidugli**, Dipartimento di Ingegneria Civile, Università di Roma "Tor Vergata", Via di Tor Vergata 110, 00133 Roma, Italy
- **Miroslav Šilhavý**, Mathematical Institute of the AV ČR, itná 25, 115 67 Prague 1, Czech Republic
- **André M. Sonnet**, Dipartimento di Matematica, Istituto Nazionale di Fisica della Materia, Università di Pavia, Via Ferrata 1, 27100 Pavia, Italy
- **Giorgio Vergara Caffarelli**, Dipartimento di Metodi e Modelli Matematici per le Scienze Applicate, Università di Roma "La Sapienza", Via Scarpa 16, 00161 Roma, Italy

- **Epifanio G. Virga**, Dipartimento di Matematica, Istituto Nazionale di Fisica della Materia, Università di Pavia, Via Ferrata 1, 27100 Pavia, Italy
- **Krzysztof Wilmanski**, Weierstrass Institute for Applied Analysis and Stochastics, Berlin, Germany
- **Ying Zhang**, Department of Materials Science, Xiamen University, Xiamen, Fujian 361005, China
- **Henryk Zorski**, Institute of Fundamental Technological Research, Polish Academy of Sciences, 00-049 Warsaw, Świętokrzyska 21, Poland

Contents

Heat Conduction in a Non-Inertial Frame

Elvira Barbera, Ingo Müller

Abstract. Stationary heat conduction is considered in a monatomic gas between two co-axial cylinders which are at rest in a non-inertial frame. Instead of the usual Navier–Stokes–Fourier theory we employ the more reliable 13-moment theory of extended thermodynamics. Three surprising observations are made:

- There is a tangential heat flux between the cylinders.
- No rigid rotation of the heat conducting gas is possible.
- There is a normal pressure field in the axial direction.

1 Introduction

In the hierarchy of moment equations in the kinetic theory of monatomic gas – and in extended thermodynamics – all the equations except the first contain inertial terms [1, 2]. This observation is borne out by the set (1.1) which represents the equations for the first 13 moments in a non-inertial frame[1],

$$\begin{aligned}
&\frac{\partial F}{\partial t} + F^k_{;k} = 0,\\
&\frac{\partial F^i}{\partial t} + F^{ik}_{;k} - F\underset{0}{i}^i - 2F^k W^i_k = 0,\\
&\frac{\partial F^{ij}}{\partial t} + F^{ijk}_{;k} - 2F^{(i}\underset{0}{i}^{j)} - 4F^{k(i}W^{j)}_k = \frac{1}{\tau}(F_E^{<ij>} - F^{<ij>}),\\
&\frac{\partial F^{ji}_j}{\partial t} + F^{jik}_{j;k} - 3F^{(i}_j\underset{0}{i}^{j)} - 6F^{k(i}_j W^{j)}_k = \frac{1}{\tau}(F^{ji}_{Ej} - F^{ji}_j).
\end{aligned} \tag{1.1}$$

Here $\underset{0}{i}^i$ denotes the velocity-independent inertial accelerations and W^i_k is the matrix of the angular velocity of the frame. The right-hand sides of (1.1) represent the productions due to collisions, simplified according to the BGK approach (see, e.g., [3]). Further, τ is a relaxation time of the order of magnitude of a mean time of free flight of the atoms of the gas.

We expect the inertial contributions to provide some unusual phenomena. Let us look, for instance, at the gas-filled gap between two cylinders (see Fig. 1.1). If the inner cylinder is heated we expect an inward radial temperature gradient and an outward radial heat flux; and that is indeed all there is to it in an inertial frame. However, if the cylinders are placed on a turntable – a rotating frame – with their

[1] Upper and lower indices represent contra- and covariant components respectively while the semicolon indicates covariant differentiation. Symmetrization is understood with respect to bracketed indices. Angular brackets denote trace-less tensors.

axes parallel to the angular velocity of the frame, we expect a tangential component of the heat flux as well as a radial component in a 13-moment theory – in contrast to customary Navier–Stokes–Fourier theory.

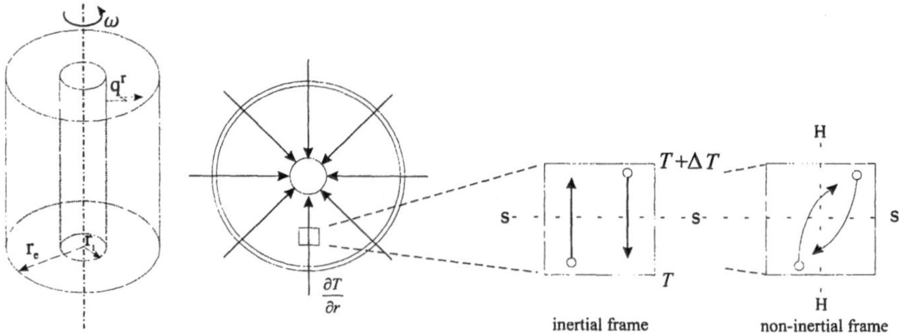

Fig. 1.1. On the origin of the radial and tangential components of the heat flux

This expectation is motivated by the following argument from the elementary kinetic theory of gases. In Fig. 1.1 we focus attention on a small volume element of the order of magnitude of a mean free path. Atoms from below will carry less energy upward than atoms from above will carry downward. Thus, if a pair of atoms passes through the middle layer, a downward transport of energy is effected, i.e., a heat flux in the radial direction. In a non-inertial frame the atoms do not fly in straight paths, rather their orbits are bent by the Coriolis force so that, along with the radial heat flux, there is a tangential one upon the passage of each pair of particles. Fig. 1.1 illustrates this effect.

This is interesting enough. However, a careful evaluation of the system (1.1) will provide further unusual phenomena in the rotating gas cylinder:

- A heat conducting gas in the cylindrical gap cannot perform a rigid rotation.
- There are normal stress differences which tend to destroy the planar character of the velocity field.

We proceed to prove these phenomena and describe them quantitatively.

2 Field equations

2.1 General field equations

The moments $F^{i_1 i_2,\dots,i_n}$ in (1.1) may be decomposed into internal parts, i.e., velocity independent parts, and other parts that depend on the velocity explicitly. This decomposition is dictated by the Galilean invariance of the system (1.1). It reads:

$$
\begin{aligned}
F &= \rho \\
F_i &= \qquad\quad \rho v_i \\
F_{ij} &= \rho_{ij} \qquad\qquad\quad + \rho v_i v_j \\
F_{ijk} &= \rho_{ijk} + 3\rho_{(ij}v_{k)} \qquad\qquad + \rho v_i v_j v_k \\
F_{ikll} &= \rho_{ikll} + 4\rho_{(ikl}v_{l)} + 6\rho_{(ik}v_l v_{l)} \qquad\quad + \rho v_i v_k v^2,
\end{aligned}
\tag{2.1}
$$

where the ρ's are the internal moments. Some of them are easily interpreted. Thus ρ is the mass-density, ρ_{ij} is the pressure tensor with $p = \frac{1}{3}\rho_i^i$ being the pressure, and $\frac{1}{2}\rho_{ij}^j$ is the heat flux.

Elimination of the F's between (1.1) and (2.1) and subsequent elimination of the velocity provides the equations:

$$
\begin{aligned}
&\frac{d\rho}{dt} + \rho v^k_{;k} = 0 \\
&\rho^{ik}_{;k} + \left\{\rho\left(\frac{dv^i}{dt} - \underset{0}{i}^i - 2W^i_k v^k\right)\right\} = 0 \\
&\frac{d\rho^{ij}}{dt} - 2W^{(i}_k \rho^{j)k} + \rho^{ij} v^k_{;k} \rho^{ijk}_{;k} + \left\{2\rho^{k(i}(v^{j)}_{;k} - W^{j)}_k)\right\} = -\frac{1}{\tau}\rho^{\langle ij\rangle} \\
&\frac{d\rho^{ij}_j}{dt} - W^i_k \rho^{kj}_j + \rho^{ij}_j v^k_{;k} + \rho^{ijk}_{j;k} \\
&+ \left\{3\rho^{(i}_{\;\;j}\left(\frac{dv^{j)}}{dt} - \underset{0}{i}^{j)} - 2W^{j)}_k v^k\right) + \rho^{k(i}_{\;\;j}(v^{j)}_{;k} - W^{j)}_k)\right\} = -\frac{1}{\tau}\rho^{ij}_j .
\end{aligned}
\tag{2.2}
$$

This system is not closed because of the occurrence of the fluxes $\rho^{\langle ij\rangle k}$ and ρ^{ijk}_j. We close it by calculating these fluxes from the 13-moment Grad distribution function [4]:

$$
f_G = f_E\left(1 + \frac{1}{2\frac{k}{m}p\theta}\rho_{\langle ij\rangle}C^i C^j - \frac{1}{2\frac{k}{m}p\theta}\rho^j_{ij}C^i\left(1 - \frac{1}{5\frac{k}{m}\theta}g_{pq}C^p C^q\right)\right).
\tag{2.3}
$$

Here f_E is the Maxwell distribution and θ is the kinetic temperature of the gas, a measure for the mean kinetic energy of the atoms while C^i denotes the components of the velocity of the atoms in the local rest frame of the gas. With (2.3) the two internal moments $\rho^{<ij>k}$ and ρ^{ijk}_j become:

$$
\begin{aligned}
\rho^{\langle ij\rangle k} &= \frac{1}{5}\left(\rho^{in}_n g^{jk} + \rho^{jn}_n g^{ik} - \frac{2}{3}\rho^{kn}_n g^{ij}\right) \\
\rho^{ijk}_j &= 5\frac{k}{m}p\theta g^{ik} + 7\frac{k}{m}\theta\rho^{\langle ik\rangle}.
\end{aligned}
\tag{2.4}
$$

We shall be interested in heat conduction in a gas between two co-axial cylinders. Therefore cylindrical coordinates (r, ϑ, z) are appropriate and we remind the reader of the metric tensor and of the Cristoffel symbols in these coordinates:

$$g^{ik} = \begin{pmatrix} 1 & 0 & 0 \\ 0 & \frac{1}{r^2} & 0 \\ 0 & 0 & 1 \end{pmatrix} \quad \text{and } \Gamma^1_{22} = -r, \quad \Gamma^2_{21} = \Gamma^2_{12} = \frac{1}{r}, \quad \Gamma^m_{kn} = 0 \quad \text{otherwise.} \tag{2.5}$$

The bulk of the paper will refer to a non-inertial frame with a constant angular velocity ω and without translational motion with respect to an inertial frame. In that case we have:

$$\underset{0}{i}^i = \begin{pmatrix} r\omega^2 \\ 0 \\ 0 \end{pmatrix}, \quad W^{ij} = \begin{pmatrix} 0 & r\omega & 0 \\ -\frac{1}{r}\omega & 0 & 0 \\ 0 & 0 & 0 \end{pmatrix}. \tag{2.6}$$

2.2 Heat conduction in a gas at rest in an inertial frame

Müller and Ruggeri [5] have used Eqs. (2.2), (2.4) to calculate the stationary fields of the gas at rest in an inertial frame between two co-axial cylinders with inner and outer radii r_i and r_e. The gas was heated by the heat flux q_i^r at r_i and the temperature was kept at the value θ_e at r_e. The solution turns out to be:

$$p = p_i = \text{const}, \quad \mathbf{v} = 0, \quad \theta = \theta_e - \frac{q_i^r r_i}{5\frac{k}{m}\tau p} \ln\left(\frac{\frac{28}{25}\frac{\tau}{p} q_i^r r_i + r^2}{\frac{28}{25}\frac{\tau}{p} q_i^r r_i + r_e^2}\right),$$

$$\rho_j^{ij} = \begin{bmatrix} 2q_i^r \frac{r_i}{r} \\ 0 \\ 0 \end{bmatrix}, \quad \rho^{\langle kl \rangle} = \begin{bmatrix} \frac{4}{5}\tau q_i^r \frac{r_i}{r^2} & 0 & 0 \\ 0 & -\frac{4}{5}\tau q_i^r \frac{r_i}{r^4} & 0 \\ 0 & 0 & 0 \end{bmatrix}. \tag{2.7}$$

Thus even this simple case of the 13-moment theory offers a surprise:

- The temperature at the inner cylinder remains bounded as r_i tends to zero.
- There are normal components of the deviatoric stress $\rho^{<kl>}$.

Recall that the normal stresses would be zero in a Navier–Stokes–Fourier solution, since the gas is at rest, while the temperature field would read

$$\theta^{NSF} = \theta_e - \frac{2}{5}\frac{q_i^r r_i}{\frac{k}{m}\tau p} \ln\left(\frac{r}{r_e}\right). \tag{2.8}$$

For a fixed heating $2\pi r_i q_i^r$ per unit length of the inner cylinder the temperature field (2.8) has a logarithmic singularity at r_i when r_i tends to zero.

2.3 A heat-conducting gas is incapable of a rigid rotation

After the heuristic argument for a tangential component of the heat flux, which was presented in the introduction, naturally we are interested in the effect of the inertial terms on the stationary heat conduction between the two cylinders. Therefore we have made the semi-inverse ansatz:

$$p = p(r), \quad \mathbf{v} = 0, \quad \theta = \theta(r),$$
$$\rho_j^{ij} = 2\begin{bmatrix} q^r(r) \\ q^\vartheta(r) \\ 0 \end{bmatrix}, \quad \rho^{<ij>} = \begin{bmatrix} \sigma^{rr}(r) & \sigma^{r\vartheta}(r) & 0 \\ \sigma^{r\vartheta}(r) & \sigma^{\vartheta\vartheta}(r) & 0 \\ 0 & 0 & \sigma^{zz}(r) \end{bmatrix}, \tag{2.9}$$

only to find out that there is no non-trivial solution of the system (2.2), (2.4) for this case. Nor is there a non-trivial solution if – for $\mathbf{v} = 0$ – we also set $\sigma^{r\vartheta} = 0$ or $\sigma^{r\vartheta}$ *and* q^ϑ equal to zero.

Therefore we conclude that the heat conducting gas cannot be at rest in a non-inertial frame. Or, in other words, a heat conducting gas is incapable of a rigid rotation.

3 General planar solution

3.1 A new semi-inverse ansatz and the corresponding system of differential equations

Since heat conduction in a rigidly rotating gas is impossible, we generalise the semi-inverse ansatz (2.9) by assuming that the velocity field can have the form

$$\mathbf{v} = \begin{bmatrix} v^r(r) \\ v^\vartheta(r) \\ 0 \end{bmatrix}. \tag{3.1}$$

Actually the balance of mass immediately requires that v^r is zero, because there is no in- or out-flux of mass through the cylinders. Thus only *one* new field is added to the fields (2.9), viz. v^ϑ. This modest addition, however, makes for a formidable system of coupled equations which can no longer be solved analytically. The system is given in (3.2), where we have now employed physical rather than co- or contravariant components. The physical coordinates are characterised by the indices in square brackets:

$$\frac{d(p + \sigma[rr])}{dr} - \frac{1}{r}(\sigma[\vartheta\vartheta] - \sigma[rr]) = \frac{p}{\frac{k}{m}\theta}\frac{1}{r}(r\omega + v[\vartheta])^2,$$

$$\sigma[r\vartheta] = \frac{D}{r^2},$$

$$q\,[r] + \sigma\,[r\vartheta]\,v\,[\vartheta] = \frac{C}{r},$$

$$-\frac{5}{2}\frac{1}{\tau}\sigma\,[rr] = -10\omega\sigma\,[r\vartheta] + \frac{4}{3}\frac{dq\,[r]}{dr} - \frac{5}{3}\sigma\,[r\vartheta]\frac{dv\,[\vartheta]}{dr} - \frac{2}{3}\frac{q\,[r]}{r} - \frac{25}{3}\frac{\sigma\,[r\vartheta]\,v\,[\vartheta]}{r},$$

$$-\frac{5}{2}\frac{1}{\tau}\sigma\,[r\vartheta] = \frac{5}{r}\left(\sigma\,[rr] - \sigma\,[\vartheta\vartheta]\right)\left(r\omega + v\,[\vartheta]\right) + \frac{dq\,[\vartheta]}{dr} - \frac{q\,[\vartheta]}{r} + \frac{5}{2}\left(p + \sigma\,[rr]\right)\left(\frac{dv\,[\vartheta]}{dr} - \frac{v\,[\vartheta]}{r}\right),$$

$$-\frac{5}{2}\frac{1}{\tau}\sigma\,[\vartheta\vartheta] = 10\omega\sigma\,[r\vartheta] + \frac{2}{r}q\,[r] + 4\sigma\,[r\vartheta]\frac{dv\,[\vartheta]}{dr} + \frac{6}{r}\sigma\,[r\vartheta]\,v\,[\vartheta], \quad (3.2)$$

$$-\frac{2}{\tau}q\,[r] = -4\omega q\,[\vartheta] - \frac{2}{r}\sigma\,[rr]\left(r\omega + v\,[\vartheta]\right)^2 + \frac{k}{m}\left(5p + 7\sigma\,[rr]\right)\frac{d\theta}{dr} + 2\frac{k}{m}\theta\left(\frac{d\left(\sigma\,[rr]\right)}{dr} + \frac{1}{r}\left(\sigma\,[rr] - \sigma\,[\vartheta\vartheta]\right)\right) + \frac{4}{5}q\,[\vartheta]\frac{dv\,[\vartheta]}{dr} - \frac{24}{5}\frac{q\,[\vartheta]\,v\,[\vartheta]}{r},$$

$$-\frac{2}{\tau}q\,[\vartheta] = \frac{4}{r}q\,[r]\left(r\omega + v\,[\vartheta]\right) - \frac{2}{r}\sigma\,[r\vartheta]\left(r\omega + v\,[\vartheta]\right)^2 + 7\frac{k}{m}\sigma\,[r\vartheta]\frac{d\theta}{dr} + \frac{14}{5}q\,[r]\left(\frac{dv\,[\vartheta]}{dr} - \frac{v\,[\vartheta]}{r}\right),$$

where C and D are constants of integration.

By use of Eqs. $(3.2)_{2,3,6}$ we may eliminate $\sigma\,[r\vartheta]$, $q\,[r]$ and $\sigma\,[\vartheta\vartheta]$ from the system (3.2). Thus rearrangement of the remaining five equations will provide a system of the generic form

$$\frac{d\mathbf{u}}{dr} = \mathbf{A}\left(\mathbf{u}, r; C, D, \omega, \tau\right) \quad \text{where } \mathbf{u} = \left(p, v\,[\vartheta], \theta, \sigma\,[rr], q\,[\vartheta]\right). \quad (3.3)$$

This will be solved by stepwise integration after appropriate boundary values are chosen.

We shall fix $\tau = 10^{-5}$ s, appropriate to a pressure of about 10^2 Pa, and we let ω assume the large value $10^4 \frac{1}{s}$ in order to emphasise the rotational effects. The cylinders have radii $r_i = 10^{-3}$ m and $r_e = 10^{-2}$ m. Thus, in summary, we set

$$r_i = 10^{-3}\,\text{m}, \quad r_e = 10^{-2}\,\text{m}, \quad \tau = 10^{-5}\,\text{s}, \quad \omega = 10^4 \frac{1}{\text{s}}. \quad (3.4)$$

We do not introduce non-dimensional quantities, rather we give all numbers in SI units.

3.2 Boundary values

It turns out that the system (3.3) has singular points at $r = 0$ and at $r = \infty$. These singularities require that the numerical solution must start its stepwise construction at the outer cylinder.

Therefore we provide the following boundary values:

$$\begin{aligned} \text{at } r_i: \quad & \text{velocity } v[\vartheta] = 0, \\ & \text{radial heat flux } q[r] = 10^4 \frac{\mathrm{W}}{\mathrm{m}^2}, \\ \text{at } r_e: \quad & \text{velocity } v[\vartheta] = 0, \\ & \text{temperature } \theta, \\ & \text{normal pressure } p + \sigma[rr], \, p + \sigma[zz], \\ & \text{tangential heat flux } q[\vartheta]. \end{aligned} \tag{3.5}$$

We envisage both cylinders to be at rest and the natural boundary condition for $v[\vartheta]$ at r_i and r_e is therefore the no-slip condition $v[\vartheta] = 0$. At r_e this condition is directly applied, while at r_i we ensure it by shooting. The shooting parameter is the value of $\sigma[r\vartheta]$ at r_e, or, by $(3.2)_2$, the constant D.

The constant C is no problem. We think of the inner wall as heated by prescribed heating per unit axial length $2\pi r_i q[r]_i$. According to $(3.2)_3$, $(3.4)_1$ and $(3.5)_{1,2}$ we have

$$C = 10 \frac{\mathrm{W}}{\mathrm{m}}. \tag{3.6}$$

The temperature θ_e may be guaranteed by a heat bath.

More problematic are the shear pressures. We have to realise that in practice it is impossible to assign p and $\sigma[rr]$ separately. What we may be able to assign at r_e are the normal pressures $(3.5)_5$. From these, by $(3.2)_{2,3,4,6}$ we obtain a linear system for the boundary values p_e and $\sigma[rr]_e$, viz.:

$$\begin{aligned} p_e + \sigma[rr]_e &= \{p + \sigma[rr]\}_e, \\ p_e + \frac{1}{3}\sigma[rr]_e &= \{p + \sigma[zz]\}_e + \frac{4}{3}\tau\omega\frac{D}{r_e^2} + \frac{4}{15}\tau\frac{C}{r_e^2}. \end{aligned} \tag{3.7}$$

Note that in this way the boundary values p_e and $\sigma[rr]_e$ depend on the shooting parameter D. The numerical results will all be calculated for the boundary values

$$\{p + \sigma[rr]\}_e = 110\,\mathrm{Pa} \quad \text{and} \quad \{p + \sigma[zz]\}_e = 110\,\mathrm{Pa}. \tag{3.8}$$

There is also a question mark about the assignability and controllability of the tangential heat flux $q[\vartheta]$ at r_e, or, indeed, anywhere. This is a very serious difficulty and we do not know how to solve it. However there is an indication that the system, in an unknown manner, – mysterious to us – knows which boundary value for $q[\vartheta]$ is permissible.

If we solve the system of equations for a set of boundary values p_e, $v[\vartheta]_i = 0$, $q[r]_i$, $v[\vartheta]_e = 0$, θ_e, $\sigma[rr]_e$, – which according to the above arguments, *can* be assigned, – *and, for widely different values of* $q[\vartheta]_e$ between $-2000\frac{\mathrm{W}}{\mathrm{m}^2}$ and $+2000\frac{\mathrm{W}}{\mathrm{m}^2}$, the corresponding function $q[\vartheta](r)$ will sweep into essentially one single function within a very thin boundary layer (see Fig. 3.1, left). Thus it seems that in the case shown in the figure the "correct" boundary value is approximately equal to $-200\frac{\mathrm{W}}{\mathrm{m}^2}$. Incidentally that is the value of $q[\vartheta]$ which we expect from the heuristic argument for $q[\vartheta]$, presented in the introduction, and which corresponds – for the data (3.4) and (3.5) – to the leading term $q[\vartheta] = -2\tau\omega q[r]$ in $(3.2)_8$.

While from the above it is suggestive to take the value $q[\vartheta]_e \approx -200\frac{\mathrm{W}}{\mathrm{m}^2}$ there is not proof that we *have to* take it. This matter is currently under investigation. However, in the paper we proceed with this boundary value.

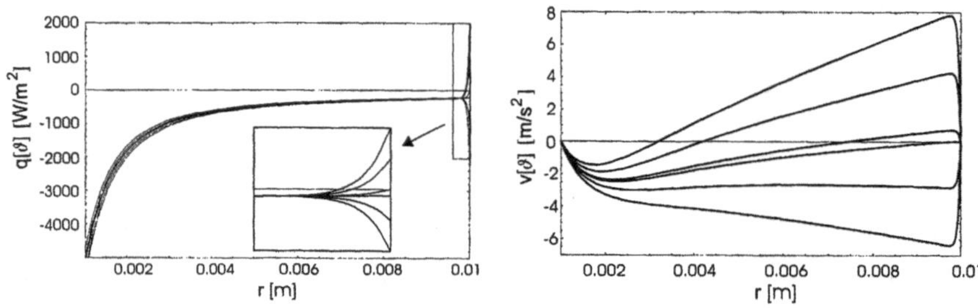

Fig. 3.1. $q[\vartheta]$ as a function of r for different boundary values **(left)**; $v[\vartheta](r)$ for different boundary values of $q[\vartheta]_e$ **(right)**

It is true that the different data $q[\vartheta]_e$ create minimally different functions $q[\vartheta](r)$, – except in the boundary layer; but the corresponding curves for $v[\vartheta](r)$ differ considerably with the data $q[\vartheta]_e$ (see Fig. 3.1, right). Specifically *the* data, for which $q[\vartheta](r)$ shows a boundary layer behaviour, also show such a layer in $v[\vartheta](r)$.

We proceed to show the complete solution of our problem for the boundary data implied by (3.5), (3.8) and $q[\vartheta]_e \approx -200\frac{\mathrm{W}}{\mathrm{m}^2}$, i.e., the value of $q[\vartheta]_e$ for which neither $q[\vartheta]$ nor $v[\vartheta]$ exhibit a boundary layer.

3.3 Solutions

Figures 3.2, 3.3 and 3.4 represent the solution of the system (3.2) for the boundary values discussed in the previous paragraph. The individual functions are arranged as follows.

Figure 3.2 shows the fields $T(r)$, $q[r](r)$ and $p(r)$. These are in some sense the "main" functions to be determined in a heat-conducting gas on a turntable. By this we mean that these three functions are also non-constant, if we consider the gas as a Navier–Stokes–Fourier fluid. In order to exhibit the effect of 13 moments as opposed to Navier–Stokes–Fourier we have printed both in the graphs of Fig. 3.2 – dashed lines for Navier–Stokes–Fourier solutions.

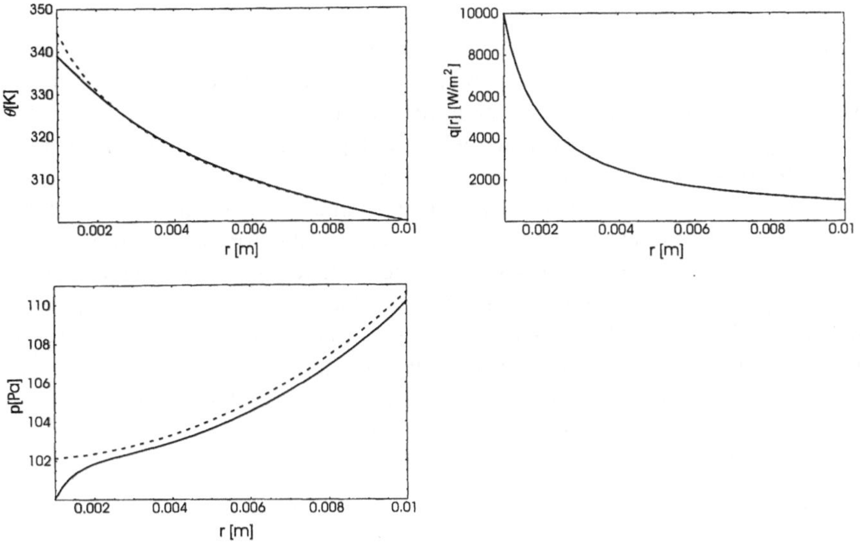

Fig. 3.2. Temperature (**top left**), heat flux (**top right**) and pressure (**bottom**). *Bold lines*: 13 moments; *Dashed lines*: Navier–Stokes–Fourier. For the heat flux $q\,[r]$ the two lines coincide on the scale of the graphs

The next set of graphs, in Fig. 3.3, represent two effects due to the inertial terms: non-zero functions for $q\,[\vartheta]$ and $v\,[\vartheta]$. We conclude that the heat flux has a tangential component, i.e., a component perpendicular to the temperature gradient. This phenomenon was already predicted by Müller [6] in 1972 by an argument from the kinetic theory of gases. Also we conclude that there is a tangential component of the velocity, so that the gas does not rotate rigidly with the cylinders except at the boundaries, because there is no slip. To our knowledge this phenomenon has never been observed nor predicted from calculations and it may be considered as the most important single result of this work.

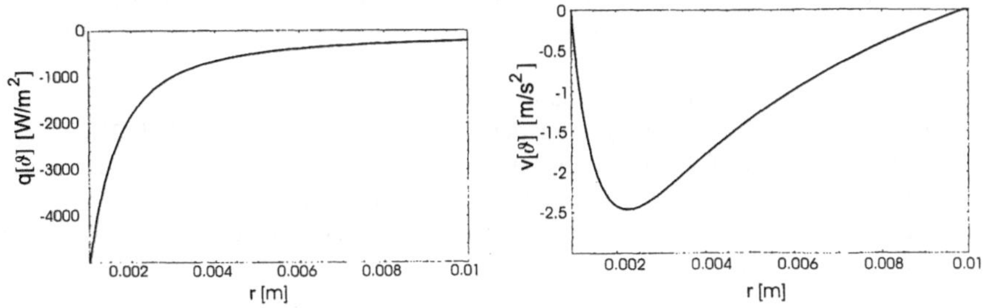

Fig. 3.3. Tangential component of $q\,[\vartheta]$ (**left**) and $v\,[\vartheta]$ (**right**)

The remaining results – represented in Fig. 3.4 – concern the deviatoric components of the pressure tensor which all vanish in a Navier–Stokes–Fourier theory of the rotating heat-conducting gas. Here, in the 13-moment theory, they do not vanish. Note that there is no obvious relation between the shear pressure $\sigma\,[r\vartheta]$ and the velocity gradient $\frac{dv[\vartheta]}{dr}$; in particular where $v\,[\vartheta]$ has a minimum the shear pressure does not vanish as it would in a Navier–Stokes–Fourier fluid.

We have unequal normal stresses – much as in a non-Newtonian fluid, albeit for an entirely different reason. Of particular interest is the normal pressure $p + \sigma\,[zz]$ which is quite non-uniform mostly due to the centrifugal force which tends to drive the gas upwards at the outer cylinder. We conclude from this that our semi-inverse ansatz, which implies $v\,[z] = 0$, can only be realised by applying a stress field on the horizontal surfaces. Note that in the neighbourhood of the *inner cylinder* the normal pressure turns sharply up; once again the analogy to non-Newtonian fluids – and in particular to the Weissenberg effect – catches the eye.

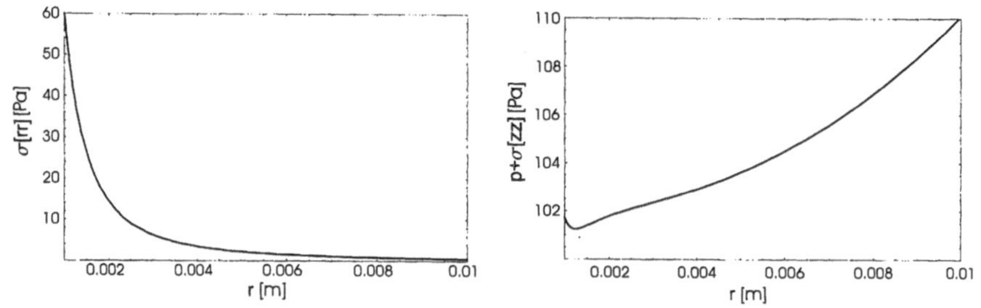

Fig. 3.4. Normal stresses $\sigma\,[rr]$ (**left**) and $p + \sigma\,[zz]$ (**right**)

References

[1] Müller, I., Ruggeri, T. (1993): Extended thermodynamics. (Springer Tracts in Natural Philosophy, vol. 37). Springer, New York

[2] Müller, I., Ruggeri, T. (1998): Rational extended thermodynamics. Second ed. (Springer Tracts in Natural Philosophy, vol. 37). Springer, New York

[3] Bhatnagar, P.L., Gross, E.P., Krook, M. (1954): A model for collision processes in gases. I. Small amplitude processes in charged and neutral one-component systems. Phys. Rev. **94**, 511–525

[4] Grad, H. (1958) Principles of the kinetic theory of gases. In: Flügge, S. (ed.) Handbuch der Physik, Bd. 12. Thermodynamik der Gase. Springer, Berlin, pp. 205–294

[5] Müller, I., Ruggeri, T. (2000): Stationary heat conduction in radially symmetric situations – an application of extended thermodynamics. Submitted

[6] Müller, I. (1972): On the frame dependence of stress and heat flux. Arch. Rational Mech. Anal. **45**, 241–250

Mediated Interactions of Proteins in Lipid Membranes

Paolo Biscari

Abstract. The mediated interactions of proteins embedded in lipid membranes has been widely studied in the Monge gauge approximation, which consists in treating the membrane shape as a linear perturbation of a reference shape (planar or spherical). Here I report the results of a two-dimensional study in which the closed a curve representing the membrane shape can be analytically determined. The nonlinear effects of a closed geometry determine new qualitative features such as new, nontrivial equilibrium configurations of the proteins along the membrane.

1 Introduction

Lipid membranes are aggregates of amphipathic molecules, that is, molecules having a hydrophilic head and a hydrophobic tail. Living in an aqueous environment, these molecules tend to form closed bilayers (*vesicles*), where the hydrophobic tails are in contact with each other, and both hide from the surrounding water behind the hydrophilic heads. From the mathematical point of view, the vesicles can be modelled by closed, compact surfaces, since their thickness (which is of the order of 10^{-9} m [9], that is, double the length of the molecules) is much smaller than their lateral dimensions. Furthermore, and because of their amphipathic nature, the molecules making up the vesicle are not likely to leave it, so that the total area of the membrane is fixed. To determine the shape of the membrane, either isolated or subject to external fields, it is then necessary to minimize (under the constraint of fixed area) the free energy functional, whose elastic part depends on the principal curvatures of the surface representing the membrane.

When a lipid membrane embeds a protein (which we can model as a rigid rod), the direct interaction at the membrane-protein contact points influences the membrane shape, fixing the angle between the normal to the membrane and the protein axis to be equal to a preferred value, which depends on the protein conformation [4]. The membrane-protein interaction gives rise to deep modifications in the membrane shape which may even give rise [3] to the expulsion or the complete absorption of the protein. Thus, when two or more proteins are embedded by the same membrane, the deformations they induce in the membrane create an indirect interaction between them: they attract or repel each other through a *mediated force*, that depends on their distance as well as on their conformation. This interaction has been widely studied under the assumption that the presence of the proteins determines only small variations on the shape conformation, which remains almost planar [5, 7, 8, 13] or spherical [6]. Nevertheless, two-dimensional studies [2, 3] have now shown that even the smallest proteins may induce dramatic changes in the membrane shape.

In this paper I illustrate some features of the mediated force that can be derived from a two-dimensional model, where the shape of the embedding membrane (now represented by a closed curve) can be analytically determined. The results reported in this paper have been obtained mainly in collaboration with Fulvio Bisi and Riccardo Rosso (see [1–3]); I acknowledge their contributions, and I refer to the above mentioned papers for details omitted here.

2 Equilibrium shapes

In the absence of external fields, the free energy associated with the curve γ that represents a planar lipid membrane depends only on its curvature. Using the classic model of Helfrich [10], here I consider

$$\mathcal{F}[\gamma] := \frac{\kappa}{2} \int_\gamma \left(\sigma(s) - \sigma_0\right)^2 ds, \tag{2.1}$$

where κ and σ_0 are the *bending rigidity* and the *spontaneous curvature* [5] of the membrane, respectively: thus, the preferred shape for a free membrane is a circle of radius σ_0^{-1}. Furthermore, to take into account the inextensibility constraint, we have to replace $\mathcal{F}$ by an effective free energy, which also contains a Lagrange multiplier function λ:

$$\mathcal{F}_{\rm eff}[\,\gamma\,] := \int_\gamma \left\{ \frac{\kappa}{2}\left(\sigma(s) - \sigma_0\right)^2 + \lambda(s)\,\left|\gamma'(s)\right| \right\} ds. \tag{2.2}$$

The geometric quantities which characterize the ith protein are its length a_i and the contact angle $\varphi_i \in \left(-\frac{\pi}{2}, \frac{\pi}{2}\right)$ that the tangent unit vector $\boldsymbol{t}$ to the curve γ determines with the protein axis at the contact points. Thus, if we introduce the angle ϑ, between $\boldsymbol{t}$ and a fixed unit vector $\boldsymbol{e}_x$ (which we will choose to be parallel to one of the proteins), it turns out that ϑ has jump discontinuities across the proteins, with $\Delta\vartheta_i = 2\varphi_i$.

Figure 1 illustrates the geometric setting in the case of two embedded proteins; in it, the contact angle is positive in the first protein (inclusions with $\varphi_i > 0$ are usually labelled as *inner*), and negative in the second (which is thus an *outer* inclusion).

In the two-dimensional model, if a membrane embeds n proteins, the curve γ that describes its shape splits into n connected pieces $\{\gamma_1, \ldots, \gamma_n\}$, and it is possible to prove [11, 12, 3, 2] that the Euler-Lagrange equations associated to the functional (2.2) can be analytically solved, and that the curvature $\sigma = \vartheta'(s)$ along every γ_i is given by

$$\sigma_i^2(\vartheta) = \frac{1}{L^2}\left(\tilde{\lambda}_i + \mu_i \cos\vartheta + \nu_i \sin\vartheta\right). \tag{2.3}$$

In (2.3), $\tilde{\lambda}_i$, μ_i and ν_i are real numbers; furthermore, it turns out [2] that the Lagrange multiplier function $\lambda(s)$ is piecewise constant, with the value that it assumes on every γ_i given by

$$\lambda_i := \lambda(\gamma_i) = \frac{\kappa}{2L^2}\left(\tilde{\lambda}_i - (\sigma_0 L)^2\right).$$

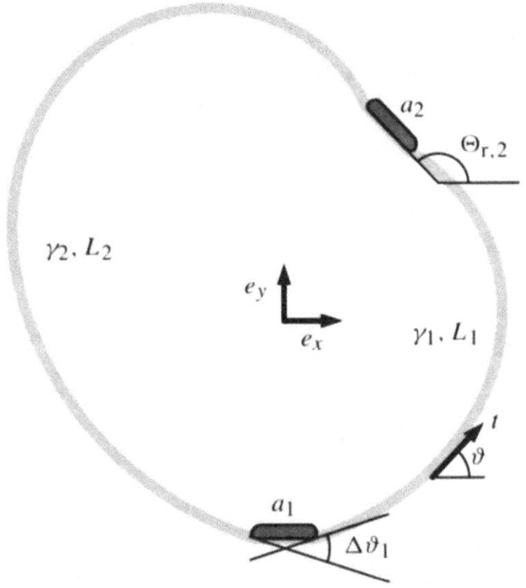

Fig. 1. Geometrical setting for a membrane embedding two proteins. The angles ϑ and $\Theta_{r,2}$ (relative angle between the second and the first inclusion) are measured with respect to the direction of the unit vector $\boldsymbol{e}_x$, parallel to the first inclusion

In the presence of inclusions, the equilibrium equation for the membrane is not sufficient by itself to ensure the equilibrium of the whole system: we must also take into account the balance equations of the forces and torques acting on every single protein. The balance of forces [2] requires that $\mu_i \equiv \mu$ and $\nu_i \equiv \nu$ for all $i = 1, \ldots, n$, while the balance of torques determines the relative angles between the inclusions (such as $\Theta_{r,2}$ in Fig. 1).

To complete the determination of the equilibrium configuration we must lastly impose the constraints on the total length L_i of every γ_i (these equations determine the $\tilde{\lambda}_i$'s) and the closure conditions $\Delta x_{\text{tot}} = \Delta y_{\text{tot}} = 0$, which determine the common values μ and ν of the μ_i's and the ν_i's.

3 Mediated force

By following the steps described above, it is possible to determine the equilibrium configuration for any particular set of values of the distances L_i between neighbouring proteins (clearly, the sum of the lengths L_i of all the pieces γ_i of γ must coincide with the total length L of the membrane), to which corresponds a value of the free energy $\mathcal{F}(L_1, \ldots, L_n)$. It is then natural to define the *mediated force* acting on the ith protein as minus the rate of increase of the free energy when only the position of that protein is varied, all the other proteins kept fixed.

Definition 3.1 (Mediated force). Let $\{L_i,\ i = 1, \ldots, n\}$ be the lengths of the curves $\{\gamma_i,\ i = 1, \ldots, n\}$ that connect n embedded proteins inside a planar membrane (with the assumption that the curve γ_i connects the ith and the $(i+1)$th proteins, except for γ_n, which connects the last inclusion with the first one). The *mediated force* acting on the ith inclusion is defined as:

$$F_{m,i} := -\lim_{\epsilon\to 0} \frac{\mathcal{F}(L_1,\dots,L_{i-1}+\epsilon,L_i-\epsilon,\dots,L_n)-\mathcal{F}(L_1,\dots,L_{i-1},L_i,\dots,L_n)}{\epsilon} = \frac{\partial\mathcal{F}}{\partial L_i}-\frac{\partial\mathcal{F}}{\partial L_{i-1}}.$$

We remark that, when $F_{m,i} > 0$, the ith protein is pushed forwards (in the direction defined by the arc-length along the curve), while, if $F_{m,i} < 0$, it is pulled backwards.

Theorem 3.2 (Determination of the mediated force). *In the conditions of Definition 3.1, we have that*

$$\frac{\partial\mathcal{F}}{\partial L_i} = -\lambda_i, \tag{3.1}$$

so that the above defined mediated force on the ith inclusion is simply given by

$$F_{m,i} = \lambda_{i-1} - \lambda_i.$$

Theorem 3.2 gives a physical meaning to λ_i: it represents the tendency of γ_i to increase its length, pushing away the proteins that are at its ends. Clearly, the sum of the mediated forces on all the proteins is null since the system is assumed to be isolated. Finally, all the mediated forces vanish only when all the λ_i's are equal.

Proof of Theorem 3.2. I begin by giving the free energy functional in (2.1) an expression which is easier to work with. Taking into account the fact that $\sigma(s) = \vartheta'(s)$ and replacing the curvature in (2.1) by its expression (2.3), we have:

$$\begin{aligned}\mathcal{F} &= \frac{\kappa}{2}\int_\gamma (\sigma(s)-\sigma_0)^2\,ds = \frac{\kappa}{2}\left(\int_\gamma \sigma^2\,ds - 2\sigma_0\int_\gamma \sigma\,ds + \sigma_0^2\int_\gamma ds\right)\\ &= \frac{\kappa}{2L}\left(\sum_{i=1}^n\int_{\gamma_i}\left(\tilde{\lambda}_i+\mu\cos\vartheta+\nu\sin\vartheta\right)\frac{ds_i}{L} - 2(\sigma_0 L)\,\Delta\vartheta_{\rm tot} + (\sigma_0 L)^2\right)\\ &= \frac{\kappa}{2L}\left(\sum_{i=1}^n\tilde{\lambda}_i\,\frac{L_i}{L}+\mu\,\frac{\Delta x_{\rm tot}}{L}+\nu\,\frac{\Delta y_{\rm tot}}{L} - 2(\sigma_0 L)\,\Delta\vartheta_{\rm tot} + (\sigma_0 L)^2\right),\end{aligned} \tag{3.2}$$

where, because of the jumps of the tangent to γ across the proteins,

$$\Delta\vartheta_{\rm tot} = 2\pi - 2\sum_{i=1}^n \varphi_i,$$

and, from the closure condition on γ and the definition of the relative angles $\Theta_{r,i}$,

$$\Delta x_{\rm tot} = -\sum_{i=1}^n a_i\cos\Theta_{r,i} \quad\text{and}\quad \Delta y_{\rm tot} = -\sum_{i=1}^n a_i\sin\Theta_{r,i},$$

assuming that a_i denotes the length of the ith inclusion and that we define the relative angle of the first protein with respect to itself to be null ($\Theta_{r,1} = 0$). In any case, I emphasize the fact that the last two terms in (3.2) depend only on constitutive parameters, and not on the distances L_i, so that they do not contribute to the mediated force.

When we differentiate in (3.2) with respect to L_i, we have to consider not only the explicit dependence, but also the implicit dependence through the parameters $\tilde{\lambda}_i$, μ, ν, and $\vartheta_{r,i}$, since all of them depend on L_i.

The derivatives of these parameters with respect to L_i can be obtained by differentiating the equations that determine them (that is, as I have said above, the n constraint equations on the lengths of the γ_i's, the closure conditions on γ, and the balance of torques on the proteins). Thus, for example, from the equation fixing the length of γ_i, which is

$$\int_{\vartheta_{r,i}+\varphi_i}^{\vartheta_{r,i+1}-\varphi_{i+1}} \frac{d\vartheta}{\sqrt{\lambda_i + \mu\cos\vartheta + \nu\sin\vartheta}} = \frac{L_i}{L},$$

and with a prime denoting the derivative with respect to L_i, we obtain

$$-\frac{1}{2}\int_{\vartheta_{r,i}+\varphi_i}^{\vartheta_{r,i+1}-\varphi_{i+1}} \frac{\lambda_i' + \mu'\cos\vartheta + \nu'\sin\vartheta}{\left(\lambda_i + \mu\cos\vartheta + \nu\sin\vartheta\right)^{\frac{3}{2}}}\,d\vartheta + \frac{\vartheta_{r,i+1}'}{\sigma\left(\vartheta_{r,i+1} - \varphi_{i+1}\right)} - \frac{\vartheta_{r,i}'}{\sigma\left(\vartheta_{r,i} + \varphi_i\right)} = \frac{1}{L}. \tag{3.3}$$

At this point it is a long but straightforward task to rearrange the terms in the derivative of $\mathcal{F}$ to obtain exactly the combinations like that in the left-hand side of (3.3); then, by replacing them by the right-hand side, (3.1) follows immediately. □

4 Equilibrium of two proteins

In this final section I apply the results above to the particular case of two proteins embedded by the same membrane. In particular, I focus attention on the configurations in which the mediated forces vanish. All the results here are proved in detail in [1].

We consider two proteins with lengths $a_1, a_2 \in (0, L)$ and contact angles $\varphi_1, \varphi_2 \in (-\frac{\pi}{2}, \frac{\pi}{2})$. Varying all the characteristic parameters of the proteins it is possible to prove that there are at most four different types of configurations in which the proteins do not feel any mediated force. In the following, I briefly describe each of these equilibrium configurations.

4.1 Antipodal equilibrium configuration

For any choice of the protein constitutive parameters there is a symmetrical equilibrium configuration in which the proteins are poles apart along the membrane, that is, in which $L_1 = \frac{1}{2}L$ and $\Theta_{r,2} = \pi$ (see Fig. 2a). Depending on the lengths and the contact angles of the proteins, this equilibrium configuration can be either stable or unstable with respect to small variations of the distance between the proteins. In the former case (which, for example, arises if the proteins are identical), the proteins repel each other towards the antipodal configuration if they are slightly moved from it; in the latter, they attract each other if the antipodal configuration is perturbed.

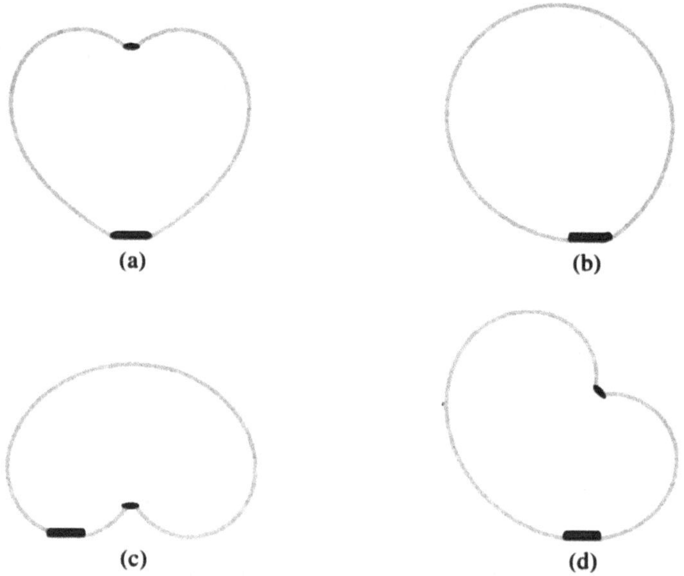

Fig. 2. Equilibrium configurations for two proteins. In the four configurations the lengths of the proteins are equal to $0.05L$ and $0.01L$, where L is the length of the membrane. In the antipodal configuration **(a)**, which is an inner-outer configuration, $\varphi_1 = 30°$ and $\varphi_2 = -45°$, and the distance between the inclusions is $L_1 = \frac{1}{2}L$. In the contact configuration **(b)**, an inner-inner configuration, $\varphi_1 = 5°$ and $\varphi_2 = 15°$; the distance between the inclusions is negligible. When these proteins approach each other, they become equivalent to an asymmetrical effective protein of length $a_{\text{eff}} = 0.0595L$, rotated of an angle $\Theta_{r,\text{eff}} = 10.35°$ with respect to $\boldsymbol{e}_x$, and with effective contact angles respectively equal to 31.7° and 8.3°. In the parallel configuration **(c)**, again an inner-outer configuration, $\varphi_1 = 10°$ and $\varphi_2 = -60°$; the distance between the proteins is $L_1 = 0.0829L$. Finally, in the asymmetric configuration **(d)**, which is still an inner-outer configuration, $\varphi_1 = 8.52°$, $\varphi_2 = -54°$, and the distance between the proteins is $L_1 = 0.3438L$

4.2 Contact equilibrium configuration

When $L_1 \ll L$, the proteins approach each other until they become almost equivalent to an effective (and eventually asymmetrical) single protein, whose length and contact angles can be related to those of the actual proteins (see Fig. 2b). This allows us to easily determine the shape of the membrane by having resort to the methods developed in [3] for a single embedded protein. We remark that, since L_1 cannot become negative, this is a boundary equilibrium configuration, and thus the equilibrium condition becomes $\lambda_1 - \lambda_2 \leq 0$.

The contact configuration is not an equilibrium one for all values of the characteristic parameters. In particular, if the proteins are identical, the contact configuration is an attractor for the proteins if both are outer proteins ($\varphi_1 = \varphi_2 < 0$), whereas the proteins repel each other at small distances if they are inner ($\varphi_1 = \varphi_2 > 0$). In the intermediate case $\varphi_1 = \varphi_2 = 0$, we have an unstable equilibrium configuration.

4.3 Parallel equilibrium configuration

For any values of a_1 and a_2, if $\varphi_1 + \varphi_2 \leq 0$, there is an equilibrium configuration in which the proteins lie parallel and equally oriented (that is, with $\Theta_{r,2} = 0$) (see Fig. 2c). In the particular case of identical proteins with negative contact angles, it is always unstable, but it may become stable if one of the contact angles is positive and the other negative, with a sufficiently large absolute value.

4.4 Asymmetric equilibrium configuration

For some particular values of the protein characteristic parameters (typically characterized by the fact that one of the characteristic angles is sufficiently small in absolute value), there is a fourth equilibrium configuration not characterized by any symmetry property (see Fig. 2d). The existence of this equilibrium may already be predicted by the stability studies whose results have been reported above: for example, if $\varphi_1 + \varphi_2 > 0$ (which implies that no parallel equilibrium configuration exists), a stable equilibrium configuration (local minimum of the free energy) at a certain intermediate distance must exist when both the contact and the antipodal equilibrium configurations are unstable.

References

[1] Biscari, P., Bisi, F. (2001): Characterization of membrane-mediated forces between proteins. Quad. 459/P, Dipartimento di Matematica, Politecnico di Milano, Milano

[2] Biscari, P., Bisi, F. and Rosso, R. (2002): Curvature effects on membrane-mediated interactions of proteins. J. Math. Biol. To appear

[3] Biscari, P., Rosso, R. (2001): Inclusions embedded in lipid membranes. J. Phys. A **34**, 439–459

[4] Dan, N., Berman, A., Pincus, P., Safran, S.A. (1994): Membrane-induced interactions between inclusions. J. Physique II **4**, 1713–1725

[5] Dan, N., Safran, S.A. (1998): Effect of lipid characteristics on the structure of transmembrane proteins. Biophys. J. **75**, 1410–1414

[6] Dommersnes, P.G., Fournier, J.-B., Galatola, P. (1998): Long-range elastic forces between membrane inclusions in spherical vesicles. Europhys. Lett. **42**, 233–238

[7] Goulian, M., Bruinsma, R., Pincus, R. (1993): Long-range forces in heterogeneous fluid membranes. Europhys. Lett. **22**, 145–150
[8] Golestanian, R., Goulian, M., Kardar, M. (1996): Fluctuation-induced interactions between rods on membranes and interfaces. Europhys. Lett. **33**, 241–245
[9] Lipowsky, R., Döbereiner, H.-G. (1998): Vesicles in contact with nanoparticles and colloids. Europhys. Lett. **43**, 219–225
[10] Petrov, A.G. (1999): The lyotropic state of matter: Molecular physics and living matter physics. Gordon and Breach, Amsterdam
[11] Rosso, R., Virga, E.G. (1998): Adhesion by curvature of lipid tubules. Contin. Mech. Thermodyn. **10**, 359–367
[12] Rosso, R., Virga, E.G. (1998): Adhesion of lipid tubules in an assembly. European J. Appl. Math. **9**, 485–506; Erratum. ibid. **10** (1999), 221
[13] Weikl, T.R., Kozlov, M.M., Helfrich, W. (1998): Interaction of conical membrane inclusions: Effect of lateral tension. Phys. Rev. E **57**, 6988–6995 (1998)

Second-Order Surface Potentials in Finite Elasticity

Sandra Carillo, Paolo Podio-Guidugli, Giorgio Vergara Caffarelli

Abstract. A study of elastic, shape-dependent body-environment interactions is presented in which the contact loads that the environment applies on the body at a given deformation are modelled by a surface potential whose density depends not only on the first deformation gradient but also on the second. The cases of tangential and simple surface potentials are given special attention, as well as the case of surface potentials admitting an energetically equivalent replacement by a volume potential which is a second-order null Lagrangian. Throughout the paper the body-environment boundary is assumed to be a smooth surface, except in the last section, where a body with a line discontinuity of the boundary normal is considered.

1 Introduction

The primary objective of the constitutive theory of continuous material bodies is to characterize their stress response to deformation. The most studied case is that of *simple* material bodies, for which the stress is determined by the history of the first deformation gradient; for the elastic subcase, only the current value of the first deformation gradient matters.

The characterization of body-environment interactions is a much less studied chapter of the classical constitutive theory. The standard view in the mechanics of solids is to picture these interactions by the assignment of a pair of vector fields over the reference shape of the body, the distance and contact *dead loads*. Consideration of *live loads*, that is, of *shape-dependent body-environment interactions*, is rare (see [1] and the literature cited there).

No doubt, a dead boundary can be coupled with a simple material body of arbitrary response. In general, however, the constitutive characterizations of bodies and environments should not be regarded as separable problems.[1] This paper, which builds upon previous work by two of us [2], is an attempt to exemplify such an integrated approach for the basic case of *elastic material bodies with live environments of elastic type*. To concentrate on the main issue, the posing of interaction conditions at the body-environment boundary, we deliberately ignore distance interactions.

The formulation of conditions to hold at a body's boundary is a simplified manner of accounting for the complexities that a more detailed picture of the role of the environment would entail.

Suppose that the environment is viewed as a body in itself. Then, the modelling of interactions at the common boundary reduces to laying down suitable *transmission*

[1] There are interesting applications where even the notion of a sharp interface between a body and its environment is inappropriate, for reasons that are either physical or mathematical, or both.

conditions between the body of interest and its environmental body. The expression of such transmission conditions is not especially difficult, provided that both bodies belong, in a sense that should be made precise, to the "same" constitutive class. We will try to clarify this last point.

If both the body and its environmental body are simple, then the transmission conditions have the form of *continuity conditions* on the displacement vector and on the traction vector,[2] with the latter continuity condition stated in terms of first deformation gradients, inner and outer. If the displacement vector is required to be continuous, then the tangential part of the first deformation gradient must also be continuous. To be consistent with these facts, a boundary condition of Neumann type for the given body should amount to setting the traction vector equal to a live applied load which depends only on the tangential deformation gradient (this is indeed the case with uniform pressure loading; see [2] and Footnote 6 in Sect. 4 below). Traction boundary conditions involving deformation gradients higher than the first (such as membrane loading [2], which depends on the second tangential derivatives of deformation) suggest consideration of environmental bodies that are *not* made of simple materials; the same is true for traction boundary conditions where live loads depending on the first normal derivative of the deformation are specified.

Simple materials are also called materials of grade 1, while *materials of grade* N [4] are those whose stress response to an admissible deformation history depends on the first N spatial gradients of that history: the higher the grade the less local the stress response. In particular, an elastic material of grade N can be characterized by means of a constitutive mapping $\sigma({}_1\mathbf{F}, {}_2\mathbf{F}, \dots, {}_N\mathbf{F})$ delivering the density of energy stored per unit material volume for a given list $({}_1\mathbf{F}, {}_2\mathbf{F}, \dots, {}_N\mathbf{F})$ of successive deformation gradients, from the first to the Nth. In this case the mechanical response to deformation consists in a list $({}_1\mathbf{S}, {}_2\mathbf{S}, \dots, {}_N\mathbf{S})$ of *i-stresses* [5]

$$ {}_i\mathbf{S} := \sigma_{{}_i\mathbf{F}} \tag{1} $$

(here $\sigma_{{}_i\mathbf{F}}$ denotes the derivative of the mapping σ with respect to its ith argument).

How can a material body of grade N be "put in contact" with a material body of grade M at a common boundary?

Assume, without any substantial loss of generality, that $M \geq N$. Aside from kinematical continuity conditions on the displacement vector (and, perhaps, one or more of its successive gradients), M continuity conditions suggest themselves, one for each of the i-tractions ${}_i\mathbf{Sn}$ (here $\mathbf{n}$ is the normal at the common boundary). Among these conditions, those for $i = N+1, \dots, M$ require that the corresponding i-tractions in the grade-M material body be null. [3]

[2] The requirements on the displacement and traction vectors can be algebraically complementary or not, according to the type of contact one wishes to model; a discussion can be found in [3].

[3] There seems to be little doubt that two materials of different grade belong to the "same" constitutive class, in that writing these transmission conditions is always possible and meaningful. This would not be generally the case for two material bodies of different *microstructure* [6], although there are ways to regard the grade itself as a type of microstructure.

Guided by our understanding of the situation of two material bodies of arbitrary grades in mutual contact, we propose the following variational characterization of equilibria for an *elastic body-environment pair of grade* (N, M): all those deformations f at which the first variation of the functional

$$f \mapsto \Big(\int_{\Omega} \sigma({}_1\mathbf{F}, {}_2\mathbf{F}, \ldots, {}_N\mathbf{F}) dV - \int_{\partial\Omega} \tau({}_0\mathbf{F}, {}_1\mathbf{F}, {}_2\mathbf{F}, \ldots, {}_M\mathbf{F}) dA \Big) \tag{2}$$

takes the value 0. In this functional, the surface integral is meant to approximate the energetic contribution of the environment; the first argument of the density function τ is the zeroth deformation gradient, that is, the position vector with respect to a fixed origin of the typical point of $f(\partial\Omega)$. Of the M *natural boundary conditions* imposed by the stationarity requirement, those for $i = N + 1, \ldots, M$ has significance only for the corresponding live applied tractions, which must all be null; for reasons that will be clearer later, we shall call them *tangency conditions*, whereas we shall give no special name to the remaining N conditions.

Our paper has the following organization. In the next section, after some notational preliminaries, we recapitulate the few elements of differential geometry of surfaces we need to manipulate interaction mappings defined over surfaces; a key tool we introduce in Sect. 2.4 is a decomposition of the surface trace of the first and second deformation gradients into "tangential" and "normal parts". In Sect. 3 we formulate the general variational equilibrium problem for an elastic body-environment pair of grade (N, M), and motivate our discussion by means of some examples for the cases $M = 1$ and $M = 2$. Section 4 contains the central results of our work, the tangency and boundary conditions for body-environment pairs of grade (1,2). Comparison of these results with both the standard case (1,0) and the case (1,1) examined in [2] suggests a number of developments: (i) for the case (1,2), the study of *tangential surface potentials*, whose density depends only on the tangential parts of the first and second deformation gradients (Sect. 6); (ii) the study of the cases (2,2) and (3,2), the former, in particular, for *simple surface potentials* such that the differential order of the associated boundary operator is 3 (Sect. 7); (iii) the case of second-order surface potentials replaceable by *null Lagrangians*, that is, by energetically equivalent second-order volume potentials (Sect. 8). In all these cases, for simplicity, the body-environment boundary is taken to be smooth; in the last section a representative case is treated of a body with a line discontinuity of the boundary normal and hence with special tangency conditions on that line.

2 Preliminaries

Our notation is direct. We let the order of a tensorial quantity be indicated by the font which we use: specifically, we denote fourth-order tensors by Blackboard bold letters (e.g., $\mathbb{T}$); third-order tensors by bold-face uppercase Greek letters; second-order tensors and vectors by bold-face Roman letters, uppercase and lowercase, respectively; functionals and scalar fields by Greek letters, uppercase and lowercase, respectively.

We regard tensorial quantities of order higher than one as linear transformations over spaces of tensors of lower order: $\mathbf{L}[\mathbf{l}]$ and $\Lambda[\mathbf{L}]$ are vectors, while $\Lambda[\mathbf{l}]$ is a second-order tensor; the square-brackets are meant to signal a linear transformation: in particular, according to convenience, we regard third-order tensors as linear transformations either of second-order tensors into vectors or of vectors into second-order tensors.

Simple *juxtaposition* of two second-order tensors $\mathbf{A}$ and $\mathbf{B}$ is defined by $(\mathbf{AB})[\mathbf{l}] := \mathbf{A}[\mathbf{B}[\mathbf{l}]]$ for all vectors $\mathbf{l}$. By a centered dot we denote the *inner product* of two tensorial quantities of the same order: $\mathbf{a} \cdot \mathbf{b}$, $\mathbf{A} \cdot \mathbf{B}$, etc. Juxtaposition with an interposed $\otimes$ signifies *dyadic product*: $\mathbf{a} \otimes \mathbf{b}$ is a second-order tensor, while $\mathbf{a} \otimes \mathbf{A}$, $\mathbf{A} \otimes \mathbf{a}$ and $\mathbf{a} \otimes \mathbf{b} \otimes \mathbf{c}$ are third-order tensors; note that $(\mathbf{a} \otimes \mathbf{B})\mathbf{C} = (\mathbf{B} \cdot \mathbf{C})\mathbf{a}$, $(\mathbf{B} \otimes \mathbf{a})\mathbf{c} = (\mathbf{a} \cdot \mathbf{c})\mathbf{B}$, and $(\mathbf{a} \otimes \mathbf{B})\mathbf{c} = \mathbf{a} \otimes (\mathbf{Bc})$. The appropriate notion of inner product allows us to define *transposition* of tensors: e.g., $\mathbf{L}^{\mathrm{T}} \cdot \mathbf{a} \otimes \mathbf{b} := \mathbf{L} \cdot \mathbf{b} \otimes \mathbf{a}$ for all vectors $\mathbf{a}$ and $\mathbf{b}$; and $\Lambda^{\mathrm{T}} \cdot \mathbf{A} \otimes \mathbf{a} := \Lambda \cdot \mathbf{a} \otimes \mathbf{A}$ for all choices of a vector $\mathbf{a}$ and a tensor $\mathbf{A}$; in particular, $(\mathbf{a} \otimes \mathbf{b})\mathbf{C} = \mathbf{a} \otimes (\mathbf{C}^{\mathrm{T}}\mathbf{b})$. The symmetric and skew-symmetric subspaces of the space of second-order tensors are denoted by Sym and Skw, respectively (if $\mathbf{L} \in \mathrm{Sym}$, then $\mathbf{L} = \mathbf{L}^{\mathrm{T}}$; and if $\mathbf{L} \in \mathrm{Skw}$, then $\mathbf{L} = -\mathbf{L}^{\mathrm{T}}$). Finally, the subspace of third-order tensors satisfying $\Lambda[\mathbf{a} \otimes \mathbf{b}] = \Lambda[\mathbf{b} \otimes \mathbf{a}]$ for all vectors $\mathbf{a}, \mathbf{b}$ is denoted by Σym.

Remark. When components are used, (i) linear action translates into complete contraction with respect to all indices of the quantity on the right: $(\mathbf{L}[\mathbf{l}])^i = L^{ij} l_j$, $(\Lambda[\mathbf{L}])^i = \Lambda^{ijk} L_{jk}$, etc.; (ii) simple juxtaposition implies a single index contraction, e.g., $(\mathbf{AB})^i{}_j = A^{ih} B_{hj}$; (iii) as to inner products, $\mathbf{A} \cdot \mathbf{B} = A^{ij} B_{ij}$, $\Lambda \cdot \Sigma = \Lambda^{ijk} \Sigma_{ijk}$, etc.; (iv) since $(\mathbf{a} \otimes \mathbf{b})\mathbf{c} := (\mathbf{b} \cdot \mathbf{c})\mathbf{a}$, for all vectors $\mathbf{c}$, defines the dyadic product of $(\mathbf{a}, \mathbf{b})$, then $(\mathbf{a} \otimes \mathbf{b})^{ij} = a^i b^j$.

2.1 Gradient and divergence in generalized coordinates

Given a smooth scalar-valued field φ depending on a list of generalized coordinates ζ^i $(i = 1, 2, 3)$, its gradient is the construct

$$\nabla\varphi = \varphi_{,i}\, \mathbf{g}^i, \quad \varphi_{,i} = \frac{\partial \varphi}{\partial \zeta^i}, \tag{3}$$

where $\mathbf{g}^i$ is the ith vector of the contravariant basis. Building on this definition, for the gradient of a tensor field Ψ of any order greater than or equal to 1 we have

$$\nabla\Psi = \Psi_{,i} \otimes \mathbf{g}^i. \tag{4}$$

For $\mathbf{g}_i$ the ith vector of the covariant basis and $\mathbf{G} := \mathbf{g}^i \otimes \mathbf{g}_i$ the metric tensor, the divergence of a vector field $\mathbf{v}$ is defined to be

$$\mathrm{Div}\mathbf{v} = \nabla\mathbf{v} \cdot \mathbf{G} = \mathbf{v}_{,i} \cdot \mathbf{g}^i. \tag{5}$$

Likewise, for Ψ a tensor field of order greater than 1,

$$\mathrm{Div}\Psi = \nabla\Psi[\mathbf{G}] = \Psi_{,i}\,[\mathbf{g}^i]. \tag{6}$$

Finally, by iteration of definition (4), the *second gradient* of a vector field $\mathbf{v}$ is

$$\nabla^{(2)}\mathbf{v} = (\mathbf{v},_i \otimes \mathbf{g}^i),_j \otimes \mathbf{g}^j. \tag{7}$$

By definition, the second gradient must have the following symmetry:

$$(\nabla^{(2)}\mathbf{v})[\mathbf{a} \otimes \mathbf{b}] = (\nabla^{(2)}\mathbf{v})[\mathbf{b} \otimes \mathbf{a}], \text{ for all vectors } \mathbf{a}, \mathbf{b}, \tag{8}$$

that is to say, $\nabla^{(2)}\mathbf{v} \in \Sigma\text{ym}$.

2.2 Elements of surface geometry in normal coordinates

We denote by $\mathbf{n}$ the unit normal at a point of a regular surface $\mathcal{S}$ parametrized by the coordinates ζ^α $(\alpha = 1, 2)$. The typical point in a tubular neighborhood $\mathcal{N}_\mathcal{S}$ of $\mathcal{S}$ is

$$p = \hat{p}(x, \zeta) = x + \zeta\mathbf{n}(x),$$

where x is a typical point of $\mathcal{S}$ and $\zeta \equiv \zeta^3$ takes values in an open interval of 0, possibly depending on x so as to guarantee that the mapping $p \leftrightarrow (x, \zeta)$ is one-to-one. Thus, in $\mathcal{N}_\mathcal{S}$, the covariant bases at, respectively, p and $x = \hat{p}(x, 0)$ are:

$$\mathbf{g}_\alpha := p,_\alpha = \mathbf{e}_\alpha + \zeta\mathbf{n},_\alpha, \quad \mathbf{e}_\alpha := x,_\alpha, \quad \mathbf{g}_3 := p,_3 = \mathbf{n}. \tag{9}$$

As to the contravariant base at p,

$$\mathbf{g}^\alpha := (-1)^\alpha \mathbf{N}\mathbf{g}_{\alpha+1} \text{ (mod2, } \alpha \text{ unsummed)}, \quad \mathbf{g}^3 := \mathbf{n}, \tag{10}$$

where

$$\mathbf{N} := -\mathbf{g}^1 \otimes \mathbf{g}^2 + \mathbf{g}^2 \otimes \mathbf{g}^1; \tag{11}$$

note that, for

$$\nu := \mathbf{g}_1 \times \mathbf{g}_2 \cdot \mathbf{n}, \tag{12}$$

the *volume form* at p, $\widetilde{\mathbf{N}} := \nu\mathbf{N}$ is the skew tensor uniquely associated to the normal unit vector $\mathbf{n}$. Likewise, at x, the contravariant base is defined to be

$$\nu_0\mathbf{e}^\alpha := (-1)^\alpha \mathbf{n} \times \mathbf{e}_{\alpha+1} \text{ (mod2, } \alpha \text{ unsummed)}, \quad \mathbf{e}^3 := \mathbf{n}, \tag{13}$$

where

$$\nu_0 := \mathbf{e}_1 \times \mathbf{e}_2 \cdot \mathbf{n}. \tag{14}$$

It follows from (9), (12) and (14) that

$$\nu = \nu_0(1 - 2\zeta H + \zeta^2 K); \tag{15}$$

the *Weingarten tensor*

$$\mathbf{W} := -\mathbf{n},_\alpha \otimes \mathbf{e}^\alpha \tag{16}$$

summarizes the second-order differential characters of the surface $\mathcal{S}$, in particular, its *mean curvature* $H := \frac{1}{2}\mathrm{tr}\mathbf{W}$ and its *Gaussian curvature* $K := \det \mathbf{W}$.

We now list three consequences of (9)–(16) that will be relevant to some of our later developments. The first is that the surface metric tensor $\mathbf{E} := \mathbf{e}^\alpha \otimes \mathbf{e}_\alpha$ satisfies

$$\mathbf{E} = \mathbf{G}|_{\zeta=0} - \mathbf{n} \otimes \mathbf{n}, \tag{17}$$

the second that

$$\mathbf{g}^\alpha|_{\zeta=0} = \mathbf{e}^\alpha, \quad \mathbf{g}^\alpha,_\beta|_{\zeta=0} = \mathbf{e}^\alpha,_\beta\,. \tag{18}$$

Now, since (9) and (10) imply that $\mathbf{g}^\alpha \cdot \mathbf{g}_\beta = \delta^\alpha_\beta$, we easily find that

$$\mathbf{g}^\alpha,_3 = -(\mathbf{g}^\beta \otimes \mathbf{n},_\beta)\mathbf{g}^\alpha, \tag{19}$$

and hence, by the first part of (18) and (16), that

$$\mathbf{g}^\alpha,_3|_{\zeta=0} = \mathbf{W}^{\mathrm{T}}\mathbf{e}^\alpha. \tag{20}$$

2.3 Surface gradient and surface divergence

Let Ψ be a tensor field of any order greater than or equal to 1, defined in a tubular neighborhood $\mathcal{N}_\mathcal{S}$ of the surface $\mathcal{S}$. From (4) and (9) we have that

$$\nabla\Psi = \Psi,_\alpha \otimes \mathbf{g}^\alpha + \Psi,_3 \otimes \mathbf{n}. \tag{21}$$

This leads us to define the *surface gradient* of Ψ as

$${}^s\nabla\Psi = \Psi,_\alpha \otimes \mathbf{e}^\alpha; \tag{22}$$

here we have made use of $(18)_1$ and, for brevity, we have written $\Psi,_\alpha$ in place of $\Psi,_\alpha|_{\zeta=0}$. Needless to say, (22) can be taken as the formal definition of surface gradient when Ψ is only defined over $\mathcal{S}$; in particular, we see from (16) that

$$\mathbf{W} = -{}^s\nabla\mathbf{n}.$$

As to the *surface divergence* operator, for a vector field we have

$${}^s\mathrm{Div}\mathbf{v} = {}^s\nabla\mathbf{v} \cdot \mathbf{E} = \mathbf{v},_\alpha \cdot \mathbf{e}^\alpha; \tag{23}$$

for a tensor field of order greater than 1,

$${}^s\mathrm{Div}\Psi = {}^s\nabla\Psi[\mathbf{E}] = \Psi,_\alpha[\mathbf{e}^\alpha]. \tag{24}$$

2.4 Tangential and normal parts of gradients

We find it expressive and useful to split the gradient at a point of $\mathcal{S}$ of a vector field $\mathbf{v}$ defined over $\mathcal{N}_{\mathcal{S}}$ into its "tangential" and "normal" parts, defined as follows:

$$\nabla\mathbf{v}|_{\zeta=0} = (\nabla\mathbf{v})_t + (\nabla\mathbf{v})_n, \tag{25}$$

with

$$(\nabla\mathbf{v})_t := {}^s\nabla\mathbf{v}, \tag{26}$$

$$(\nabla\mathbf{v})_n := \partial_{\mathbf{n}}\mathbf{v}\otimes\mathbf{n}, \quad \partial_{\mathbf{n}}\mathbf{v} = (\nabla\mathbf{v}|_{\zeta=0})\mathbf{n} = \mathbf{v},_3|_{\zeta=0}. \tag{27}$$

As to the second gradient of $\mathbf{v}$, relation (7) yields

$$\begin{aligned}\nabla^{(2)}\mathbf{v} = {} & \mathbf{v},_{\alpha\beta}\otimes\mathbf{g}^\alpha\otimes\mathbf{g}^\beta + \mathbf{v},_{\alpha 3}\otimes(\mathbf{g}^\alpha\otimes\mathbf{n}+\mathbf{n}\otimes\mathbf{g}^\alpha) + \mathbf{v},_{33}\otimes\mathbf{n}\otimes\mathbf{n} \\ & + \mathbf{v},_\alpha\otimes(\mathbf{g}^\alpha,_\beta\otimes\mathbf{g}^\beta + \mathbf{g}^\alpha,_3\otimes\mathbf{n}) + \mathbf{v},_3\otimes\mathbf{n},_\alpha\otimes\mathbf{g}^\alpha.\end{aligned} \tag{28}$$

But, due to (19),

$$\mathbf{v},_\alpha\otimes\mathbf{g}^\alpha,_3\otimes\mathbf{n} = -\mathbf{v},_\alpha\otimes(\mathbf{g}^\alpha\cdot\mathbf{n},_\beta)\mathbf{g}^\beta\otimes\mathbf{n} = -(\mathbf{v},_\alpha\otimes\mathbf{g}^\alpha)(\mathbf{n},_\beta\otimes\mathbf{g}^\beta)\otimes\mathbf{n},$$

whence, with (16),

$$(\mathbf{v},_\alpha\otimes\mathbf{g}^\alpha,_3\otimes\mathbf{n})|_{\zeta=0} = ({}^s\nabla\mathbf{v})\mathbf{W}\otimes\mathbf{n}. \tag{29}$$

In view of (18) and (29), relation (28) implies that

$$\begin{aligned}\nabla^{(2)}\mathbf{v}|_{\zeta=0} = {} & {}^s\nabla^{(2)}\mathbf{v} + ({}^s\nabla\mathbf{v})\mathbf{W}\otimes\mathbf{n} + (\partial_{\mathbf{n}}\mathbf{v}),_\alpha\otimes(\mathbf{e}^\alpha\otimes\mathbf{n}+\mathbf{n}\otimes\mathbf{e}^\alpha) \\ & - \partial_{\mathbf{n}}\mathbf{v}\otimes\mathbf{W} + \partial^{(2)}_{\mathbf{n}}\mathbf{v}\otimes\mathbf{n}\otimes\mathbf{n}, \quad {}^s\nabla^{(2)}\mathbf{v} = (\mathbf{v},_\alpha\otimes\mathbf{e}^\alpha),_\beta\otimes\mathbf{e}^\beta.\end{aligned} \tag{30}$$

Just as we did for the first, we split the second gradient at a point of $\mathcal{S}$ into its tangential and normal parts:

$$\nabla^{(2)}\mathbf{v}|_{\zeta=0} = (\nabla^{(2)}\mathbf{v})_t + (\nabla^{(2)}\mathbf{v})_n, \tag{31}$$

where

$$(\nabla^{(2)}\mathbf{v})_t := {}^s\nabla^{(2)}\mathbf{v} + ({}^s\nabla\mathbf{v})\mathbf{W}\otimes\mathbf{n}, \tag{32}$$

$$(\nabla^{(2)}\mathbf{v})_n := (\partial_{\mathbf{n}}\mathbf{v}),_\alpha\otimes(\mathbf{e}^\alpha\otimes\mathbf{n}+\mathbf{n}\otimes\mathbf{e}^\alpha) - \partial_{\mathbf{n}}\mathbf{v}\otimes\mathbf{W} + \partial^{(2)}_{\mathbf{n}}\mathbf{v}\otimes\mathbf{n}\otimes\mathbf{n}. \tag{33}$$

Relations (25)–(27) and (31)–(33) will play a central role in our study of second-order surface potentials.

2.5 Surface-divergence identities

Let the curve Γ be the (possibly empty) smooth boundary of the surface $\mathcal{S}$, with Γ so oriented as to leave $\mathcal{S}$ on the left, and let $\mathbf{t}$ denote the unit tangent to Γ. Then, for $\mathbf{a}$ a vector field over $\mathcal{S}$, the following *surface-divergence identity* holds:

$$\int_{\mathcal{S}} {}^s\mathrm{Div}\mathbf{a}\, dA = -\int_{\mathcal{S}} 2H(\mathbf{a}\cdot\mathbf{n})dA + \int_{\Gamma} \mathbf{a}\cdot\mathbf{t}\times\mathbf{n}\, ds. \tag{34}$$

In the right-hand side of this relation, the first integral vanishes whenever $\mathbf{a}$ is *tangential* (i.e., $\mathbf{a}\cdot\mathbf{n}\equiv 0$ over $\mathcal{S}$), the second whenever $\mathcal{S}$ is closed. In the former instance, (34) implies that

$$\int_{\mathcal{S}} \mathbf{A}\cdot{}^s\nabla\mathbf{v} = -\int_{\mathcal{S}} \mathbf{v}\cdot{}^s\mathrm{Div}\mathbf{A}\, dA + \int_{\Gamma} \mathbf{v}\cdot\mathbf{A}(\mathbf{t}\times\mathbf{n})ds; \tag{35}$$

in the latter case, that

$$\int_{\mathcal{S}} \mathbf{A}\cdot{}^s\nabla\mathbf{v} = -\int_{\mathcal{S}} \mathbf{v}\cdot({}^s\mathrm{Div}\mathbf{A} + 2H\mathbf{A}\mathbf{n})dA. \tag{36}$$

We shall make use of these last two relations later on.

3 Variational formulation of equilibrium problems

When for simplicity distance interactions are ignored, the energy

$$E\{f\} = \Sigma\{f\} - T\{f\} \tag{37}$$

associated to the deformation f of an elastic body from its reference placement Ω consists of the *total stored energy*

$$\Sigma\{f\} = \int_{\Omega} \sigma(x, \mathsf{F}(x))dV, \tag{38}$$

and the *surface-interaction potential*

$$T\{f\} = \int_{\partial\Omega} \tau(x, \mathsf{G}(x))dA, \tag{39}$$

where $\mathsf{F}(x)$ and $\mathsf{G}(x)$ are two lists of values of successive deformation gradients, evaluated at the point x. Formally, the energy functional is required to be stationary at elastic equilibria:

$$\delta E\{f\}[\mathbf{v}] = 0 \quad \text{for all } \mathbf{v}\in\mathcal{V}, \tag{40}$$

where $\mathcal{V}$ denotes the space of all admissible variations.

Some hypotheses are needed to make this stationarity requirement precise; here we do not list the most general ones, but rather those that allow for an easy and quick exemplification of the type of results we are seeking. First, for the time being, we take Ω to be a connected, open and bounded subset of $\mathbb{R}^3$, whose C^3-boundary $\partial\Omega$ is a globally regular surface without self-intersections and with everywhere well-defined outer normal $\mathbf{n}$ (more general boundaries will be considered in Sect. 8). Under these assumptions, the surface $\partial\Omega$ has a *tubular neighborhood* $\mathcal{N}_{\partial\Omega}$, each point p of which is uniquely associated with a triplet $(\zeta^1, \zeta^2, \zeta)$, where ζ^α $(\alpha = 1, 2)$ are the generalized coordinates of the minimal-distance point x on $\partial\Omega$, and where $\zeta = (p - x) \cdot \mathbf{n}(x)$ is the signed distance of p from $\partial\Omega$ (cf. Sect. 2.2). Secondly, we confine attention to deformations f of class $C^4(\overline{\Omega})$; and we assume that the constitutive mappings $\sigma(x, \cdot)$ and $\tau(x, \cdot)$ are smooth in $\overline{\Omega}$. Finally, we choose $\mathcal{V}$ to be C^∞, implicitly restricting attention to Neumann-type conditions over the entire boundary.

In this paper, we consider two assignments of the stored-energy mapping $\sigma(x, \mathsf{F})$, namely, the classical case when the list F reduces to a single second-order tensor $\mathbf{F}$, and $\mathsf{F}(x) = \nabla f(x)$; the case of *second-grade elastic materials*, when $\mathsf{F} = (\mathbf{F}, \mathbf{\Phi})$ and $\mathsf{F}(x) = (\nabla f(x), \nabla^{(2)} f(x))$; and, very briefly, the case of *third-grade elastic materials*, when $\mathsf{F} = (\mathbf{F}, \mathbf{\Phi}, \mathbb{F})$ and $\mathsf{F}(x) = (\nabla f(x), \nabla^{(2)} f(x), \nabla^{(3)} f(x))$. As to the surface-interaction mapping, we take $\mathsf{G} = (\mathbf{f}, \mathbf{F}, \mathbf{\Phi})$, so that

$$\tau(x, \mathsf{G}) = \tau(x, \mathbf{f}, \mathbf{F}, \mathbf{\Phi}) \tag{41}$$

and

$$T\{f\} = \int_{\partial\Omega} \tau(x, f(x) - o, \nabla f(x), \nabla^{(2)} f(x)) dA.^4 \tag{42}$$

For each fixed $x \in \Omega$ and, respectively, for each fixed $x \in \partial\Omega$, the constitutive mappings $\sigma(x, \cdot)$ and $\tau(x, \cdot)$ are considered to be defined for all vectors $\mathbf{f}$, all second-order tensors $\mathbf{F}$ with positive determinant and all third-order tensors in Σym.

Remark. When the surface interaction is *dead*, that is, shape-independent, its density is linear in the deformation:

$$\tau(x, f(x) - o) = \mathbf{s}_o(x) \cdot (f(x) - o),$$

with $\mathbf{s}_o$ the (dead) surface load. Such a simple model cannot account for the shape-dependent, *live* surface interactions encountered in the cases of *a body wrapped by an elastic membrane*, of *a body immersed in a fluid* of *a membrane and the frame over which the membrane is stretched.* The first two examples motivated the study of live loads with first-order surface potentials [2, 1]; the third example requires that surface potentials also depending on the second gradient of deformation are treated, the subject of our present study.

---

[4] Here the point o is arbitrarily fixed once and for all. Recall the notation used in the Introduction in connection with grade-N materials and note that hereafter we write $\mathbf{f}$, $\mathbf{F}$, $\mathbf{\Phi}$, and $\mathbb{F}$, for the gradients ${}_0\mathbf{F}$, ${}_1\mathbf{F}$, ${}_2\mathbf{F}$, and ${}_3\mathbf{F}$, respectively.

4 Elastic body-environment pairs of grade (1,2)

In this section we consider standard elastic materials, whose total stored energy is

$$\Sigma\{f\} = \int_{\Omega} \sigma(x, \nabla f(x)) dV, \tag{43}$$

and take the surface-interaction potential as in (42). This is the basic case we study; variants, of both lesser and greater generality, will be considered later on.

4.1 Field equation and tangency conditions

Under the present assumptions, we deduce from the stationarity requirement (40) the following Euler–Lagrange equation:

$$\int_{\Omega} \sigma_{\mathbf{F}} \cdot \nabla \mathbf{v} dV = \int_{\partial\Omega} (\tau_{\mathbf{f}} \cdot \mathbf{v} + \tau_{\mathbf{F}} \cdot \nabla \mathbf{v} + \tau_{\Phi} \cdot \nabla^{(2)} \mathbf{v}) dA, \tag{44}$$

to hold for all $\mathbf{v} \in \mathcal{V}$.

Now, as is well-known, an application of the divergence theorem to the right-hand side yields

$$\int_{\Omega} \sigma_{\mathbf{F}} \cdot \nabla \mathbf{v} dV = -\int_{\Omega} \text{Div} \sigma_{\mathbf{F}} \cdot \mathbf{v} dV + \int_{\partial\Omega} \sigma_{\mathbf{F}}[\mathbf{n}] \cdot \mathbf{v} dA, \tag{45}$$

which implies the standard field equation

$$\text{Div} \sigma_{\mathbf{F}} (x, \nabla f(x)) = 0; \quad x \in \Omega. \tag{46}$$

On the other hand, with the use of the decomposition (25) for the first gradient and the identity (36), we find that

$$\int_{\partial\Omega} \tau_{\mathbf{F}} \cdot \nabla \mathbf{v} dA = \int_{\partial\Omega} \Big(-({}^s\text{Div} \tau_{\mathbf{F}} + 2H \tau_{\mathbf{F}}[\mathbf{n}]) \cdot \mathbf{v} + \tau_{\mathbf{F}}[\mathbf{n}] \cdot \partial_{\mathbf{n}} \mathbf{v} \Big) dA.^5 \tag{47}$$

Similarly, with the use of the decomposition (31)–(33) for the second gradient, we arrive at

$$\begin{aligned} \int_{\partial\Omega} \tau_{\Phi} \cdot \nabla^{(2)} \mathbf{v} dA = \int_{\partial\Omega} \Big(& \tau_{\Phi} \cdot {}^s\nabla^{(2)} \mathbf{v} - (\tau_{\Phi}[\mathbf{n}]^s \nabla \mathbf{n}) \cdot {}^s\nabla \mathbf{v} \\ & + \big(-2^s\text{Div}(\tau_{\Phi}[\mathbf{n}]) - 4H \tau_{\Phi}[\mathbf{n} \otimes \mathbf{n}] + \tau_{\Phi}[{}^s\nabla \mathbf{n}]\big) \cdot \partial_{\mathbf{n}} \mathbf{v} + \tau_{\Phi}[\mathbf{n} \otimes \mathbf{n}] \cdot \partial_{\mathbf{n}}^{(2)} \mathbf{v} \Big) dA. \end{aligned} \tag{48}$$

Since both the second and first normal derivatives of the variation $\mathbf{v}$ are arbitrary, relations (44)–(48) yield two *tangency conditions*:

$$\mathbf{0} = \tau_{\Phi}[\mathbf{n} \otimes \mathbf{n}], \tag{49}$$

$$\mathbf{0} = \tau_{\mathbf{F}}[\mathbf{n}] - 2^s\text{Div}(\tau_{\Phi}[\mathbf{n}]) + \tau_{\Phi}[{}^s\nabla \mathbf{n}], \tag{50}$$

which are to hold at each point $x \in \partial\Omega$.

[5] Here we have made use of the fact that the third-order tensor $\tau_{\Phi} \in \Sigma\text{ym}$.

Remark. In view of the identity

$$({}^s\mathrm{Div}\tau_\Phi)[\mathbf{n}] = {}^s\mathrm{Div}(\tau_\Phi[\mathbf{n}]) - \tau_\Phi[{}^s\nabla\mathbf{n}], \tag{51}$$

the second tangency condition can be given the alternative form

$$\mathbf{0} = \tau_\mathbf{F}[\mathbf{n}] - 2({}^s\mathrm{Div}\tau_\Phi)[\mathbf{n}] - \tau_\Phi[{}^s\nabla\mathbf{n}]. \tag{52}$$

4.2 Natural boundary condition

What remains of the stationarity condition (44) is

$$\begin{aligned}\int_{\partial\Omega}\sigma_\mathbf{F}[\mathbf{n}]\cdot\mathbf{v}dA = \int_{\partial\Omega}\Big(&(\tau_\mathbf{f} - {}^s\mathrm{Div}\tau_\mathbf{F} - 2H\tau_\mathbf{F}[\mathbf{n}])\cdot\mathbf{v}\\ &- (\tau_\Phi[\mathbf{n}]{}^s\nabla\mathbf{n})\cdot{}^s\nabla\mathbf{v} + \tau_\Phi\cdot{}^s\nabla^{(2)}\mathbf{v}\Big)dA = 0,\end{aligned} \tag{53}$$

to hold for all $\mathbf{v}\in\mathcal{V}$. With the use of (36), since the tensor $\tau_\mathcal{F}[\mathbf{n}]{}^s\nabla\mathbf{n}$ is tangential because ${}^s\nabla\mathbf{n}$ is, we find that

$$-\int_{\partial\Omega}(\tau_\Phi[\mathbf{n}]{}^s\nabla\mathbf{n})\cdot{}^s\nabla\mathbf{v}dA = \int_{\partial\Omega}{}^s\mathrm{Div}(\tau_\Phi[\mathbf{n}]{}^s\nabla\mathbf{n})\cdot\mathbf{v}dA; \tag{54}$$

moreover, by (34), (36), (49) and (51), we also find that

$$\begin{aligned}\int_{\partial\Omega}\tau_\Phi\cdot{}^s\nabla^{(2)}\mathbf{v}dA = \int_{\partial\Omega}\Big(&2H(2({}^s\mathrm{Div}\tau_\Phi)[\mathbf{n}] + \tau_\Phi[{}^s\nabla\mathbf{n}])\\ &+ 2\tau_\Phi[{}^s\nabla H\otimes\mathbf{n}] + {}^s\mathrm{Div}({}^s\mathrm{Div}\tau_\Phi)\Big)\cdot\mathbf{v}dA.\end{aligned} \tag{55}$$

Thus, after localization and with the use of the second tangency condition in the form (52), relations (53)–(55) yield the *natural boundary condition*

$$\sigma_\mathbf{F}[\mathbf{n}] = \tau_\mathbf{f} - {}^s\mathrm{Div}\tau_\mathbf{F} + {}^s\mathrm{Div}(\tau_\Phi[\mathbf{n}]{}^s\nabla\mathbf{n}) + 2\tau_\Phi[{}^s\nabla H\otimes\mathbf{n}] + {}^s\mathrm{Div}({}^s\mathrm{Div}\tau_\Phi), \tag{56}$$

to hold at each point of $\partial\Omega$; this condition prescribes that, at the boundary, the *stress vector* $\mathbf{Sn}$ *equals the surface-load vector* $\mathbf{s}_o$:

$$\mathbf{Sn} = \mathbf{s}_o, \tag{57}$$

with $\mathbf{S} = \sigma_\mathbf{F}$ the Piola *stress tensor* and with

$$\mathbf{s}_o = \tau_\mathbf{f} - {}^s\mathrm{Div}\tau_\mathbf{F} + {}^s\mathrm{Div}(\tau_\Phi[\mathbf{n}]{}^s\nabla\mathbf{n}) + 2\tau_\Phi[{}^s\nabla H\otimes\mathbf{n}] + {}^s\mathrm{Div}({}^s\mathrm{Div}\tau_\Phi). \tag{58}$$

5 Lower-order surface potentials

For first-order surface interactions, *ceteris paribus*, the tangency condition (49) evaporates, while (50), the other tangency condition, reduces to

$$\mathbf{0} = \tau_{\mathbf{F}}[\mathbf{n}]; \tag{59}$$

the boundary condition (56) reduces to

$$\sigma_{\mathbf{F}}[\mathbf{n}] = \tau_{\mathbf{f}} - {}^{s}\mathrm{Div}\tau_{\mathbf{F}}. \tag{60}$$

The tangency condition (59) is satisfied by the surface-interaction mappings appropriate for a body immersed in a fluid (*pressure loading*) or wrapped by an elastic membrane (*membrane loading*) [2]. In each of these cases the surface potential and the associated surface-load vector

$$\mathbf{s}_{\mathrm{o}} = \tau_{\mathbf{f}} - {}^{s}\mathrm{Div}\tau_{\mathbf{F}}$$

are called *tangential*, because they depend on tangential derivatives only. These derivatives are of the first order in the former case, of the second order in the latter; thus, while pressure loading is *simple* in the sense of Spector [7, 8, 2], membrane loading is not. In addition, pressure loading – but not membrane loading – has an associated first-order *null Lagrangian*, i.e., a volume potential

$$N\{f\} = \int_{\Omega} \nu(f(x) - o, \nabla f(x))dV$$

that can effectively replace the (tangential and simple) surface potential

$$T_N\{f\} = \int_{\partial\Omega} \tau(\mathbf{n}(x), f(x) - o, {}^{s}\nabla f(x))dA$$

because

$$T_N\{f\} = N\{f\} \tag{61}$$

for all admissible deformations f (see [2] and the literature quoted there).[6] We shall study tangential surface interactions of the second-order in the next section, and surface interactions associated to second-order null Lagrangians in Sect. 8.

[6] For pressure loading, it can be shown [2] that there are two constitutive mappings, the scalar-valued *pressure function* $\pi = \hat{\pi}(\mathbf{f})$ and the vector-valued function

$$\mathbf{w}(\mathbf{f}, \mathbf{F}) = \bar{\pi}(\mathbf{f})(\mathbf{F}^*)^{\mathrm{T}}\mathbf{f}, \quad \bar{\pi}(\mathbf{v}) = \int_0^1 \hat{\pi}(\gamma\mathbf{v})\gamma^2 d\gamma$$

(here $\mathbf{F}^*$ is the cofactor of $\mathbf{F}$), in terms of which the densities of the associated volume and surface potentials are, respectively,

$$\nu(x) = \mathrm{Div}\mathbf{w}(f(x) - x, \nabla f(x)), \quad x \in \Omega,$$

A remarkable feature of the boundary condition (56) which we derived for general second-order surface interactions is that the geometrical properties of the boundary now enter that condition explicitly: they did not in either the case of dead (zeroth order) interactions or the case of live interactions of the first order [2].

Another remarkable feature of condition (56) is the following. In the case of dead loads, the stored-energy mapping σ alone determines both the field operator

$$\mathbf{f} := -\mathrm{Div}\mathbf{S}$$

and the boundary operator

$$\mathbf{b} := \mathbf{S}\mathbf{n} - \mathbf{s}_o,$$

the former turning out to be of the second order, the latter of the first. When the surface load is live, the relative surface-interaction mapping τ contributes to determine the boundary operator, in a manner that becomes decisive as its order increases: the classical variational circumstance that the order of the field operator is twice the order of the boundary operator is preserved for first-order surface interactions only when they are simple. In the case of second-order interactions, the boundary operator $\mathbf{b}$ is of order 3 or 4, according to whether the highest-order term coming from ${}^s\mathrm{Div}({}^s\mathrm{Div}\tau_\Phi)$ is present in (56) or not; in any case, the principal part of $\mathbf{b}$ is determined by the surface-interaction mapping τ alone. This last feature changes for elastic materials of grade higher than 1, a case we study briefly in Sect. 7, where we also introduce a notion of simplicity for second-order surface interactions that is modelled on the corresponding notion for first-order interactions.

6 Tangential surface potentials

Given a region Ω with boundary $\partial\Omega$ and a mapping τ as in (41), we call the resulting surface-interaction potential of type (42) *tangential* for Ω if

$$\int_{\partial\Omega} \tau(x, f(x) - o, \nabla f(x), \nabla^{(2)} f(x)) dA$$
$$= \int_{\partial\Omega} \tau(x, f(x) - o, (\nabla f)_t(x), (\nabla^{(2)} f)_t(x)) dA \quad (62)$$

for all deformations f.

---

and

$$\tau(x) = \mathbf{w}(f(x) - x, \nabla f(x)) \cdot \mathbf{n}(x), \quad x \in \partial\Omega.$$

The divergence lemma and the well-known fact that $(\nabla f)^*$ depends only on tangential derivatives imply that condition (61) is satisfied. Note that

$$\nu(\mathbf{f}, \mathbf{F}) = \mathbf{w}_\mathbf{f}(\mathbf{f}, \mathbf{F}) \cdot \mathbf{F}^T = \bar{\pi}(\mathbf{f}) \det \mathbf{F};$$

hence, in general, we cannot expect volume densities of null Lagrangians to be translation-invariant.

Proposition 1. *A surface-interaction mapping τ yields a tangential potential for a body Ω if and only if it satisfies the tangency conditions (49)–(50) at each point of $\partial\Omega$.*

Proof. It is enough to show that

$$\int_{\partial\Omega}(\tau_{\mathbf{F}}\cdot\nabla\mathbf{v}+\tau_{\Phi}\cdot\nabla^{(2)}\mathbf{v})dA=\int_{\partial\Omega}(\tau_{\mathbf{F}}\cdot(\nabla\mathbf{v})_t+\tau_{\Phi}\cdot(\nabla^{(2)}\mathbf{v})_t)dA,$$

for each deformation f and for all variations $\mathbf{v}$; or rather, equivalently in view of relations (25) and (31), to show that

$$\int_{\partial\Omega}(\tau_{\mathbf{F}}\cdot(\nabla\mathbf{v})_n+\tau_{\Phi}\cdot(\nabla^{(2)}\mathbf{v})_n)dA=0, \tag{63}$$

if (49)–(50) hold. Substitution of (27) and (33) into (63), and the arbitrariness in the choice of $\partial_{\mathbf{n}}\mathbf{v}$ and $\partial_{\mathbf{n}}^{(2)}\mathbf{v}$, immediately yield both (49) and (50). □

Tangential potentials are mathematically convenient, in that they give rise to boundary-value problems where the only natural boundary condition has the clear physical interpretation of equating the stress and load vectors (cf. (57)).

Remark. Whenever

$$\tau_{\Phi}[\mathbf{n}]=\mathbf{0}, \tag{64}$$

condition (49) is satisfied and the only remaining tangency condition is

$$\mathbf{0}=\tau_{\mathbf{F}}[\mathbf{n}]+\tau_{\Phi}[{}^{s}\nabla\mathbf{n}]. \tag{65}$$

7 Higher-grade material bodies and simple surface potentials

We now choose a grade-2 elastic material with stored-energy density $\sigma=\sigma(x,\mathbf{F},\Phi)$, and accordingly write (38) as

$$\Sigma\{f\}=\int_{\Omega}\sigma(x,\nabla f(x),\nabla^{(2)}f(x))dV. \tag{66}$$

The consequential changes in the stationarity analysis of Sect. 4 are:

- the field equation (46) becomes

$$\mathrm{Div}\widetilde{\mathbf{S}}=\mathbf{0}, \tag{67}$$

  with

$$\widetilde{\mathbf{S}}:=\sigma_{\mathbf{F}}-\mathrm{Div}\sigma_{\Phi}; \tag{68}$$

– the tangency condition (50) now becomes the boundary condition for the 2-stress $\sigma_{2\mathbf{F}} \equiv \sigma_\Phi$ (cf. (1)):

$$\sigma_\Phi[\mathbf{n}\otimes\mathbf{n}] = \tau_\mathbf{F}[\mathbf{n}] - 2^s\mathrm{Div}(\tau_\Phi[\mathbf{n}]) + \tau_\Phi[{}^s\nabla\mathbf{n}]; \tag{69}$$

– the boundary condition (56) becomes

$$\begin{aligned}\widetilde{\mathbf{S}}\mathbf{n} = \tau_\mathbf{f} - {}^s\mathrm{Div}\tau_\mathbf{F} + {}^s\mathrm{Div}(\tau_\Phi[\mathbf{n}]^s\nabla\mathbf{n}) + 2\tau_\Phi[{}^s\nabla H\otimes\mathbf{n}]\\ + {}^s\mathrm{Div}({}^s\mathrm{Div}\tau_\Phi) + {}^s\mathrm{Div}(\sigma_\Phi[\mathbf{n}]).\end{aligned} \tag{70}$$

The tangency condition (49) does not change. Inspecting (67)–(68), we see that the field operator

$$\widetilde{\mathbf{f}} := -\mathrm{Div}(\sigma_\mathbf{F} - \mathrm{Div}\sigma_\Phi) \tag{71}$$

now has order 4. As anticipated at the end of Sect. 5, the boundary operator

$$\widetilde{\mathbf{b}} := \widetilde{\mathbf{S}}\mathbf{n} - \mathbf{s}_o \tag{72}$$

has order 3 or 4, according as the fourth-order term coming from ${}^s\mathrm{Div}({}^s\mathrm{Div}\tau_\Phi)$ occurs in $\mathbf{s}_o$ or not. In the latter alternative, borrowing the terminology used by Spector for first-order potentials, we call the corresponding second-order surface potential *simple*. Our next proposition provides an easy algebraic characterization for a class of densities of simple second-order potentials.

Proposition 2. *A second-order surface potential of density* $\tau(x, \mathbf{f}, \mathbf{F}, \Phi)$ *is simple if, for*

$$\mathbb{T}^{(pq)}_{ijhk} := \tau_{\Phi^p_{ij}\Phi^q_{hk}}, \tag{73}$$

the conditions

$$\mathbb{T}^{(pq)}_{ijhh} + 2\mathbb{T}^{(pq)}_{ihjh} + 2\mathbb{T}^{(pq)}_{hjih} + \mathbb{T}^{(pq)}_{hhij} = 0 \tag{74}$$

are satisfied.

Proof. The term in $\mathbf{s}_o$ which depends on the fourth-order derivatives of the deformation is

$$\tau_{\Phi^p_{ij}\Phi^q_{hk}}\, f^q{}_{,hkij} = \mathbb{T}^{(pq)}_{ijhk}\, f^q{}_{,hkij}\,. \tag{75}$$

Each of the fourth-order tensors $\mathbb{T}^{(pq)}$ enjoys the following minor symmetries with respect to the differentiation indices:

$$\mathbb{T}^{(pq)}_{ijhk} = \mathbb{T}^{(pq)}_{jihk} = \mathbb{T}^{(pq)}_{ijkh}\,. \tag{76}$$

Consequently, all fourth-order derivatives disappear if τ is such that conditions (74) are satisfied. □

Remark. On choosing $\sigma = \sigma(x, \mathbf{F}, \Phi, \mathbb{F})$, that is to say, on choosing a grade-3 elastic material, the tangency condition (49) takes, as expected, the form of the boundary condition for the 3-stress $\sigma_{\mathbb{F}}$, namely,

$$\sigma_{\mathbb{F}}[\mathbf{n} \otimes \mathbf{n} \otimes \mathbf{n}] = \tau_{\Phi}[\mathbf{n} \otimes \mathbf{n}]. \tag{77}$$

We leave it to the reader to derive the accompanying changes in the other boundary conditions and in the field operator; note that the elastic body-environment pair we just considered is of grade (3,2). In view of (77), we may say that, in some sense, a second-order surface potential mimics an environmental body made of an elastic material of grade 3.

8 Second-order null Lagrangians

A *second-order null Lagrangian* is a volume potential $N\{f\}$ of density $\nu(\mathbf{f}, \mathbf{F}, \Phi)$ that can be effectively replaced by a surface-interaction potential $T_N\{f\}$ of type (42), in the sense that N and T_N satisfy condition (61), namely,

$$\begin{aligned} N\{f\} &= \int_{\Omega} \nu(f(x) - o, \nabla f(x), \nabla^{(2)} f(x))dV \\ &= \int_{\partial\Omega} \tau(\mathbf{n}(x), f(x) - o, \nabla f(x), \nabla^{(2)} f(x))dA = T_N\{f\} \end{aligned} \tag{78}$$

for all admissible deformations f.

Proposition 3. *Let* $\mathbf{w}(\mathbf{f}, \mathbf{F}, \Phi)$ *be a vector-valued mapping such that the fourth-order tensor*

$$\mathbb{W} := \mathbf{w}_{\Phi}, \quad \mathbb{W}^{ij}_{\ \ hk} = (w^i)_{\Phi^j_{hk}} \tag{79}$$

satisfies, for each fixed value of the index j*, the skew-symmetry condition*

$$\mathbb{W}^{ij}_{\ \ hk} + \mathbb{W}^{kj}_{\ \ ih} + \mathbb{W}^{hj}_{\ \ ki} = 0. \tag{80}$$

Then

$$\nu(\mathbf{f}, \mathbf{F}, \Phi) = \mathbf{w}_{\mathbf{f}}(\mathbf{f}, \mathbf{F}, \Phi) \cdot \mathbf{F}^{\mathrm{T}} + \mathbf{w}_{\mathbf{F}}(\mathbf{f}, \mathbf{F}, \Phi) \cdot \Phi^{\mathrm{T}} \tag{81}$$

is a volume density of a second-order null Lagrangian whose associated surface potential has density

$$\tau(\mathbf{n}, \mathbf{f}, \mathbf{F}, \Phi) = \mathbf{w}(\mathbf{f}, \mathbf{F}, \Phi) \cdot \mathbf{n}. \tag{82}$$

Proof. The definition of $\mathbb{W}$ and the fact that $\Phi \in \Sigma\mathrm{ym}$ imply that

$$\mathbb{W}^{ij}_{\ \ hk} = \mathbb{W}^{ij}_{\ \ kh}, \tag{83}$$

for all fixed values of indices i, j.[7] It follows from (80) and (83) that

$$\nu(f(x)-o, \nabla f(x), \nabla^{(2)} f(x)) = \mathrm{Div}\mathbf{w}(f(x)-o, \nabla f(x), \nabla^{(2)} f(x)), \quad (84)$$

for all admissible deformations f. With this, the divergence lemma and (82), it is easy to conclude that (78) holds. □

Remark. The surface density (82) satisfies the tangency condition (49) automatically; in fact, since

$$\tau_\Phi = \mathbf{w}_\Phi[\mathbf{n}] = \mathbb{W}[\mathbf{n}], \quad (85)$$

that condition now reads

$$\mathbf{0} = \mathbb{W}[\mathbf{n} \otimes \mathbf{n} \otimes \mathbf{n}], \quad (86)$$

a consequence of (80).

9 Bodies with line discontinuities of the boundary normal

In this section we consider bodies whose reference configuration Ω has a boundary $\partial\Omega = \mathcal{S}^{(1)} \cup \mathcal{S}^{(2)} \cup \Gamma$, with $\mathcal{S}^{(1)}$ and $\mathcal{S}^{(2)}$ two smooth surfaces with an empty intersection, whose common boundary is the closed, simple and smooth oriented curve Γ. We treat the case when, at Γ, the normal vector $\mathbf{n}$ jumps ($\mathbf{n}^{(1)} \neq \mathbf{n}^{(2)}$) but the vector $\mathbf{n}^{(1)} \times \mathbf{n}^{(2)}$ is not null: thus, for us, Γ is never a cusp. We assume that both surfaces can be extended past their support curve Γ as two smooth surfaces that, for convenience, we continue to indicate by $\mathcal{S}^{(\alpha)}$. To handle line discontinuities of the boundary normal, we consider two surface-interaction mappings, one on each of the surfaces $\mathcal{S}^{(\alpha)}$, possibly depending in an explicit manner on the normal-vector field:

$$\tau^{(\alpha)} = \tau^{(\alpha)}(x, \mathbf{n}^{(\alpha)}, \mathbf{f}, \mathbf{F}, \Phi); \quad (87)$$

we also introduce for later use the unit vectors

$$\mathbf{m}^{(\alpha)} := \mathbf{t} \times \mathbf{n}^{(\alpha)}, \quad (88)$$

where $\mathbf{t}$ is the tangent to Γ.

Under the present circumstances, in addition to conditions on the surfaces $\mathcal{S}^{(\alpha)}$ having the same form as the tangency conditions (49)–(50) and the boundary condition (56), the stationarity condition (40) yields certain conditions to be satisfied on the discontinuity line Γ, namely,

$$\tau_\Phi^{(\alpha)}[\mathbf{n}^{(\alpha)} \otimes \mathbf{n}^{(\alpha)}] = \tau_\Phi^{(\alpha)}[\mathbf{n}^{(1)} \otimes \mathbf{n}^{(2)}] = \mathbf{0} \quad (\alpha = 1, 2) \quad \text{on } \Gamma \quad (89)$$

[7] Combination of (80) and (83) implies that the number of independent components of the tensor $\mathbb{W}$ is 24.

(with the last parts of (89) obtained by the use of Stokes theorem) and

$$\tau_{\mathbf{F}}^{(1)}[\mathbf{n}^{(2)}] + \tau_{\mathbf{F}}^{(2)}[\mathbf{n}^{(1)}] = \mathbf{0} \quad \text{on } \Gamma. \tag{90}$$

A further equilibrium condition follows from an application of the surface divergence theorem; rather than recording its complicated general form, we prefer to exemplify it in a simple but significant particular case.

To be precise, we consider surface-interaction densities (87) of the form

$$\tau^{(\alpha)} = \tau(x, a^{(\alpha)}), \tag{91}$$

where $a^{(\alpha)}$ denotes the *surface Laplacian* of f

$$^s\Delta f := {}^s\mathrm{Div}({}^s\nabla f), \tag{92}$$

evaluated at a point of $\mathcal{S}^{(\alpha)}$. Note that it follows from (92), (30) and (32) that

$$^s\Delta f = (\nabla^{(2)} f)_t[\mathbf{E}];$$

hence a surface potential of density (91) is tangential, in the sense of Sect. 6. It also follows from (92) and (32) that

$$^s\Delta f = (\nabla^{(2)} f)|_{\zeta=0}[\mathbf{E}] + 2H\nabla f|_{\zeta=0}[\mathbf{n}],$$

a relation which allows us to write (91) in the form

$$\tau^{(\alpha)} = \tau(x, \Phi[\mathbf{E}^{(\alpha)}] + 2H\mathbf{F}[\mathbf{n}^{(\alpha)}]), \tag{93}$$

a special instance of the general form (87). In particular, it follows from (93) that

$$\tau_{\Phi}^{(\alpha)} = \tau_a \otimes \mathbf{E}^{(\alpha)}, \quad \tau_{\mathbf{F}}^{(\alpha)} = 2H\tau_a \otimes \mathbf{n}^{(\alpha)}, \tag{94}$$

whence it is easy to deduce that, for these surface-interaction mappings, conditions (89) are trivially satisfied, whereas condition (90) reduces to

$$(\mathbf{n}^{(1)} \cdot \mathbf{n}^{(2)})(\tau_a^{(1)} + \tau_a^{(2)}) = \mathbf{0} \quad \text{on } \Gamma. \tag{95}$$

The further equilibrium condition we promised to exemplify is

$$\partial_{\mathbf{m}^{(1)}} \tau_a^{(1)} = \partial_{\mathbf{m}^{(2)}} \tau_a^{(2)} \quad \text{on } \Gamma. \tag{96}$$

To prove it, we set

$$T^{(\alpha)}\{f\} = \int_{\mathcal{S}^{(\alpha)}} \tau(x, \mathbf{n}^{(\alpha)}, {}^s\Delta f)dA, \tag{97}$$

whence

$$\delta T^{(\alpha)}\{f\}[\mathbf{v}] = \int_{\mathcal{S}^{(\alpha)}} \tau_a^{(\alpha)}(x, \mathbf{n}^{(\alpha)}, {}^s\Delta f) \cdot {}^s\Delta\mathbf{v}dA. \tag{98}$$

Now, firstly, recall the differential identity

$$\mathbf{u}\cdot{}^s\Delta\mathbf{v} = {}^s\mathrm{Div}(({}^s\nabla\mathbf{v})^{\mathrm{T}}[\mathbf{u}]) - {}^s\nabla\mathbf{u}\cdot{}^s\nabla\mathbf{v}; \tag{99}$$

secondly, apply the surface-divergence identity (34) to the tangential vector field $({}^s\nabla\mathbf{v})^{\mathrm{T}}[\tau_a^{(\alpha)}]$ to obtain

$$\int_{\mathcal{S}^{(\alpha)}} {}^s\mathrm{Div}(({}^s\nabla\mathbf{v})^{\mathrm{T}}[\tau_a^{(\alpha)}])dA = \int_\Gamma (-1)^\alpha \tau_a^{(\alpha)}\cdot\partial_{\mathbf{m}^{(\alpha)}}\mathbf{v}ds; \tag{100}$$

thirdly, make use of (35) to obtain

$$\int_{\mathcal{S}^{(\alpha)}} {}^s\nabla\tau_a^{(\alpha)}\cdot{}^s\nabla\mathbf{v}dA = -\int_{\mathcal{S}^{(\alpha)}} \mathbf{v}\cdot{}^s\Delta\tau_a^{(\alpha)}dA + \int_\Gamma (-1)^\alpha \mathbf{v}\cdot\partial_{\mathbf{m}^{(\alpha)}}\tau_a^{(\alpha)}ds; \tag{101}$$

finally, combine (99)–(101) to write the variational integral in (98) in the following form

$$\delta T^{(\alpha)}\{f\}[\mathbf{v}] = \int_{\mathcal{S}^{(\alpha)}} \mathbf{v}\cdot{}^s\Delta\tau_a^{(\alpha)}dA + \int_\Gamma (-1)^\alpha(\tau_a^{(\alpha)}\cdot\partial_{\mathbf{m}^{(\alpha)}}\mathbf{v} - \mathbf{v}\cdot\partial_{\mathbf{m}^{(\alpha)}}\tau_a^{(\alpha)})ds.$$

The desired conclusion follows when we consider that the choice of $\mathbf{v}$ over Γ is arbitrary in the line integral

$$\begin{aligned}&\int_\Gamma \Big(\sum_\alpha(-1)^\alpha(\tau_a^{(\alpha)}\cdot\partial_{\mathbf{m}^{(\alpha)}}\mathbf{v} - \mathbf{v}\cdot\partial_{\mathbf{m}^{(\alpha)}}\tau_a^{(\alpha)})\Big)ds\\ &\quad= \int_\Gamma \big(-\tau_a^{(1)}\cdot\partial_{\mathbf{m}^{(1)}}\mathbf{v} + \tau_a^{(2)}\cdot\partial_{\mathbf{m}^{(2)}}\mathbf{v} + \mathbf{v}\cdot(\partial_{\mathbf{m}^{(1)}}\tau_a^{(1)} - \partial_{\mathbf{m}^{(2)}}\tau_a^{(2)})\big)ds,\end{aligned} \tag{102}$$

which appears in the stationarity condition

$$\delta(E - (T^{(1)} + T^{(2)}))\{f\}[\mathbf{v}] = 0 \quad \text{for all } \mathbf{v}\in\mathcal{V}, \tag{103}$$

which characterizes the equilibrium deformations f.

Remark. In fact, since $\partial_{\mathbf{m}^{(1)}}\mathbf{v}$ and $\partial_{\mathbf{m}^{(2)}}\mathbf{v}$ can also be given arbitrary values over Γ, we also deduce from the stationarity condition (103) that

$$\tau_a^{(1)} = \tau_a^{(2)} = \mathbf{0} \quad \text{on } \Gamma, \tag{104}$$

a result that implies (95).

Acknowledgements. The support of MURST, Progetto Cofinanziato 2000 "Modelli Matematici per la Scienza dei Materiali", is gratefully acknowledged. The work of Podio-Guidugli has also been supported by TMR Contract FMRX-CT98-0229 "Pha-se Transitions in Crystalline Solids".

References

[1] Podio-Guidugli, P. (1999): Elasticity with live loads. In: Interactions between analysis and mechanics. The legacy of Gaetano Fichera (Atti Convegni Lincei, 148). Accademia Nazionale dei Lincei, Rome, pp. 127–142

[2] Podio-Guidugli, P., Vergara-Caffarelli, G. (1990): Surface interaction potentials in elasticity. Arch. Rational Mech. Anal. **109**, 343–383

[3] Podio-Guidugli, P., Vergara-Caffarelli, G., Virga, E.G. (1987): The role of ellipticity and normality assumptions in formulating live-boundary conditions in elasticity. Quart. Appl. Math. **44**, 659–664

[4] Truesdell, C., Noll, W. (1965): The non-linear field theories of mechanics. (Handbuch der Physik, Bd. III/3). Springer, Berlin

[5] Lembo, M., Podio-Guidugli, P. (2000): Second-gradient constraints in second-grade materials. In: Contributions to continuum theories. Anniversary Volume for Kryzysztof Wilmanski's 60th. (WIAS Report 18). Weierstraß-Institut für Angewandte Analysis und Stochastik, Berlin, pp. 200–206

[6] Capriz, G. (1989): Continua with microstructure. (Springer Tracts in Natural Philosophy, Vol. 35). Springer, New York

[7] Spector, S.J. (1980): On uniqueness in finite elasticity with general loading. J. Elasticity **10**, 145–161

[8] Spector, S.J. (1982): On uniqueness for the traction problem in finite elasticity. J. Elasticity **12**, 367–383

The Thermistor Problem with Thomson's Effect

Giovanni Cimatti

Abstract. The electrical heating of a conductor is studied with Thomson's effect taken into account. A theorem of existence and uniqueness is proved by using a transformation which permits the relevant nonlinear elliptic boundary value problem to be reduced to the mixed problem for the Laplacian.

1 Introduction

In this paper we prove a theorem of existence and uniqueness and of non-existence for the nonlinear boundary value problem modelling the electrical heating of a conductor; the equations Thomson's effect take into account. Let $\mathbf{J}$ and $\mathbf{q}$ denote the current density and the heat flux. The basic equations [1] read

$$\mathbf{J} = -\sigma(u)(\nabla\varphi + \alpha(u)\nabla u), \tag{1.1}$$

$$\mathbf{q} = -\kappa(u)\nabla u + (\varphi + u\alpha(u))\mathbf{J}, \tag{1.2}$$

where φ is the electric potential and u the temperature. In (1.1) and (1.2) the electric and thermal conductivities $\sigma(u) > 0$, $\kappa(u) > 0$ and Thomson's coefficient $\alpha(u)$ are given functions of the temperature.

Energy and charge conservation, under steady conditions, require

$$\nabla \cdot \mathbf{J} = 0, \tag{1.3}$$

$$\nabla \cdot \mathbf{q} = 0. \tag{1.4}$$

Introducing the auxiliary functions, we have

$$v = \varphi + A(u), \quad A(u) = \int_0^u \alpha(t)dt, \tag{1.5}$$

$$\nu(u) = u\alpha(u) - A(u), \tag{1.6}$$

Equations (1.1) and (1.2) become

$$\mathbf{J} = -\sigma(u)\nabla v, \tag{1.7}$$

$$\mathbf{q} = -\kappa(u)\nabla u - \sigma(u)(v + \nu(u))\nabla v. \tag{1.8}$$

Let Ω be an open, bounded and connected subset of $\mathbf{R}^3$ with a C^2-boundary S, consisting of three parts, S_0, S_1 and S_2, such that

$$S_0 S_1 = \emptyset, \ \overset{\circ}{S}_j \overset{\circ}{S}_i = \emptyset, \quad i, j = 1, 2, 3, \quad S = S_0 \cup S_1 \cup S_2,$$

where $\overset{\circ}{S}$ denotes the interior (relative to S) of S_i. The body is insulated on S_2, both electrically and thermally, whereas S_0 and S_1 represent the electrodes to which a difference of potential is applied. These requirements lead to the boundary conditions:

$$\varphi = 0 \quad \text{on } S_0, \quad \varphi = \bar{\varphi} \quad \text{on } S_1, \tag{1.9}$$

$$u = 0 \quad \text{on } S_0 \cup S_1, \tag{1.10}$$

$$\frac{\partial \varphi}{\partial n} = 0 \text{ on } S_2, \tag{1.11}$$

$$\frac{\partial u}{\partial n} = 0 \text{ on } S_2, \tag{1.12}$$

where $\bar{\varphi}$ is a given positive constant. Substituting (1.7) and (1.8) into (1.3) and (1.4) and expressing the boundary conditions in terms of v and u, we arrive at problem P_{uv}:

$$\nabla \cdot (\sigma(u)\nabla v) = 0, \tag{1.13}$$

$$\nabla \cdot [\kappa(u)\nabla u + \sigma(u)(v + \nu(u))\nabla v] = 0, \tag{1.14}$$

$$v = 0 \text{ on } S_0, \quad v = \bar{v} \text{ on } S_1, \tag{1.15}$$

$$\frac{\partial v}{\partial n} = 0 \text{ on } S_2, \quad \frac{\partial u}{\partial n} = 0 \text{ on } S_2, \tag{1.16}$$

where, by (1.5), $\bar{v} = \bar{\varphi}$. The energy equation (1.14) can be written alternatively as

$$-\nabla \cdot (\kappa(u)\nabla u) = \sigma(u)|\nabla v|^2 + u\alpha'(u)\sigma(u)\nabla u \cdot \nabla v. \tag{1.17}$$

If $(u(x), v(x))$ is a regular solution of P_{uv} the maximum principle implies that

$$0 < v(x) < \bar{\varphi} \quad x \in \Omega, \tag{1.18}$$

$$u(x) > 0 \quad x \in \Omega. \tag{1.19}$$

We remark that the case $\bar{\varphi} < 0$ can be reduced immediately to the present one, since φ is defined apart from an arbitrary constant and the roles of S_0 and S_1 can be exchanged. We want to prove that, quite surprisingly, problem P_{uv} can be reduced to a nearly complete integration. More precisely, if the compatibility condition (2.3) given below is satisfied, the problem P_{uv} has either a unique solution which can be written in terms of the linear mixed problem

$$\Delta\xi = 0 \text{ in } \Omega, \tag{1.20}$$

$$\xi = 0 \text{ on } S_0, \tag{1.21}$$

$$\xi = k \text{ on } S_1, \tag{1.22}$$

$$\frac{\partial \xi}{\partial n} = 0 \text{ on } S_2, \tag{1.23}$$

or no solution.

2 The main result

The transformation which linearizes problem P_{uv} is based on the following elementary result.

Lemma 1. *The differentials forms*

$$\sigma(u)dv, \tag{2.1}$$

$$\kappa(u)du + \sigma(u)(v + \nu(u))dv \tag{2.2}$$

have a global common integrating factor if and only if

$$\frac{\sigma(u)\nu'(u)}{\kappa(u)} = \tau = \text{constant}. \tag{2.3}$$

Proof. Let $(I(u, v))^{-1}$ be a common integrating factor. The two differential forms

$$\frac{\sigma(u)}{I(u,v)}dv, \qquad \frac{\kappa(u)}{I(u,v)}du + \frac{\sigma(u)(v + \nu(u))}{I(u,v)}dv$$

are both integrable; therefore

$$\frac{\partial}{\partial u}\left(\frac{\sigma(u)}{I(u,v)}\right) = 0, \tag{2.4}$$

$$\frac{\partial}{\partial v}\left(\frac{\kappa(u)}{I(u,v)}\right) = \frac{\partial}{\partial u}\left(\frac{\sigma(u)(v + \nu(u))}{I(u,v)}\right). \tag{2.5}$$

Hence

$$\frac{\sigma(u)}{I(u,v)} = B(v), \tag{2.6}$$

where $B(v)$ is an arbitrary function. Inserting (2.6) in (2.5) we get, after separation of variables,

$$\frac{\sigma(u)}{\kappa(u)}\nu'(u) = \frac{B'(v)}{B(v)}. \tag{2.7}$$

Then (2.3) follows and we have

$$B'(v) = \tau B(v). \qquad \square \tag{2.8}$$

Let $\tau = 0$. Since $\sigma(u) > 0$, $\kappa(u) > 0$ and $\nu(0) = 0$, we have, by (2.3), $\nu(u) = 0$. From (2.8) it follows that $B_0(v) = C$ =constant, which we normalize by setting

$$B_0(v) = 1. \tag{2.9}$$

Thus (2.6) gives

$$I(u, v) = \sigma(u). \tag{2.10}$$

The primitives θ and ψ of

$$\frac{1}{\sigma(u)}\sigma(u)dv = d\theta \tag{2.11}$$

$$\frac{\kappa(u)}{\sigma(u)}du + vdv = d\psi \tag{2.12}$$

are given by

$$\theta_0(u, v) = v, \tag{2.13}$$

$$\psi_0(u, v) = F(u) + \frac{v^2}{2}, \tag{2.14}$$

where we define

$$F(u) = \int_0^u \frac{\kappa(t)}{\sigma(t)}dt. \tag{2.15}$$

Let $\tau \neq 0$. From (2.3) we have

$$\nu(u) = \tau F(u). \tag{2.16}$$

Recalling (2.9) we see that Eq. (2.8) implies that

$$B_\tau(v) = e^{\tau v}. \tag{2.17}$$

In this case,

$$I(u, v) = \frac{\sigma(u)}{e^{\tau v}}, \tag{2.18}$$

and the primitive $\theta_\tau(u, v)$ of the differential form

$$(I(u, v))^{-1}\sigma(u)dv = d\theta_\tau \tag{2.19}$$

is given by

$$\theta_\tau(u, v) = \frac{e^{\tau v} - 1}{\tau}, \tag{2.20}$$

where the constant of integration has been chosen to agree with θ_0 as $\tau \to 0$. The primitive $\psi_\tau(u, v)$ of

$$\frac{\kappa(u)}{I(u, v)}du + \frac{\sigma(u)(v + \nu(u))}{I(u, v)}dv \tag{2.21}$$

is also easily computed and given by

$$\psi_\tau(u, v) = e^{\tau v}F(u) + \left(\frac{v}{\tau} - \frac{1}{\tau^2}\right)e^{\tau v} + \frac{1}{\tau^2}. \tag{2.22}$$

Again the constant of integration has been chosen to be compatible with (2.14).

Remark 1. Hereafter we assume that (2.3), i.e.,

$$\frac{\sigma(u)\nu'(u)}{\kappa(u)} = \tau,$$

holds. If $\tau = 0$ then Thomson's effect is absent and we are dealing with the usual thermistor problem which is well-studied and understood [2, 3]. Condition (2.3) is less restrictive than it might appear since, for a large class of conductors, the Wiedemann–Franz law, i.e.,

$$\frac{\sigma(u)}{\kappa(u)} = \frac{k}{u}, \tag{2.23}$$

applies. On the other hand, A. Tait [4] has found, with accurate measurements, that $\alpha(u)$ depends linearly on the temperature; thus, if (2.23) holds, (2.3) is satisfied.

Lemma 1 permits us to write problem P_{uv} in a more symmetric form. To this end let $(u(x), v(x))$ be a solution. By Lemma 1 we have

$$\kappa(u)\nabla u + (\sigma(u)(v + \nu(u))\nabla v = I(u, v)\nabla\psi \tag{2.24}$$

and

$$\sigma(u)\nabla v = I(u, v)\nabla\theta; \tag{2.25}$$

hence, by (1.13) and (1.14), we obtain

$$\nabla \cdot (I(u, v)\nabla\psi) = 0, \tag{2.26}$$

$$\nabla \cdot (I(u, v)\nabla\theta) = 0. \tag{2.27}$$

In terms of θ and ψ the boundary conditions become

$$\theta = 0 \text{ on } S_0, \quad \psi = 0 \text{ on } S_0, \tag{2.28}$$

$$\theta|_{S_1} = \bar{\theta} = \begin{cases} \frac{e^{\tau\bar{v}}-1}{\tau}, & \text{if } \tau \neq 0 \\ \bar{v}, & \text{if } \tau = 0 \end{cases} \tag{2.29}$$

$$\psi|_{S_1} = \bar{\psi} = \begin{cases} \left(\frac{\bar{v}}{\tau} - \frac{1}{\tau^2}\right)e^{\tau\bar{v}} + \frac{1}{\tau^2}, & \text{if } \tau \neq 0 \\ \frac{\bar{v}^2}{2}, & \tau = 0 \end{cases} \tag{2.30}$$

$$\frac{\partial\theta}{\partial n} = 0, \quad \frac{\partial\psi}{\partial n} = 0 \text{ on } S_2. \tag{2.31}$$

Since θ and ψ satisfy the same equation and $\bar{\theta}$ and $\bar{\psi}$ are constants, between θ and ψ the following functional relation holds:

$$\psi = \frac{\bar{\psi}}{\bar{\theta}}\theta. \tag{2.32}$$

Using (2.32) we see that Eq. (2.22) becomes, by (2.20),

$$\frac{\bar{\psi}}{\bar{\theta}}\frac{e^{\tau v}-1}{\tau}=e^{\tau v}F(u)+\left(\frac{v}{\tau}-\frac{1}{\tau^2}\right)e^{\tau v}+\frac{1}{\tau^2} \tag{2.33}$$

when $\tau \neq 0$ and

$$\frac{\bar{\psi}}{\bar{\theta}}v=F(u)+\frac{v^2}{2} \tag{2.34}$$

if $\tau = 0$. Solving (2.33) and (2.34) with respect to $F(u)$ we obtain

$$F(u)=H_\tau(v,\bar{v}), \tag{2.35}$$

where

$$H_\tau(v,\bar{v})=\frac{\bar{\psi}}{\bar{\theta}}\frac{1-e^{-\tau v}}{\tau}-\frac{v}{\tau}+\frac{1}{\tau^2}-\frac{e^{-\tau v}}{\tau^2} \tag{2.36}$$

when $\tau \neq 0$ and

$$H_0(v,\bar{v})=\frac{1}{2}v\bar{v}-\frac{1}{2}v^2 \tag{2.37}$$

when $\tau = 0$. If (2.35) can be solved with respect to u, we have

$$u=F^{-1}(H_\tau(v,\bar{v})) \tag{2.38}$$

and the equation

$$\nabla\cdot(\sigma(u)\nabla v)=0$$

can be written entirely in terms of v, i.e.,

$$\nabla\cdot[\sigma(F^{-1}(H_\tau(v,\bar{v})))\nabla v]=0 \tag{2.39}$$

with the boundary conditions

$$v=0 \ \text{ on } \ S_0,\qquad v=\bar{v} \ \text{ on } \ S_1,\qquad \frac{\partial v}{\partial n}=0 \ \text{ on } \ S_2. \tag{2.40}$$

Problem (2.39), (2.40) can be reduced to a linear problem by defining

$$L(v)=\int_0^v \sigma(F^{-1}(H_\tau(t,\bar{v})))\,dt. \tag{2.41}$$

If $\xi(x)$ is a solution to the problem

$$\Delta\xi=0 \qquad \text{in } \ \Omega, \tag{2.42}$$

$$\xi=0 \qquad \text{on } \ S_0, \tag{2.43}$$

$$\xi = L(\bar{v}) \quad \text{on } S_1, \tag{2.44}$$

$$\frac{\partial \xi}{\partial n} = 0 \qquad \text{on } S_2, \tag{2.45}$$

then $(u(x), v(x))$ is given in terms of $\xi(x)$ by

$$u(x) = F^{-1}(H_\tau(L^{-1}(\xi(x)), \bar{v})), \tag{2.46}$$

$$v(x) = L^{-1}(\xi(x)). \tag{2.47}$$

Clearly, the method works if F^{-1} and L^{-1} exist. We make precise the conditions under which this is true in the next section.

3 Existence, non-existence and uniqueness of solution

Let

$$\sigma(t) \in C^0(\mathbf{R}^1_+), \quad \sigma(t) > 0 \text{ for all } t > 0, \tag{3.1}$$

$$\kappa(t) \in C^0(\mathbf{R}^1_+), \quad \kappa(t) > 0 \text{ for all } t > 0, \tag{3.2}$$

$$\alpha(t) \in C^1(\mathbf{R}^1_+). \tag{3.3}$$

Define

$$\nu(t) = t\alpha(t) - \int_0^t \alpha(\tau)d\tau \tag{3.4}$$

and

$$F(t) = \int_0^t \frac{\kappa(\tau)}{\sigma(\tau)} d\tau. \tag{3.5}$$

Then the following result holds.

Theorem 1. *Suppose that (3.1), (3.2), (3.3) hold and that*

$$\frac{\sigma(t)\nu'(t)}{\kappa(t)} = \tau. \tag{3.6}$$

If

$$\lim_{t\to\infty} F(t) = \infty, \tag{3.7}$$

then problem P_{uv} has one and only one solution. Let

$$H^* = \max_{v\in[0,\bar{v}]} H_\tau(v, \bar{v}) \tag{3.8}$$

where $H_\tau(v, \bar{v})$ is defined in (2.36) and (2.37). If

$$\lim_{t\to\infty} F(t) = \ell < \infty, \tag{3.9}$$

then problem P_{uv} *has no solution when*

$$\ell \geq H^* \tag{3.10}$$

and exactly one solution if

$$\ell < H^*. \tag{3.11}$$

Proof. The following properties of the function $H_\tau(v, \bar{v})$ are easily verified by direct calculations:

$$H_\tau(0, \bar{v}) = 0, \quad H_\tau(\bar{v}, \bar{v}) = 0, \tag{3.12}$$

$$\frac{\partial^2 H_\tau}{\partial v^2}(v, \bar{v}) < 0, \quad v \in [0, \bar{v}]. \tag{3.13}$$

It follows that

$$H_\tau(v, \bar{v}) > 0, \quad v \in (0, \bar{v}). \tag{3.14}$$

If either (3.7) or (3.9) and (3.11) hold, the equation

$$F(u) = H_\tau(v, \bar{v}), \quad v \in [0, \bar{v}], \tag{3.15}$$

can be solved uniquely with respect to u since $F'(u) > 0$ for $u > 0$; thus

$$u = F^{-1}(H_\tau(v, \bar{v})). \tag{3.16}$$

In particular, when $v \in (0, \bar{v})$, we have

$$u > 0. \tag{3.17}$$

Therefore, the function

$$L(v) = \int_0^v \sigma(F^{-1}(H_\tau(t, \bar{v})))d\tau \tag{3.18}$$

is well-defined and maps one–to–one and onto $[0, \bar{v}]$ in $[0, L(\bar{v})]$. Let

$$\bar{\xi} = L(\bar{v}) > 0 \tag{3.19}$$

and $\xi(x)$ be the solution of the problem

$$\Delta\xi = 0 \quad \text{in } \Omega, \tag{3.20}$$

$$\xi = 0 \quad \text{on } S_0, \tag{3.21}$$

$$\xi = \bar{\xi} \quad \text{on } S_1, \tag{3.22}$$

$$\frac{\partial \xi}{\partial n} = 0 \quad \text{on } S_2. \tag{3.23}$$

By the maximum principle we have

$$0 < \xi(x) < \bar{\xi}, \quad x \in \Omega. \tag{3.24}$$

Define

$$v(x) = L^{-1}(\xi(x)), \quad x \in \Omega, \tag{3.25}$$

$$\theta(x) = \begin{cases} \frac{e^{\tau v(x)}-1}{\tau}, & \tau \neq 0 \\ v(x), & \tau = 0, \end{cases}, \tag{3.26}$$

$$\bar{\theta} = \begin{cases} \frac{e^{\tau \bar{v}}-1}{\tau,} & \tau \neq 0 \\ \bar{v}, & \tau = 0, \end{cases} \tag{3.27}$$

$$\bar{\psi} = \begin{cases} \left(\frac{\bar{v}}{\tau} - \frac{1}{\tau^2}\right) e^{\tau \bar{v}} + \frac{1}{\tau^2}, & \tau \neq 0 \\ \frac{\bar{v}^2}{2}, & \tau = 0, \end{cases} \tag{3.28}$$

$$\psi(x) = \frac{\bar{\psi}}{\bar{\theta}} \theta(x), \tag{3.29}$$

$$u(x) = F^{-1}(H_\tau(v(x), \bar{v})). \tag{3.30}$$

We claim that $u(x)$, $v(x)$ and $\theta(x)$, given respectively by (3.30), (3.25) and (3.26), satisfy the equation

$$\nabla \cdot [I(u(x), v(x)) \nabla \theta(x)] = 0, \tag{3.31}$$

where

$$I(u, v) = \frac{\sigma(u)}{e^{\tau v}}. \tag{3.32}$$

For, we have

$$I(u, v) \nabla \theta = \sigma(u) \nabla v = \sigma(F^{-1}(H_\tau(v; \bar{v}))) \nabla v = \nabla \xi \tag{3.33}$$

and (3.31) follows from (3.20). On the other hand, by (3.29) we also obtain

$$\nabla \cdot [I(u, v) \nabla \psi] = 0. \tag{3.34}$$

Now (3.34) is precisely (1.14) since $\nu(u) = \tau F(u)$ and, by (3.29) and (3.30), we get

$$\psi = \frac{e^{\tau v}}{\tau} \nu(u) + \left(\frac{v}{\tau} - \frac{1}{\tau^2}\right) e^{\tau v} + \frac{1}{\tau^2}. \tag{3.35}$$

Hence

$$I(u, v) \nabla \psi = \kappa(u) \nabla u + \sigma(u)(v + \nu(u)) \nabla v$$

and (1.14) follows from (3.34). By (3.33) Eq. (1.13) is also satisfied. The boundary conditions for $v(x)$ are satisfied since, by (3.25), we have $v = 0$ on S_0, $v = \bar{v}$ on S_1 and $\frac{\partial v}{\partial n} = 0$ on S_2. Moreover, by (3.30), we obtain

$$u|_{S_0} = F^{-1}(H_\tau(v|_{S_0}, \bar{v})) = F^{-1}(H_\tau(0, \bar{v})) = F^{-1}(0) = 0,$$
$$u|_{S_1} = F^{-1}(H_\tau(v|_{S_1}, \bar{v})) = F^{-1}(H_\tau(\bar{v}, \bar{v})) = F^{-1}(0) = 0,$$

and $\frac{\partial u}{\partial n} = 0$ on S_2.

We claim that $(u(x), v(x))$ is the unique solution. Let $(U(x),\ V(x))$ be a second solution and define

$$\Theta(x) = \begin{cases} \frac{e^{\tau V(x)}-1}{\tau}, & \tau \neq 0 \\ V(x), & \tau = 0 \end{cases} \tag{3.36}$$

$$\Psi(x) = \begin{cases} e^{\tau V(x)} F(U(x)) + \left(\frac{V(x)}{\tau} - \frac{1}{\tau^2}\right) e^{\tau V(x)} + \frac{1}{\tau^2}, & \tau \neq 0 \\ F(U(x)) + \frac{V^2(x)}{2}, & \tau = 0. \end{cases} \tag{3.37}$$

We have

$$\nabla \cdot [I(U, V)\nabla\Theta] = 0 \quad \text{in } \Omega, \tag{3.38}$$

$$\Theta = 0 \text{ on } S_0,\ \Theta = \bar{\theta} \text{ on } S_1,\ \frac{\partial \Theta}{\partial n} = 0 \text{ on } S_2, \tag{3.39}$$

$$\nabla \cdot [I(U, V)\nabla\Psi] = 0 \quad \text{in } \Omega, \tag{3.40}$$

$$\Psi = 0 \text{ on } S_0,\ \Psi = \bar{\psi} \text{ on } S_1,\ \frac{\partial \Psi}{\partial n} = 0 \text{ on } S_2. \tag{3.41}$$

Hence Θ and Ψ are related by the same functional relation as θ and ψ, i.e.,

$$\Psi = \frac{\bar{\psi}}{\bar{\theta}}\Theta. \tag{3.42}$$

Therefore, we obtain

$$F(U) = H_\tau(V, \bar{v}), \tag{3.43}$$

and the equation $\nabla \cdot (\sigma(U)\nabla V) = 0$ can be written as

$$\nabla \cdot [\sigma(F^{-1}(H_\tau(V, \bar{v})))\nabla V] = 0. \tag{3.44}$$

The function

$$\Xi = L(V) = \int_0^V \sigma(F^{-1}(H_\tau(t, \bar{v})))\, dt \tag{3.45}$$

satisfies

$$\Delta\Xi = 0 \qquad \text{in } \Omega, \tag{3.46}$$

$$\Xi = 0 \quad \text{on } S_0, \tag{3.47}$$

$$\Xi = L(\bar{v}) \quad \text{on } S_1, \tag{3.48}$$

$$\frac{\partial \Xi}{\partial n} = 0 \quad \text{on } S_2. \tag{3.49}$$

On the other hand, the solution of problem (3.46)–(3.49) is unique. Hence $\xi = \Xi$ and this implies that

$$U(x) = F^{-1}(H_\tau(V(x), \bar{v})) = F^{-1}(H_\tau(v(x), \bar{v})) = u(x), \tag{3.50}$$

$$V(x) = L^{-1}(\Xi(x)) = L^{-1}(\xi(x)) = v(x). \tag{3.51}$$

Suppose that (3.9) and (3.10) hold. We claim that problem P_{uv} has no solution. Suppose the contrary, and let $(u(x), v(x))$ be a solution. We again have the functional relation

$$F(u(x)) = H_\tau(v(x), \bar{v}))\ , \quad x \in \bar{\Omega}. \tag{3.52}$$

Let $v^* \in (0, \bar{v})$ satisfy

$$H_\tau(v^*, \bar{v}) = \max_{v \in [0,\bar{v}]} H_\tau(v, \bar{v}) = H^*;$$

since $v = 0$ on S_0 and $v = \bar{v}$ on S_1 there exists $x^* \in \Omega$ such that $v(x^*) = v^*$. By (3.52) we have

$$H^* > F(u(x^*)) = H_\tau(v(x^*), \bar{v}) = H_\tau(v^*, \bar{v}) = H^*,$$

and no solution to problem P_{uv} can exist. □

Remark 2. If $\tau = 0$ we have $v^* = \frac{\bar{v}}{2}$ and $H^* = \frac{1}{8}\bar{v}^2$ and the conditions (3.10) and (3.11) become particularly simple.

References

[1] Landau, L., Lifchitz, E. (1969): Électrodynamique de milieux continus. Éditions Mir, Moscow

[2] Cimatti, G. (1989): Remark on existence and uniqueness for the thermistor problem under mixed boundary conditions. Quart. Appl. Math. **47**, 117–121

[3] Howison, S.D., Rodrigues, J.F., Shillor, M. (1993): Stationary solutions to the thermistor problem. J. Math. Anal. Appl. **174**, 573–588

[4] Tait, P.G. (1870, 1871): Laboratory notes: On thermo-electricity. Proc. Roy. Soc. Edinburgh **7**, 308–311; 597–602

Elastic Symmetry Restrictions from Structural Gradients

Stephen C. Cowin

Abstract. In a continuum model for a material with structural gradient inhomogenieties there are restrictions on the type of elastic symmetry. It is shown that, if the material symmetry and the structural gradient are determined on the same scale for a linearly elastic material, then the only linear elastic symmetries possible for gradient materials are the trigonal, monoclinic and triclinic symmetries. The key point is that, on the same scale, the normal to a plane of symmetry and a material structural gradient are incompatible unless they are perpendicular. This incompatibility restricts the type of linear elastic symmetries possible for gradient materials to the trigonal, monoclinic and triclinic symmetries.

1 Introduction

It is argued that, if the material symmetry and the structural gradient are determined on the same scale for a linearly elastic material, then the only linear elastic symmetries possible for gradient materials are the trigonal, monoclinic and triclinic symmetries. In order to develop this point some recent results in the representation of the anisotropic symmetries for linear elasticity are reviewed. In the next section materials with structural gradients are described. In the three sections that immediately follow that section the notation for a generalized Hooke's law is reviewed, the concept of the plane of mirror symmetry is described, and the classification of the linear elastic material symmetries by the normals to their planes of mirror symmetry is summarized. The key point is made in the section on the symmetry of elastic materials with large structural gradients. That point is that, on the same scale, the normal to a plane of symmetry and a material structural gradient are incompatible unless they are perpendicular. This incompatibility restricts the type of linear elastic symmetries possible for gradient materials to the trigonal, monoclinic and triclinic symmetries. After a section on trigonal symmetry, the final section on continuum modeling considerations describes criteria for determining when the problem of this incompatibility must be addressed and when it may be neglected. In a material with a structural gradient, if a representative volume element (RVE) may be selected so that it is large enough to adequately average over the microstructure and small enough to insure that the structural gradient across the RVE is negligible, then it is not necessary to restrict the material symmetry to accommodate the gradient. However, in a material with a structural gradient, if an RVE cannot be selected such that the structural gradient across an adequately sized RVE is negligible, then it is not necessary to restrict the material symmetry to accommodate the structural gradient.

2 Material structural gradients

The concern here is with elastic materials that have a structural gradient. A material containing a structural gradient, such as increasing/decreasing porosity, is said to be a gradient material. Figure 1 is an illustration of a material with a layered structural gradient. Spheres of varying diameters and one material type are layered in a material of another type. As a special case, the spheres may be voids. Figure 2 is an illustration of a material with an unlayered structural gradient. Spheres of varying diameters and one material type are graded in a size distribution in a material of another type. Again, as a special case, the spheres may be voids. Gradient materials may be man-made, but are often natural. Examples of natural materials with structural gradients include cancellous bone and the growth rings of trees.

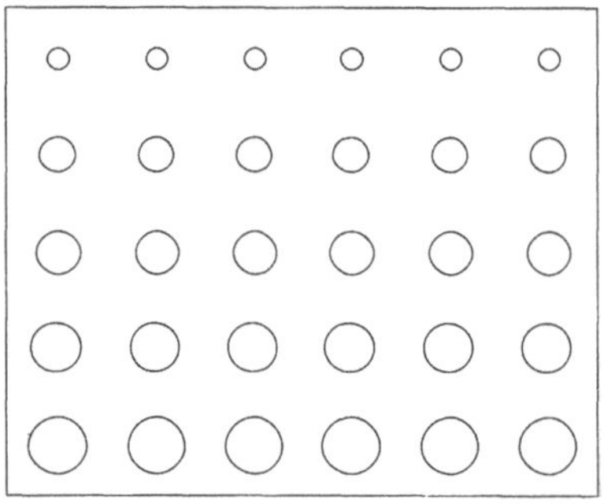

Fig. 1. An illustration of a material with a layered structural gradient. Spheres of varying diameters and one material type are layered in a material of another type. As a special case, the spheres may be voids

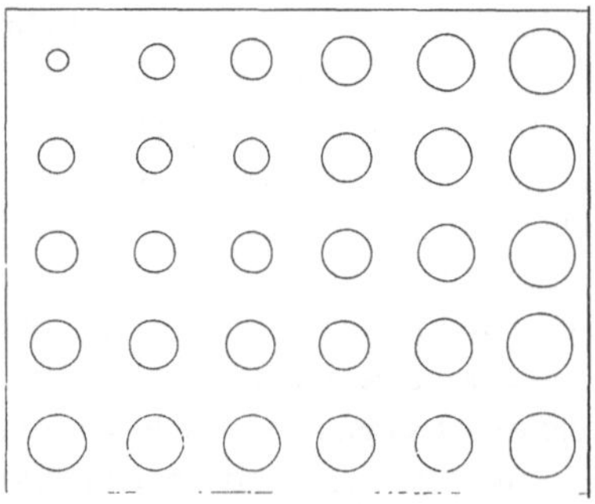

Fig. 2. An illustration of a material with a unlayered structural gradient. Spheres of varying diameters and one material type are graded in a size distribution in a material of another type. As a special case, the spheres may be voids

3 The generalized Hooke's law

The traditional notation for the anisotropic Hooke's law is due to Voigt [1]. The Voigt notation employs a fourth rank tensor in a three-dimensional space to represent the elastic coefficients. The components of the second rank Cartesian stress and strain tensors are denoted by T_{ij} and E_{ij}, respectively, and the components of the fourth rank tensors of elastic coefficients and compliance by C_{ijkm} and S_{ijkm}, respectively. In the Cartesian tensorial notation of Voigt, Hooke's law is written either as the stress-strain relations

$$T_{ij} = C_{ijkm} E_{km}, \tag{1}$$

or the strain-stress relations

$$E_{ij} = S_{ijkm} T_{km}. \tag{2}$$

There are three important symmetry restrictions on the tensors with components C_{ijkm} and S_{ijkm}. These restrictions, which require that components with the subscripts $_{ijkm}$, $_{jikm}$, and $_{kmij}$ be equal, follow from the symmetry of the stress tensor, the symmetry of the strain tensor, and the requirement that no work be produced by the elastic material in a closed loading cycle, respectively.

The Cartesian tensor transformation law for these two fourth rank tensors is of interest in the considerations that follow. Two reference bases or coordinate systems will be employed in three-dimensional space; these two systems have equivalent bases in six-dimensional space. The first basis is called the Latin basis or coordinate system because Latin letters are used to indicate indices associated with the system; it has base vectors $\mathbf{e}_i$, $i = 1, 2, 3$. The second basis is exactly like the first but a slightly different notation is employed in order to distinguish it from the first. For the other system primed indices are employed and the base vectors are $\mathbf{e}_{i'}$, $i' = 1', 2', 3'$. The orthogonal transformation from the primed to the unprimed system in three dimensions is represented by the matrix $\mathbf{Q}$ with components $Q_{ij'} = \mathbf{e}_i \cdot \mathbf{e}'_j$. The components of the fourth rank tensor of elastic coefficients in these two different coordinate systems, C_{ijkm} and $C_{i'j'k'm'}$, are related by

$$\begin{aligned} C_{ijkm} &= Q_{ii'} Q_{jj'} Q_{kk'} Q_{mm'} C_{i'j'k'm'} \quad \text{and} \\ C_{i'j'k'm'} &= Q_{ii'} Q_{jj'} Q_{kk'} Q_{mm'} C_{ijkm'} \end{aligned} \tag{3}$$

where $Q_{ii'}$ represents the components of the orthogonal transformation between the two different Cartesian coordinate systems. Equations (3) also apply to the components of the compliance tensor S_{ijkm}; one need only replace C by S in these equations.

Voigt [1] also employed a matrix notation for the stress-strain and strain-stress relations. In this notation the stress-strain relations (1) become

$$\begin{bmatrix} T_{11} \\ T_{22} \\ T_{33} \\ T_{23} \\ T_{13} \\ T_{12} \end{bmatrix} = \begin{bmatrix} c_{11} & c_{12} & c_{13} & c_{14} & c_{15} & c_{16} \\ c_{12} & c_{22} & c_{23} & c_{24} & c_{25} & c_{26} \\ c_{13} & c_{23} & c_{33} & c_{34} & c_{35} & c_{36} \\ c_{14} & c_{24} & c_{34} & c_{44} & c_{45} & c_{46} \\ c_{15} & c_{25} & c_{35} & c_{45} & c_{55} & c_{56} \\ c_{16} & c_{26} & c_{36} & c_{46} & c_{56} & c_{66} \end{bmatrix} \begin{bmatrix} E_{11} \\ E_{22} \\ E_{33} \\ 2E_{23} \\ 2E_{13} \\ 2E_{12} \end{bmatrix} \tag{4}$$

which is written as a linear transformation in six dimensions,

$$\tilde{\mathbf{T}} = \mathbf{c}\, \tilde{\mathbf{E}}, \tag{5}$$

where the formal and asymmetric column vectors representing the Cartesian components of the second rank symmetric tensors of stress and strain are given by

$$\tilde{\mathbf{T}} = \begin{bmatrix} T_{11} \\ T_{22} \\ T_{33} \\ T_{23} \\ T_{13} \\ T_{12} \end{bmatrix}, \quad \tilde{\mathbf{E}} = \begin{bmatrix} E_{11} \\ E_{22} \\ E_{33} \\ 2E_{23} \\ 2E_{13} \\ 2E_{12} \end{bmatrix}. \tag{6}$$

The strain-stress relations, i.e., the inverse of (4), are

$$\begin{bmatrix} E_{11} \\ E_{22} \\ E_{33} \\ 2E_{23} \\ 2E_{13} \\ 2E_{12} \end{bmatrix} = \begin{bmatrix} s_{11} & s_{12} & s_{13} & s_{14} & s_{15} & s_{16} \\ s_{12} & s_{22} & s_{23} & s_{24} & s_{25} & s_{26} \\ s_{13} & s_{23} & s_{33} & s_{34} & s_{35} & s_{36} \\ s_{14} & s_{24} & s_{34} & s_{44} & s_{45} & s_{46} \\ s_{15} & s_{25} & s_{35} & s_{45} & s_{55} & s_{56} \\ s_{16} & s_{26} & s_{36} & s_{46} & s_{56} & s_{66} \end{bmatrix} \begin{bmatrix} T_{11} \\ T_{22} \\ T_{33} \\ T_{23} \\ T_{13} \\ T_{12} \end{bmatrix} \tag{7}$$

in Voigt's notation and

$$\tilde{\mathbf{E}} = \mathbf{s}\, \tilde{\mathbf{T}}, \tag{8}$$

where **s** represents the elastic compliance coefficients in this notation,

$$\mathbf{s} = \mathbf{c}^{-1} \tag{9}$$

The relationship between the Voigt matrix notation, **c** and **s**, and the Voigt Cartesian tensor notation, C_{ijkm} and S_{ijkm}, for the elastic constants is somewhat awkward. The components of **c** and C_{ijkm} are related by replacing the six-dimensional indices 1, 2, 3, 4, 5 and 6 by the pairs of three-dimensional indices 1, 2 and 3: thus 1, 2, 3, 4, 5 and 6 become 11, 22, 33, 23 or 32, 13 or 31, 12 or 21, respectively. The members of the paired indices 23 or 32, 13 or 31, 12 or 21 are equivalent because of the symmetry of the tensors of stress and strain. The relationship of the components of the symmetric matrix **s** to the components S_{ijkm} is not the same as, nor as simple as, the relationship of the components of the symmetric matrix components of **c** to the components C_{ijkm}. This is because, although **s** is the inverse of **c**, neither **s** nor **c** is a matrix of tensor components. The relationship between the components of **s** and S_{ijkm} involves factors of 1, 2 and 4. The elements in the upper left hand 3-by-3 sub-matrix of **s** have a proportionality factor of one, e.g., $S_{1111} = s_{11}$; the elements in the upper right-hand (and lower left-hand) 3-by-3 sub-matrix have a proportionality factor of two, e.g., $2S_{2311} = s_{41}$; the elements in the lower right-hand 3-by-3 sub-matrix have a proportionality factor of four, e.g., $4S_{1212} = s_{66}$; see [2].

4 Planes of mirror symmetry

The concept of a plane of mirror or reflective symmetry will be used to classify the eight types of anisotropy possible in linear elasticity and to impose restrictions on the elasticity tensor C_{ijkm} appropriate to these anisotropy types. Recall that the elasticity tensor C_{ijkm} transforms according to the rule (3) when the coordinate system is changed. In order to apply the mirror or reflective symmetry restrictions using (3), it is necessary to have a representation for the orthogonal transformation **Q** for the case of a plane of reflective symmetry. A plane of mirror symmetry and an orthogonal transformation representing a plane of mirror symmetry are described in this section.

Recall that two objects are geometrically congruent if they can be superposed upon one another so that they coincide. The two tetrahedra at the top of Fig. 3 are congruent.

Congruence of two shapes is a necessary but not sufficient condition for mirror symmetry. A pair of congruent geometric objects is said to have mirror symmetry with respect to a plane if for each point of either object there is a point of the other object such that the pair of points is symmetric with respect to the plane. The two congruent tetrahedra at the bottom of Fig. 3 have the special relationship of mirror symmetry with respect to the plane whose end view is indicated by an m. Each congruent geometric object is said to be the reflection of the other. The plane with respect to which two objects have mirror symmetry is called their *plane of reflective symmetry*. A material is said to have a *plane of reflective symmetry* or a *mirror plane* at a point in the material if the structure of the material has mirror symmetry with respect to a plane passing through the point.

In order to construct a formula for an orthogonal transformation **Q** representing a plane of reflective symmetry, let **a** be a unit vector representing the normal to a plane of reflective symmetry and let **b** be any vector perpendicular to **a**, then $\mathbf{a} \cdot \mathbf{b} = \mathbf{0}$ for

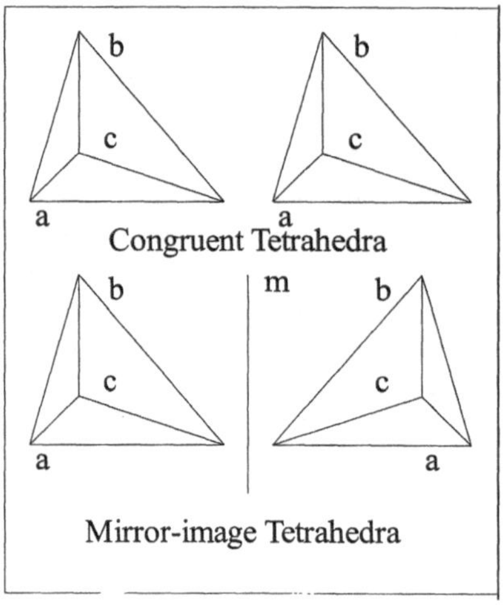

Fig. 3. Mirror symmetry illustration

all **b**. An orthogonal transformation **R** with the properties

$$\mathbf{R}^{(\mathbf{a})}\,\mathbf{a} = -\mathbf{a}, \quad \mathbf{R}^{(\mathbf{a})}\,\mathbf{b} = \mathbf{b}, \tag{10}$$

represents a plane of reflective, or mirror, symmetry. The orthogonal transformation (10) carries every vector parallel to the vector **a**, the normal to the plane of mirror symmetry, into the direction $-\mathbf{a}$ and it carries every vector **b** parallel to the plane into itself. It is easy to verify that the representation

$$\mathbf{R}^{(\mathbf{a})} = 1 - 2\mathbf{a} \otimes \mathbf{a}, \quad R_{ij}^{(a)} = \delta_{ij} - 2a_i a_j. \tag{11}$$

5 The eight elastic material symmetries

It has been shown that all the linear elastic symmetries may be classified on the basis of the normals to their planes of mirror or reflective symmetry [3]. It has also recently been shown that the classification of the eight material symmetries of linear elasticity upon the basis of the number and orientation of their planes of material symmetry is equivalent to the classification by the crystallographic groups [4]. Here we follow the approach of [4] and [5] in which the allowable symmetries are constructed systematically by increasing the number of planes of symmetry possessed by the material. Within this framework elastic materials fall into eight classes. In ascending order of symmetry, they are the triclinic, monoclinic, trigonal, orthotropic, tetragonal, cubic, transversely isotropic and isotropic classes, the first seven comprising the possible types of anisotropic elastic materials. Figure 4 illustrates the relationship between the various crystalline symmetries. Figure 4 is organized so that the higher symmetries are at the right and as one moves up the table one sees crystal systems with less and less symmetry. The symmetry of each crystal system will be briefly

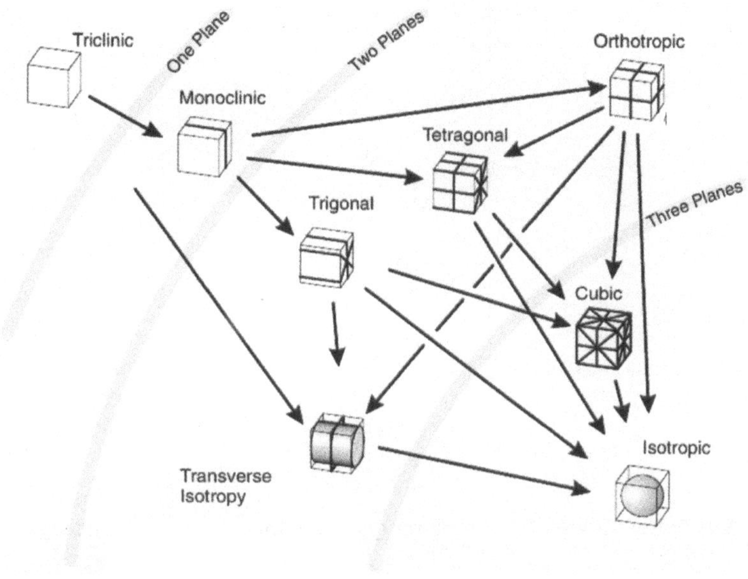

Fig. 4. The hierarchical organization of the eight material symmetries of linear elasticity. The figure is organized so that the lower symmetries are at the upper left and as one moves down and across the table to the right one encounters crystal systems with greater and greater symmetry

described. Each crystal system has one distinct elasticity tensor C_{ijkm} associated with it in a particular reference coordinate system associated with that symmetry. A reference coordinate system is selected so that there are only 18 distinct components of the elasticity tensor C_{ijkm} for triclinic symmetry, 12 for monoclinic and so that the seven constant tetragonal and trigonal symmetries are the same as the six constant tetragonal and trigonal symmetries, respectively [6]. There are, therefore, in our presentation, 7 distinct elasticity tensor C_{ijkm} reference coordinate system matrices associated with the 7 elastic anisotropic material symmetries.

The symmetry of the triclinic crystal system is almost nonexistent. Triclinic crystal systems do have some symmetry, but they do not have enough symmetry to restrict the form of the elasticity tensor C_{ijkm}. Triclinic crystal systems have no planes of mirror symmetry. The monoclinic crystal system has exactly one plane of reflective symmetry. Orthorhombic symmetry is characterized by three mutually perpendicular planes of mirror symmetry any two of which imply the existence of the third (Fig. 5).

In Fig. 5 the base vectors of the Cartesian system are indicated by $\mathbf{e}_1$, $\mathbf{e}_2$ and $\mathbf{e}_3$ while the normals to the three planes of symmetry are indicated by $\mathbf{a}_1$, $\mathbf{a}_2$ and $\mathbf{a}_3$; in the case of orthotropic symmetry the sets are coincident. Tetragonal symmetry is characterized by five planes of mirror symmetry, four of whose normals lie in the

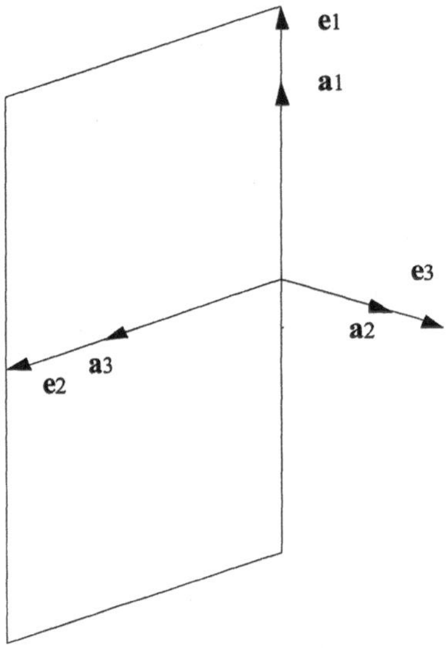

Fig. 5. Normals to planes of mirror symmetry for orthotropic symmetry

fifth plane and make angles that are multiples of $\pi/4$ with respect to one another. These planes are illustrated in Fig. 6. The base vectors of the Cartesian system are again indicated by $\mathbf{e}_1$, $\mathbf{e}_2$ and $\mathbf{e}_3$ while the normals to the five planes of symmetry are indicated by $\mathbf{a}_1$ through $\mathbf{a}_5$.

Cubic symmetry is characterized by nine planes of mirror symmetry. The nine planes of mirror symmetry ($\mathbf{a}_1$ through $\mathbf{a}_9$) include the three orthogonal planes of mirror symmetry characterizing orthorhombic symmetry (Fig. 5) and six more whose normals bisect the angles between the first three and the negatives of the first three, taken two at a time. Thus, cubic symmetry has nine planes of symmetry all intersecting at 90°, 60° or 45°. These planes are illustrated in Fig. 7.

Trigonal symmetry has three planes of mirror symmetry ($\mathbf{a}_1$ through $\mathbf{a}_3$). The normals to these three planes all lie in one plane and make angles of sixty degrees with one another. The three normals to the planes of symmetry are illustrated in Fig. 8. Since the negative of a plane of symmetry is also the normal to a plane of symmetry, it is easy to see that trigonal symmetry is a three-fold symmetry. The crystalline hexagonal symmetry is a six-fold symmetry with seven planes of mirror symmetry ($\mathbf{a}_1$ through $\mathbf{a}_7$). Six of the normals to these seven planes all lie in the seventh plane and make angles of thirty degrees with one another as illustrated in Fig. 9. The textured transversely isotropic symmetry is characterized by a single plane of isotropy. A *plane of isotropy* is a plane of mirror symmetry in which every vector is itself a normal to a plane of mirror symmetry. The elasticity tensor C_{ijkm} associated with the crystalline hexagonal symmetry and the elasticity tensor C_{ijkm} associated

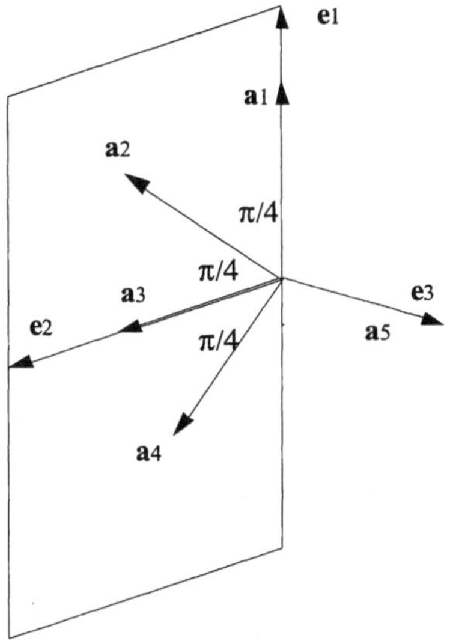

Fig. 6. Normals to planes of mirror symmetry for tetragonal symmetry

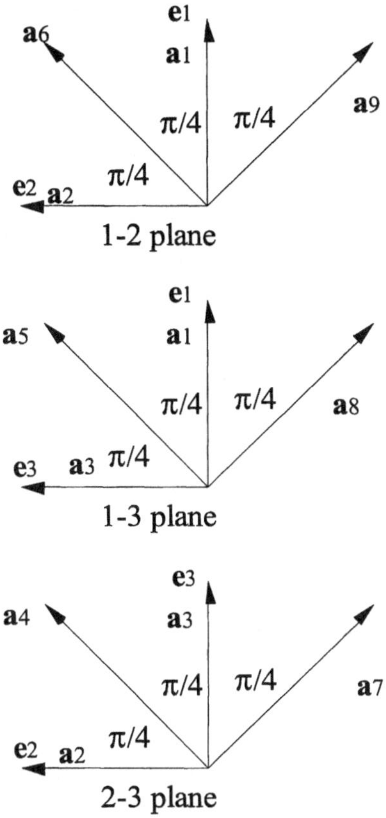

Fig. 7. The nine planes of mirror symmetry associated with cubic symmetry. These planes, shown in their three common planes, all intersect at 90° or 45°

with the textured transversely isotropic symmetry are identical and therefore the two symmetries can be considered together.

The reason that the elasticity tensor C_{ijkm} associated with these two symmetries are identical is that, if just one more plane of mirror symmetry is added in the same plane of the other three of trigonal symmetry, there are two possible outcomes, but both outcomes lead to the same elasticity tensor C_{ijkm}. Recall that for trigonal symmetry, the normals to these three planes all lie in one plane and make angles of sixty degrees with one another. If the normal to the additional plane of symmetry makes an angle of $\pi/6$ with one of the other symmetry plane normals, then 3-fold symmetry becomes a 6-fold symmetry. However, if the normal to the additional plane of symmetry does not make an angle of $\pi/6$ with the one of the other symmetry plane

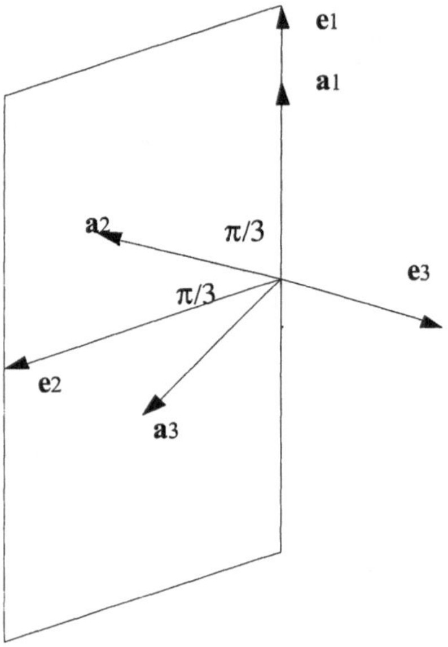

Fig. 8. The three planes of mirror symmetry associated with trigonal symmetry

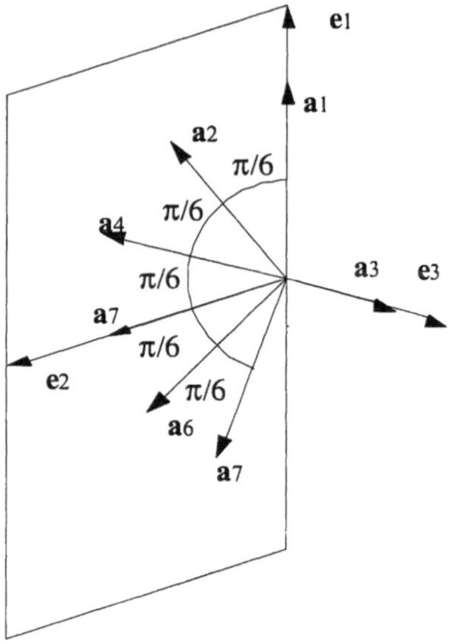

Fig. 9. An illustration of the seven planes of mirror symmetry of crystalline hexagonal symmetry

normals, then 3-fold symmetry becomes a plane of isotropy. If one more normal to a plane of symmetry, distinct from the first three normals and not making an angle of $\pi/6$ with any of the first three, is added in the same plane, then every vector in the plane can be shown to be a normal to a plane of symmetry and the plane containing these normals is itself a plane of mirror symmetry. A property of a material or a system is said to have isotropic symmetry with respect to a point when every plane passing through the point is a plane of mirror symmetry or a plane of isotropy with respect to the property in question. Isotropic symmetry is a special case of both crystalline hexagonal symmetry or the textured transversely isotropic symmetry and cubic symmetry. If just one more distinct plane of mirror symmetry is added to either of these symmetries, it becomes isotropic.

6 The symmetry of elastic materials with large structural gradients

The central argument here is that the normal to a plane of material symmetry can only be perpendicular to the direction of a uniform structural gradient. The argument is a purely geometrical one. First note that the direction of a normal to a plane of material symmetry can not be coincident with the direction of the structural gradient because the structural gradient is inconsistent with the reflective structural symmetry required by a plane of mirror symmetry. Next consider the case when the normal to a plane of material symmetry is inclined, but not perpendicular, to the direction of the structural gradient. In this case, the same situation prevails because the structural gradient is still inconsistent with the reflective structural symmetry required by a plane of mirror symmetry. The only possibility is that the normal to a plane of material symmetry is perpendicular to the direction of the structural gradient. Thus, it is concluded that the only linear elastic symmetries permitted in a material containing a structural gradient are those symmetries characterized by having all their normals to their planes of mirror or reflective symmetry perpendicular to the structural gradient. The caveat to this conclusion is that the structural gradient and the material symmetry are at the same structural scale in the material, as will be discussed in the section after next.

Employing here the classification, described above, of the linear elastic symmetries by planes of mirror symmetry, we can easily see that there are only three linear elastic symmetries satisfying the condition that they admit a direction perpendicular to all the normals to their planes of mirror or reflective symmetry. These symmetries are the trigonal, which has three planes of mirror symmetry whose normals all lie in one plane and make an angle of $2\pi/3$ with another monoclinic symmetry, which has only one plane of symmetry, and triclinic, which has no planes of symmetry. Since trigonal symmetry has the highest degree of symmetry of these three symmetries, it will be used for illustration.

7 Trigonal symmetry

Trigonal symmetry has three planes of mirror symmetry. The normals to these three planes all lie in one plane and make angles of sixty degrees with one another. Since the negative of a plane of symmetry is also the normal to a plane of symmetry, it is easy to see that trigonal symmetry is a three-fold symmetry. The normals to the three planes of mirror symmetry are all in the plane spanned by $\mathbf{e}_1$ and $\mathbf{e}_2$ (see Fig. 8).

In some texts [7], the $\mathbf{c}$ matrix with seven distinct constants,

$$\mathbf{c} = \begin{bmatrix} c_{11} & c_{12} & c_{13} & c_{14} & -c_{25} & 0 \\ c_{12} & c_{11} & c_{13} & -c_{14} & c_{25} & 0 \\ c_{13} & c_{13} & c_{33} & 0 & 0 & 0 \\ c_{14} & -c_{14} & 0 & c_{44} & 0 & 0 \\ -c_{25} & c_{25} & 0 & 0 & c_{44} & c_{14} \\ 0 & 0 & 0 & c_{25} & c_{14} & \left(\frac{1}{2}\right)(c_{11}-c_{12}) \end{bmatrix}, \tag{12}$$

is identified as a distinct type of trigonal elastic symmetry. However, as indicated in [6], a coordinate transformation reduced (12) to

$$\mathbf{c} = \begin{bmatrix} c_{11} & c_{12} & c_{13} & c_{14} & 0 & 0 \\ c_{12} & c_{11} & c_{13} & -c_{14} & 0 & 0 \\ c_{13} & c_{13} & c_{33} & 0 & 0 & 0 \\ c_{14} & -c_{14} & 0 & c_{44} & 0 & 0 \\ 0 & 0 & 0 & 0 & c_{44} & c_{14} \\ 0 & 0 & 0 & 0 & c_{14} & \left(\frac{1}{2}\right)(c_{11} - c_{12}) \end{bmatrix} \tag{13}$$

A significant point about trigonal symmetry in the present context is that the plane normal to $\mathbf{e}_3$ is not a plane of mirror symmetry. If it were a plane of mirror symmetry, then the components c_{14} and c_{25} of the elasticity matrix $\mathbf{c}$ for trigonal symmetry (12) would vanish and the material would have hexagonal symmetry (transverse isotropy). In the case of hexagonal symmetry, the existence of a structural gradient would be excluded. The plane normal to the $\mathbf{e}_3$ direction in hexagonal symmetry is a plane of isotropy, that is to say, a plane in which every vector is the normal to a plane of mirror symmetry. Clearly, the plane normal to the $\mathbf{e}_3$ direction in trigonal symmetry is not a plane of isotropy, but it does have a high degree of symmetry. It is possible to illustrate this high degree of symmetry by employing the transformation that Federov observed does not change the material symmetry. In addition, using this transformation, other representations of the $\mathbf{c}$ matrix for trigonal symmetry may be obtained. The transformation in question is an arbitrary rotation of the reference coordinate system about the $\mathbf{e}_3$ direction. The $\mathbf{Q}$ associated with the rotation of the reference coordinate system about the $\mathbf{e}_3$ direction is given by

$$\mathbf{Q} = \begin{bmatrix} \cos\alpha & -\sin\alpha & 0 \\ \sin\alpha & \cos\alpha & 0 \\ 0 & 0 & 1 \end{bmatrix}. \tag{14}$$

Employing (12) and (4) in the tensor transformation law (3) we see that the transformed matrix $\mathbf{c}$ has the form:

$$\mathbf{c} = \begin{bmatrix} c_{11} & c_{12} & c_{13} & c_{1'4'} & -c_{2'5'} & 0 \\ c_{12} & c_{11} & c_{13} & -c_{1'4'} & c_{2'5'} & 0 \\ c_{13} & c_{13} & c_{33} & 0 & 0 & 0 \\ c_{1'4'} & -c_{1'4'} & 0 & c_{44} & 0 & c_{2'5'} \\ -c_{2'5'} & c_{2'5'} & 0 & 0 & c_{44} & c_{1'4'} \\ 0 & 0 & 0 & c_{2'5'} & c_{1'4'} & \left(\frac{1}{2}\right)(c_{11} - c_{12}) \end{bmatrix} \tag{15}$$

where

$$c_{1'4'} = c_{14}\cos 3\alpha + c_{25}\sin 2\alpha, \quad c_{2'5'} = c_{25}\cos 3\alpha - c_{14}\sin 3\alpha. \tag{16}$$

This result shows that the components c_{11}, $c_{22} = c_{11}$, c_{33}, c_{44}, $c_{55} = c_{44}$, $c_{66} = (1/2)(c_{11} - c_{12})$, c_{12}, c_{13} and $c_{23} = c_{13}$ are invariant under the transformation (14); only the components c_{14} and c_{25} change. In the case of hexagonal symmetry, the components corresponding to c_{11}, $c_{22} = c_{11}$, c_{33}, c_{44}, $c_{55} = c_{44}$, $c_{66} = (1/2)(c_{11} - c_{12})$, c_{12}, c_{13} and $c_{23} = c_{13}$ are also invariant under the transformation (14).

Different representations of the **c** matrix for trigonal symmetry (12) are obtained by considering specific values of the angle of rotation α in the symmetry preserving coordinate transformation (14). The change in c_{14} and c_{25} is specified by (16). Table 1 lists the values of $c_{1'4'}$ and $c_{2'5'}$ for various values of α. The first row of this table corresponds to the representation (13) of the elasticity matrix **c** for trigonal symmetry. The other rows provide the values of $c_{1'4'}$ and $c_{2'5'}$ corresponding to other values of α. These other equivalent elasticity matrices that may be constructed from the data in Table 1 provide other representations that suggest the nature of the symmetry in the plane normal to the $\mathbf{e}_3$ direction, a symmetry that is high but short of a plane of isotropy.

Table 1. The values of the components c_{14} and c_{25} of the elasticity matrix **c** for trigonal symmetry for various values of the angle of rotation α in the symmetry preserving coordinate transformation (14)

Arctan 3α	$c_{1'4'}$	$c_{2'5'}$
$\frac{c_{25}}{c_{14}}$	$c_{14}\sqrt{1+\frac{c_{25}^2}{c_{14}^2}}$	0
$-\frac{c_{14}}{c_{25}}$	0	$c_{14}\sqrt{1+\frac{c_{25}^2}{c_{14}^2}}$
$\frac{c_{14}+c_{25}}{c_{14}-c_{25}}$	$\frac{(c_{14}-c_{25})}{\sqrt{2}}\sqrt{\frac{c_{14}^2+c_{25}^2}{(c_{14}-c_{25})^2}}$	$-\frac{(c_{14}-c_{25})}{\sqrt{2}}\sqrt{\frac{c_{14}^2+c_{25}^2}{(c_{14}-c_{25})^2}}$
$\frac{c_{25}-c_{14}}{c_{14}+c_{25}}$	$-\frac{(c_{14})+c_{25}}{\sqrt{2}}\sqrt{\frac{c_{14}^2+c_{25}^2}{(c_{14}+c_{25})^2}}$	$\frac{(c_{14}+c_{25})}{\sqrt{2}}\sqrt{\frac{c_{14}^2+c_{25}^2}{(c_{14}+c_{25})^2}}$

The elasticity matrix **c** given by (13) is associated with the rotated coordinate system corresponding to the last row of Table 1. The form of Hooke's law corresponding to this matrix is represented by the stress-strain relations

$$
\begin{aligned}
T_{11} &= c_{11}E_{11} + c_{12}E_{22} + c_{13}E_{33} + c_{14}(E_{23} - E_{13}),\\
T_{22} &= c_{12}E_{11} + c_{11}E_{22} + c_{13}E_{33} - c_{14}(E_{23} - E_{13}),\\
T_{33} &= c_{13}E_{11} + c_{13}E_{22} + c_{33}E_{33},\\
T_{23} &= c_{14}(E_{11} - E_{22}) + 2c_{44}E_{23} + 2c_{14}E_{12},\\
T_{13} &= -c_{14}(E_{11} - E_{22}) + 2C_{44}E_{23} + 2c_{14}E_{12},\\
T_{12} &= 2c_{14}(E_{13} + E_{23}) + (c_{11} - c_{12})E_{12}.
\end{aligned}
\tag{17}
$$

The form of the strain-stress relations (7) for trigonal symmetry corresponding to the first row of Table 1 are:

$$
\begin{aligned}
E_{11} &= s_{11}T_{11} + s_{12}T_{22} + s_{13}T_{33} + s_{14}T_{23}, \\
E_{22} &= s_{12}T_{11} + s_{22}T_{22} + s_{13}T_{33} - s_{14}T_{23}, \\
E_{33} &= s_{13}(T_{11} + T_{22}) + s_{33}T_{33}, \\
E_{23} &= s_{14}(T_{11} - T_{22}) + s_{44}T_{23}, \\
E_{13} &= s_{44}T_{13} + s_{14}T_{12}, \\
E_{12} &= (s_{11} - s_{12})T_{12}/2 + s_{14}T_{13}.
\end{aligned} \tag{18}
$$

The interpretation of the physical significance of the elastic compliance constants in (18) is easily done on recalling the strain-stress relations for hexagonal symmetry (transverse isotropy) in which E' and E denote the Young's moduli in the $\mathbf{e}_3$ direction and $\mathbf{e}_1/\mathbf{e}_2$ directions, respectively, ν the planar Poisson ratio, ν' the axial Poisson ratio, G the planar shear modulus; thus $s_{11} = 1/E$, $s_{12} = \nu/E$, $s_{13} = \nu'/E'$, and $s_{44} = 1/G$. After these five elastic constants familiar from hexagonal symmetry (transverse isotropy), there remains only the constant, $s_{14} = q/G$, to distinguish between trigonal symmetry and transverse isotropy; thus the strain-stress relations for trigonal materials are:

$$
\begin{aligned}
E_{11} &= (1/E)T_{11} + (\nu/E)T_{22} + (\nu'/E')T_{33} + (q/2G)T_{23}, \\
E_{22} &= (\nu/E)T_{11} + (1/E)T_{22} + (\nu'/E')T_{33} - (q/2G)T_{23}, \\
E_{33} &= (\nu'/E')(T_{11} + T_{22}) + (1/E')T_{33}, \\
E_{23} &= T_{23}/2G + (q/2G)(T_{11} - T_{22}), \\
E_{13} &= T_{13}/2G + (q/2G)T_{12}, \\
E_{12} &= (1/2E)(1 - \nu)T_{12} + (q/2G)T_{13}.
\end{aligned} \tag{19}
$$

For $q = 0$ the equations (19) represent Hooke's law for a hexagonal (transversely isotropic) elastic material. There are just five elastic constants in that case, and the $\mathbf{e}_3$ direction is the normal to a plane of symmetry. The condition for the non- existence (existence) of a plane of mirror symmetry with a normal in the $\mathbf{e}_3$ direction is $q \neq 0$ $(q = 0)$.

8 Continuum modeling considerations

The representative volume element (RVE) is a conceptual tool for forming continuum models of materials and for establishing restrictions that might be necessary for a continuum model to be applicable. An RVE for a continuum "point" is the material volume surrounding the "point", that is, a statistically homogeneous representative of the material in the neighborhood of the "point". For purposes of this discussion, the RVE is taken to be a cube of side length L_{RVE}; it could be any shape, but it is necessary that it have a characteristic length scale. An RVE is shown in Fig. 10; it is a homogenized or average image of a real material volume. At the smallest level, there

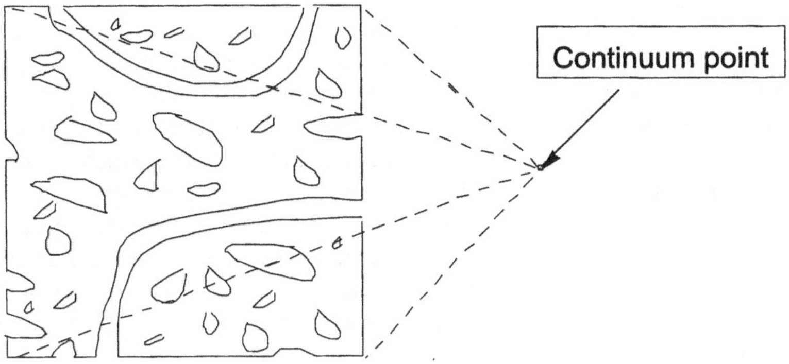

Fig. 10. The RVE for the representation of a domain of a porous medium by a continuum point

are spaces between the atoms or molecules that constitute holes in the material. Since the RVE image of the material object averages over the small holes and heterogeneous microstructures, overall it replaces a discontinuous real material object by a smooth continuum model of the object. The RVE is necessary in continuum models for all materials; the main question is how large must the length scale L_{RVE} be to obtain a reasonable continuum model. The smaller the value of L_{RVE} the better; in general the value of L_{RVE} should be much less than the characteristic dimension L_C of the problem being modeled.

The question of the size of L_{RVE} can also be posed in the following way: how large a hole is no hole? The value of L_{RVE} selected determines what the modeler has selected as too small a hole, or too small an inhomogeneity or microstructure, to influence the result the modeler is seeking. An interesting aspect of the RVE concept is that it resolves a paradox concerning stress concentrations around circular holes in linear elastic materials. The stress concentration factor associated with the hole in a circular elastic plate in a uniaxial field of otherwise uniform stress is three times the uniform stress. This means that the stress at certain points in the material on the edge of the hole is three times the stress five or six hole diameters away from the hole. The hole has a concentrating effect of magnitude 3. The paradox is that the stress concentration factor of 3 is independent of the size of the hole. Thus, no matter how small the hole, there is a stress concentration factor of 3 associated with the hole in a field of uniaxial tension or compression.

The answer to the riddle posed above is that the modeler has decided how big a hole is no hole by choosing to recognize a hole of a certain size and selecting a value of the length ratio $R_L = L_C/L_{\mathrm{RVE}}$. Since the characteristic problem length L_C is the hole radius a in the problem of a circular hole in a plate, it follows that $L_{\mathrm{RVE}} = a/R_L$. Thus, any hole whose radius is less than $L_{\mathrm{RVE}} = a/R_L$ will not appear in the continuum model although it is in the real object. The interpretation of the elastic solution to the problem of a circular elastic plate in a uniaxial field of otherwise uniform stress is that there is only one hole in the model of radius a, no holes of a size less than a and greater than $L_{\text{RVE}} = a/R_L$, and all holes in the real object of a size less than $L_{\text{RVE}} = a/R_L$ have been "homogenized" or averaged over.

The RVE also pays an important role in determining the relationship between the structural gradient and the material symmetry. In a material with a structural gradient, if an RVE may be selected so that it is large enough to adequately average over the microstructure and small enough to insure that the structural gradient across the RVE is negligible, then it is not necessary to restrict the material symmetry to accommodate the gradient. However, in a material with a structural gradient, if an RVE cannot be selected such that the structural gradient across an adequately sized RVE is negligible, then it is not necessary to restrict the material symmetry to accommodate the gradient.

If it is necessary to restrict the material symmetry to accommodate the material structural gradient, then for each RVE both an average gradient of structure (e.g., gradient of porosity) and an average structure value (e.g., porosity) are possible. In this case the average structure value (e.g., porosity) must be a potential function; the integral of the average gradient of structure around any closed path must be zero. Examples of such potential fields include the gravitational and electrostatic force fields.

References

[1] Voigt, W. (1910): Lehrbuch der Kristallphysik. Teubner, Leipzig

[2] Hearmon, R.F.S. (1961): An introduction to applied anisotropic elasticity. Oxford University Press, London

[3] Cowin, S.C., Mehrabadi, M.M. (1995): Anisotropic symmetries of linear elasticity. Appl. Mech. Rev. **48**, 247–285

[4] Chadwick, P., Vianello, M., Cowin, S.C. (2001): A proof that the number of linear anisotropic elastic symmetries is eight. J. Mech. Phys. Solids **49**, 2471–2492

[5] Cowin, S.C., Mehrabadi, M.M. (1987): On the identification of material symmetry for anisotropic elastic materials. Quart. J. Mech. Appl. Math. **40**, 451–476

[6] Fedorov, F.I. (1968): Theory of elastic waves in crystals. Plenum Press, New York

[7] Love, A.E.H. (1944): A treatise on the mathematical theory of elasticity. Dover, New York

Gaussian Curvature and Babuška's Paradox in the Theory of Plates

Cesare Davini

Abstract. Within the functional framework of the trace spaces of H^2 functions on 2-D domains with corners, in this paper we discuss the geometrical reasons for the so-called Babuška's paradox in the theory of plates and other related questions.

1 Introduction

Boundary value problems over nonsmooth domains are very special and can give rise to unexpected or curious outcomes. A classical example is the so-called Babuska's paradox pointed out in a paper over forty years ago [1]. It concerns the approximation of a circular plate, simply supported and uniformly loaded, by means of a sequence of regular polygonal plates inscribed in the circle also simply supported and uniformly loaded. Surprisingly, the solution for the polygons does not converge to that for the circle when the number of sides of the polygons tends to infinity. The difference is substantial. For instance, for a Poisson ratio $\nu = 0.3$, we calculated [13] that the normalized transverse displacement at the center of the plate is $4.347E-2$ for the circle, whereas the limit displacement for the polygonal plates is $3.199E-2$.

Various authors have tried to explain the discrepancy as an inadequacy of the mathematical model near the boundary (see, for instance, [2]); others have studied it as a peculiar feature of the mathematical model, revealing the intrinsic instability of the problem with respect to certain domain perturbations [10]. Certainly, this phenomenon and the question of the regularity of the solutions near the corners play a relevant role in the performances of various approximation schemes [3].

Working at a nonstandard discretization method for Kirchhoff-Love plates together with I. Pitacco [4], we happened to focus on the special properties of the contribution given to the strain energy

$$\mathcal{E}(u) = \frac{1}{2}\int_{\Omega}\left(|\Delta u|^2 - 2(1-\nu)\det \nabla^2 u\right)\,dx$$

of the plate by the Gaussian curvature $\det \nabla^2 u$ of the deformed middle surface. These properties have a neat geometrical nature that quite simply explains the root of Babuška's paradox and why it was to be expected. I have found no mention of this in the literature. The primary scope of this paper is to illustrate this and to discuss, in the appropriate functional framework, some properties of the class of functions used in the variational theory.

2 Preliminaries and notation

In what follows we deal with functions defined over an open bounded domain $\Omega \subset \Re^2$ simply connected and with Lipschitz boundary. We shall consider the case in which $\partial\Omega$ is the union of a finite number of disjoint arcs Γ_k, $k = 1, \dots, N$, of class C^2 up to the relative boundary. The notation is standard. $H^m(\Omega)$, with m a nonnegative integer, will denote the Sobolev space of all distributions v whose derivatives $D^l v$ belong to $L^2(\Omega)$ for all multi-indices l such that $|l| \leq m$, and $H^s(\partial\Omega)$, with s real and non-negative, the closure with respect to the appropriate norm of the space $\mathcal{D}(\partial\Omega)$ of all infinitely differentiable functions defined over $\partial\Omega$. Since $\partial\Omega$ has no boundary, there is no ambiguity in defining $H^{-s}(\partial\Omega)$ as the dual of $H^s(\partial\Omega)$. We refer to [8] for the essential details on these spaces. We also follow Grisvard in denoting by $\widetilde{H}^s(\Gamma_k)$, $s \geq 0$, the subspace of functions that belong to $H^s(\partial\Omega)$ when they are continued by zero outside Γ_k, and by $\widetilde{H}^{-s}(\partial\Omega)$ its dual.

In the next section it is shown that the integral of the Gaussian curvature relative to functions $v \in H^2(\Omega)$ reduces to a duality between the traces of the partial derivatives of v on $\partial\Omega$ and their distributional derivatives. Since $v_{,\alpha} \in H^1(\Omega)$, $\alpha = 1, 2$, it is known that for a Lipschitz domain these traces and their derivatives belong to $H^{1/2}(\partial\Omega)$ and $H^{-1/2}(\partial\Omega)$ respectively. Thus, while the degree of regularity of the boundary is of the essence when the characterization of the trace spaces of functions in $H^m(\Omega)$ with $m > 1$ is involved, as will turn out to be crucial at various stages in what follows, for the purpose of studying the integral of the Gaussian curvature it is enough to concentrate on an intrinsic definition of these spaces.

An intrinsic characterization of the spaces $H^s(\mathcal{S})$, $s \in \Re$, on a closed rectifiable simple curve $\mathcal{S}$ can be given by means of the Fourier coefficients of functions and distributions defined over $\mathcal{S}$. What follows makes this explicit in the simplest instance in which the definition of these spaces is given by using the Laplace-Beltrami operator and the theory of interpolation spaces (see [9, Remark 7.6]). Hereafter we concentrate on $H^{1/2}(\partial\Omega)$ and $H^{-1/2}(\partial\Omega)$.

Let the length of $\partial\Omega$ be 2π; we denote

$$\varphi_n(t) = \frac{1}{\sqrt{2\pi}} \exp^{\mathrm{int}}, \quad n \in \mathbb{Z},$$

the eigenfunctions of the equation $-\varphi'' = \lambda\varphi$ on $\partial\Omega$. Then, every $v \in L^2(\partial\Omega)$ can be represented in the Fourier form

$$v = \sum v_n \varphi_n \tag{1}$$

with coefficients v_n defined by

$$v_n := \int_0^{2\pi} v(t)\overline{\varphi_n(t)}\, dt.$$

In particular,

$$\|v\|^2_{L^2(\partial\Omega)} = \sum |v_n|^2 \qquad \text{(Parceval).}$$

Note that the Fourier coefficients are well-defined for the distributions because $\varphi_n \in \mathcal{D}(\partial\Omega)$. Thus, $L^2(\partial\Omega)$ can be defined as

$$L^2(\partial\Omega) = \left\{ v \in \mathcal{D}'(\partial\Omega) : \sum |v_n|^2 < +\infty, \quad v_n = \langle v, \varphi_n \rangle_{\mathcal{D}(\partial\Omega)} \right\}. \tag{2}$$

According to the definition, $H^{1/2}(\partial\Omega)$ is the interpolation space $\left[H^1(\partial\Omega), L^2(\partial\Omega)\right]_{1/2}$. It follows that $H^{1/2}(\partial\Omega)$ can be described as

$$H^{1/2}(\partial\Omega) = \left\{ u \in \mathcal{D}'(\partial\Omega) : \sum |u_n|^2 \, |n| < +\infty \right\}, \tag{3}$$

which is a Hilbert space with inner product

$$(u, v)_{H^{1/2}(\partial\Omega)} = \sum_{n \neq 0} u_n \, \bar{v}_n \, |n| + u_0 \, \bar{v}_0, \quad \text{with } u, v \in H^{1/2}(\partial\Omega). \tag{4}$$

Similarly, we define the Hilbert space

$$V = \left\{ u \in \mathcal{D}'(\partial\Omega) : \sum_{n \neq 0} \frac{|u_n|^2}{|n|} < +\infty \right\} \tag{5}$$

with inner product

$$(u, v)_V = \sum_{n \neq 0} \frac{u_n \, \bar{v}_n}{|n|} + u_0 \, \bar{v}_0, \quad \text{with } u, v \in V. \tag{6}$$

Between V and $H^{1/2}(\partial\Omega)$ there is a natural duality defined by the operation

$$\langle u, v \rangle_{V, H^{1/2}(\partial\Omega)} := \sum u_n \, \bar{v}_n. \tag{7}$$

In particular, the Cauchy–Schwarz inequality

$$\left| \sum u_n \, \bar{v}_n \right| \leq \left(\sum_{n \neq 0} \frac{|u_n|^2}{|n|} + |u_0|^2 \right)^{1/2} \left(\sum_{n \neq 0} |v_n|^2 \, |n| + |v_0|^2 \right)^{1/2} \tag{8}$$

implies that, for every $u \in V$, the operation in Eq. (7) defines a linear and continuous functional on $H^{1/2}(\partial\Omega)$, that is,

$$V \subset H^{-1/2}(\partial\Omega). \tag{9}$$

On the other hand, recalling the Riesz-Fréchet theorem, let $T : H^{-1/2}(\partial\Omega) \to H^{1/2}(\partial\Omega)$ be the isomorphism defined by the condition

$$u \in H^{-1/2}(\partial\Omega), \quad u(v) = (Tu, v)_{H^{1/2}(\partial\Omega)} \quad \forall v \in H^{1/2}(\partial\Omega),$$

or, equivalently, by Eq. (4),

$$u(v) = \sum_{n \neq 0} (Tu)_n \, \bar{v}_n \, |n| + (Tu)_0 \, \bar{v}_0 \quad \forall v \in H^{1/2}(\partial\Omega). \tag{10}$$

As $H^{1/2}(\partial\Omega) \supset \mathcal{D}(\partial\Omega)$, it follows that $H^{-1/2}(\partial\Omega) \subset \mathcal{D}'(\partial\Omega)$. Thus, the Fourier coefficients $u_n = u(\varphi_n)$ of u are well-defined and, by Eq. (10), they are related to those of Tu by

$$\begin{aligned} (Tu)_n &= \frac{u_n}{|n|}, \quad n \neq 0, \\ (Tu)_0 &= u_0. \end{aligned} \tag{11}$$

Thus, by observing that $Tu \in H^{1/2}(\partial\Omega)$, we wee that

$$\sum_{n \neq 0} \frac{|u_n|^2}{|n|} = \sum_{n \neq 0} |(Tu)_n|^2 \, |n| < +\infty.$$

Thus, $u \in V$ which proves that $H^{-1/2}(\partial\Omega)$ coincides with V.

Accordingly,

$$H^{-1/2}(\partial\Omega) = \left\{ u \in \mathcal{D}'(\partial\Omega) : \sum_{n \neq 0} \frac{|u_n|^2}{|n|} < +\infty \right\} \tag{12}$$

which is a Hilbert space with inner product

$$(u, v)_{H^{-1/2}(\partial\Omega)} = \sum_{n \neq 0} \frac{u_n \, \bar{v}_n}{|n|} + u_0 \, \bar{v}_0. \tag{13}$$

The duality pairing between $H^{-1/2}(\partial\Omega)$ and $H^{1/2}(\partial\Omega)$ is then written as

$$\langle u, v \rangle := \sum u_n \, \bar{v}_n, \tag{14}$$

with $u \in H^{-1/2}(\partial\Omega)$.

By recalling Eq. (8), it is not difficult to see that

$$\sup_{\|v\|_{H^{1/2}(\partial\Omega)}=1} |u(v)| = \left(\sum_{n \neq 0} \frac{|u_n|^2}{|n|} + |u_0|^2 \right)^{1/2} \left(\equiv \|u\|_{H^{-1/2}(\partial\Omega)} \right) \tag{15}$$

and thence that $H^{-1/2}(\partial\Omega)$ with inner product defined in Eq. (13) is endowed with the dual norm. It is also clear from Eq. (14) that the duality pairing $\langle \cdot, \cdot \rangle$ is the extension of the L^2-inner product and reduces to it whenever $u \in L^2(\partial\Omega)$.

Some straightforward consequences follow. Let t be the arc length along $\partial\Omega$ and let $(\cdot)_{,t}$ denote the distributional differentiation with respect to t.

Lemma 1. *If* $u \in H^{1/2}(\partial\Omega)$, *then* $u_{,t} \in H^{-1/2}(\partial\Omega)$ *and*

$$\|u_{,t}\|^2_{H^{-1/2}(\partial\Omega)} = \|u\|^2_{H^{1/2}(\partial\Omega)} - |u_0|^2. \tag{16}$$

Proof. The relation

$$\langle u_{,t}\,,\ \varphi_n\rangle_{\mathcal{D}(\partial\Omega)} = -(u, (\varphi_n)_{,t})_{L^2(\partial\Omega)} = i\,n(u, \varphi_n)_{L^2(\partial\Omega)}$$

yields

$$(u_{,t})_n = (i\,n)u_n. \tag{17}$$

It follows that

$$\sum_{n\neq 0} \frac{|(u_{,t})_n|^2}{|n|} = \sum_{n\neq 0} |u_n|^2\,|n| \ < +\infty.$$

So $u_{,t} \in H^{-1/2}(\partial\Omega)$. Furthermore, equality (16) holds.

Equations (16) and (17) imply

Corollary 1. *The distributional differentiation defines a continuous linear operator from* $H^{1/2}(\partial\Omega)$ *onto the subspace* $\{v \in H^{-1/2}(\partial\Omega) : \langle v, \varphi_0\rangle_{\mathcal{D}(\partial\Omega)} = 0\}$.

The continuity of the differentiation operator from spaces H^s into H^{s-1} when s and $s-1$ have the same sign is general (cf. [8]). The property established in this corollary is specific for domains without boundary and can be extended, as the definitions above, to pairs of spaces $H^s(\partial\Omega)$ and $H^{s-1}(\partial\Omega)$ for all $s \in \Re$.

Finally, we have

Theorem 1 (Green's formula).

$$\langle u_{,t}\,, v\rangle = -\langle v_{,t}\,, u\rangle \quad \forall\ u, v \in H^{1/2}(\partial\Omega). \tag{18}$$

Proof.

$$\langle u_{,t}\,, v\rangle = \sum (u_{,t})_n \bar{v}_n = \sum (in)u_n\bar{v}_n = -\sum u_n\overline{(in)v_n} = -\overline{\langle v_{,t}\,, u\rangle}.$$

For real functions the above relation reduces to Eq. (18). □

The first characterization of the trace space for H^1-functions over Lipschitz domains is due to Gagliardo [6]; that of the traces of functions of class W^m_p over non-smooth domains in $\Re^2$ was given by Yakovlev [14]. Fundamental contributions on trace spaces and on the related topic of Sobolev spaces with weighted norms were added later by various authors. The subject turns to be technical and intricate, but a general account, sufficient for our aims, can be found in the books by Nečas [11] and Grisvard [7], [8].

For later use we recall Yakovlev's theorem in a form specialized to our case.

Theorem 2 (Yakovlev, 1961). *Let $\Omega \subset \Re^2$ have boundary which is the union of disjoint arcs Γ_i, $i = 1, \dots, N$, of class C^2 up to the boundary, let $v \in H^2(\Omega)$ and let γ denote the operator that extracts the trace on $\partial\Omega$. Then, if $g_0 := \gamma v$ and $g_1 := \gamma \frac{\partial v}{\partial n}$, we have*

(i) $g_k \in H^{3/2-k}(\Gamma_i)$ *for* $k = 0, 1, i = 1, \dots N$;
(ii) $g_0 \in H^1(\partial\Omega)$;
(iii) $\psi_\alpha := (g_{0,t}\, t_\alpha + g_1 n_\alpha) \in H^{1/2}(\partial\Omega)$ $\alpha = 1, 2$, *where t_α and n_α are the Cartesian components of the unit vectors tangent and normal to the boundary, respectively.*

In addition, we have the inequality

$$\sum_{k=0}^{1} \sum_{i=0}^{N} \|g_k\|_{H^{3/2-k}(\Gamma_i)} + \sum_{\alpha=0}^{1} \|\psi_\alpha\|_{H^{1/2}(\partial\Omega)} \le C\|v\|_{H^2(\Omega)} \tag{19}$$

with constant C independent of v.

Conversely, for all g_0, g_1 satisfying the previous requirements there exists $u \in H^2(\Omega)$ such that $g_0 = \gamma u$ and $g_1 = \gamma \frac{\partial u}{\partial n}$ and the inequality

$$\|v\|_{H^2(\Omega)} \le C' \left\{ \sum_{k=0}^{1} \sum_{i=0}^{N} \|g_k\|_{H^{3/2-k}(\Gamma_i)} + \sum_{\alpha=0}^{1} \|\psi_\alpha\|_{H^{1/2}(\partial\Omega)} \right\} \tag{20}$$

holds for some constant C' independent of the pair (g_0, g_1).

In current terminology, it is said that there is a lifting between $H^2(\Omega)$ and the space of the pairs (g_0, g_1) satisfying conditions (i)–(iii).

In Yakovlev's theorem ψ_α turns to be the trace of $v_{,\alpha}$ expressed in terms of the tangential and normal derivatives of v at the boundary. Thus, (iii) is nothing but the restriction imposed by Gagliardo's theorem. It establishes, however, a compatibility condition on the assignment of the traces of v and $\frac{\partial v}{\partial n}$ near the corners of $\partial\Omega$, so that the image of $H^2(\Omega)$ under the map $\gamma : v \mapsto (\gamma v, \gamma \frac{\partial v}{\partial n})$ is not the Cartesian product of the images of γv and $\gamma \frac{\partial v}{\partial n}$ separately. This makes the treatment of nonsmooth domains special. For instance, if $g_0 \equiv 0$, then $g_1 n_\alpha \in H^{1/2}(\partial\Omega)$ and this requires that $g_1 \in \prod_{i=1}^{N} \widetilde{H}^{1/2}(\Gamma_i)$ (cf. [8, Theorems 1.4.3 and 1.4.6]. Likewise, if $g_1 \equiv 0$, then $g_{0,t} \in \prod_{i=1}^{N} \widetilde{H}^{1/2}(\Gamma_i) \bigcap \{\psi \in L^2(\partial\Omega) : \int_{\partial\Omega} \psi\, dt = 0\}$, by recalling (ii) and the embedding $H^1(\partial\Omega) \subset C^0(\partial\Omega)$.

The norm of $\widetilde{H}^{1/2}(\Gamma_i)$ is defined through that of the functions extended by zero outside Γ_i..If a superscript tilde is used to indicate the result of such an extension, it turns out that

$$\|v\|^2_{\widetilde{H}^{1/2}(\Gamma_i)} := \|\tilde{v}\|^2_{H^{1/2}(\partial\Omega)} \tag{21}$$

and there is a natural embedding of $\widetilde{H}^{1/2}(\Gamma_i)$ into $H^{1/2}(\partial\Omega)$. It follows that $\widetilde{H}^{-1/2}(\Gamma_i) \supset H^{-1/2}(\partial\Omega)$. By virtue of Eq. (21) an embedding of $\widetilde{H}^{-1/2}(\Gamma_i)$ into

$H^{-1/2}(\partial\Omega)$ can be defined by setting $\tilde{l} = l$ on $\widetilde{H}^{1/2}(\Gamma_i)$ and $\tilde{l} = 0$ on its orthogonal complement in $H^{1/2}(\partial\Omega)$, $\forall l \in \widetilde{H}^{-1/2}(\Gamma_i)$. Thus, a characterization of $\widetilde{H}^{1/2}(\Gamma_i)$ and $\widetilde{H}^{-1/2}(\Gamma_i)$ through the Fourier coefficients of these extensions is possible, although it is hardly useful for various reasons.[1] In this case the duality pairing is obviously given by Eq. (14) and all the above consequences apply.

Remark. It is worth recalling that the Fourier characterization applies to all spaces $H^s(\partial\Omega)$, with $s \in \Re$, yielding

$$H^s(\partial\Omega) = \left\{ u \in \mathcal{D}'(\partial\Omega) : \sum |u_n|^2 \, |n|^{2s} < +\infty \right\},$$

and that the duality between $H^{-s}(\partial\Omega)$ and $H^s(\partial\Omega)$ is still given by Eq. (14).

3 A representation formula for the total Gaussian curvature

For $u \in H^2(\Omega)$, we consider the functional

$$\mathcal{G}(u) := \int_\Omega \det \nabla^2 u \, dx = \int_\Omega \varepsilon_{\alpha\beta} \, u_{,1\alpha} \, u_{,2\beta} \, dx \tag{22}$$

where $\varepsilon_{\alpha\beta}$ is the two-dimensional *Ricci symbol*. By reference to the geometrical meaning of the integrand, we shall call $\mathcal{G}(u)$ the *total Gaussian curvature* relative to the function u.

The functional in Eq. (22) has a nice geometrical interpretation that makes some of its features, and in particular the reasons for Babuška's paradox, quite evident. The following representation formula provides the geometrical meaning of $\mathcal{G}(u)$ (cf. Davini and Pitacco [4]).

Theorem 3.

$$\mathcal{G}(u) = \frac{1}{2} \langle \varepsilon_{\alpha\beta} (u_{,\beta})_{,t} \, , \, u_{,\alpha} \rangle \tag{23}$$

Proof. The proof is by density. Let $\{u^{(k)}\} \subset \mathcal{D}(\bar{\Omega})$ be a sequence converging to u in $H^2(\Omega)$. If we consider the expression on the right-hand side of Eq. (22) and integrate by parts, after cancellations due to the skewsymmetry of Ricci symbols and the invariance with respect of the order of differentiation in the third order derivatives, we get

$$\int_\Omega \varepsilon_{\alpha\beta} \, u_{,1\alpha}^{(k)} \, u_{,2\beta}^{(k)} \, dx = \frac{1}{2} \int_{\partial\Omega} \varepsilon_{\alpha\beta} \, (u_{,\beta}^{(k)})_{,t} \, u_{,\alpha}^{(k)} \, dt. \tag{24}$$

Hence, since the convergence of $u^{(k)}$ to u in H^2 implies that $u_{,\alpha}^{(k)} \to u_{,\alpha}$ and $(u_{,\beta})_{,t}^{(k)} \to (u_{,\beta})_{,t}$ respectively in $H^{1/2}(\partial\Omega)$ and $H^{-1/2}(\partial\Omega)$ (cf. Corollary 1), the result follows by continuity on recalling that the duality pairing $\langle \cdot, \cdot \rangle$ is the continuous extension of the scalar product in L^2. □

---

[1] For instance, it is difficult to express the fact that supp $\tilde{v} \subset \Gamma_i$, or that the kernel of $\tilde{l}$ contains the orthogonal complement of $\widetilde{H}^{1/2}(\Gamma_i)$, by means of Fourier coefficients. Furthermore, we are not always free to choose $\tilde{l}$ in the orthogonal complement of $\widetilde{H}^{1/2}(\Gamma_i)$ in $H^{1/2}(\partial\Omega)$.

Geometrically the boundary integral in Eq. (24), and thence the pairing in Eq. (23), represents twice the area, with sign, of the region bounded by the image of the oriented curve $\partial\Omega$ under the application $G : x \mapsto \nabla v(x)$. Here G is the linearized version of the *Gauss map* and the image of $\partial\Omega$ is the *Gauss diagram*; see Fig. 1. One might use this interpretation and the duality on the right-hand side of Eq. (23)

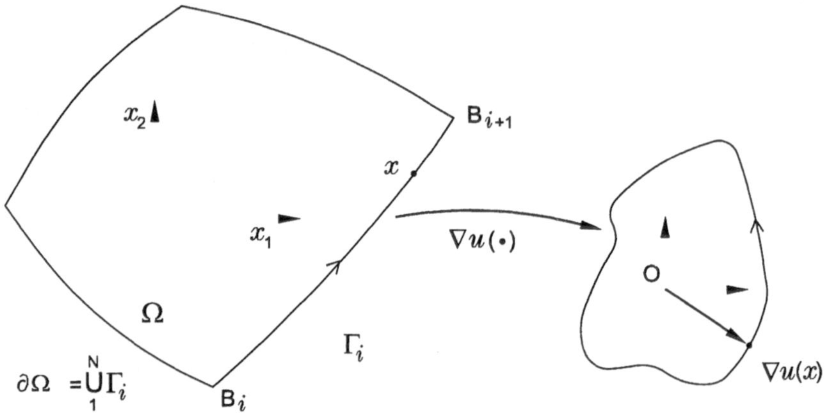

Fig. 1. Geometrical interpretation of Gaussian curvature

in order to generalize the notion of Gaussian curvature. This is how Pogorelov [12] extended that notion to the vertex of a convex polyhedron in proving the existence of solutions to Minkowski's problem. Davini and Pitacco [4] also adopted this view in their approximation scheme for the equilibrium problems of the linear theory of plates. Here it suffices to note that, for given u, one can calculate the total Gaussian curvature from the Gauss diagram of any sequence of smooth functions converging to u in H^2 and then pass to the limit, by continuity.

We focus on a case that is related to Babuška's paradox. Assume that Ω is polygonal and $u \in H^2(\Omega) \cap H_0^1(\Omega)$. Also, let $u^{(k)} \in H^2(\Omega) \cap H_0^1(\Omega)$ be a sequence of smooth functions converging to u. Since the gradients of the $u^{(k)}$ along $\partial\Omega$ are orthogonal to the sides of the polygon, when point x runs along $\partial\Omega$ its image under the Gauss map moves along rays whose directions are orthogonal to the sides of the polygon. In particular, the passage from one ray to the other occurs through the origin of the representation plane; see Fig. 2. Thus, the area of the region comprised within the Gauss diagram is zero and it follows that

$$\mathcal{G}(u) = \lim_k \mathcal{G}(u^{(k)}) = 0.$$

For a domain determined by a finite number of arcs, the representation formula (23) can be further exploited to highlight the role of the tangential and normal derivatives of the functions at the boundary. We start from smooth functions $u^{(k)}$ so that

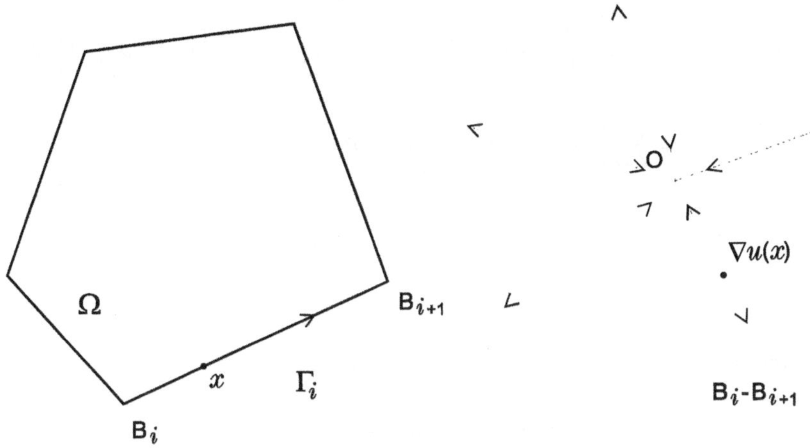

Fig. 2. Gauss diagram for functions in $H^2 \cap H_0^1$ on polygons

Eq. (24) applies:

$$\mathcal{G}(u^{(k)}) = \frac{1}{2} \sum_{i=1}^{N} \int_{\Gamma_i} \varepsilon_{\alpha\beta} \, (u,_\beta^{(k)}),_t \; u,_\alpha^{(k)} \, dt. \tag{25}$$

By writing $u,_\alpha^{(k)} = u,_n^{(k)} n_\alpha + u,_t^{(k)} t_\alpha$, applying Leibniz formula to calculate $(u,_\beta^{(k)}),_t$ and taking into account the Frenet-Serret formula, $n_\beta,_t = -\kappa t_\beta$, after some cancellations we get

$$\begin{aligned} \mathcal{G}(u^{(k)}) = & -\frac{1}{2} \sum_{i=1}^{N} \int_{\Gamma_i} \kappa (u,_n^{(k)\,2} + u,_t^{(k)\,2}) \, dt \\ & + \frac{1}{2} \sum_{i=1}^{N} \int_{\Gamma_i} \left(u,_{tt}^{(k)} \, u,_n^{(k)} - u,_{nt}^{(k)} \, u,_t^{(k)} \right) dt. \end{aligned} \tag{26}$$

Here κ is the curvature of the boundary, assumed to be positive if the center of curvature is along the outward normal to $\partial\Omega$. Thence, if we integrate the first term in the second integral by parts and manipulate the expression suitably, we find that[2]

$$\begin{aligned} \mathcal{G}(u^{(k)}) = & -\frac{1}{2} \sum_{i=1}^{N} \int_{\Gamma_i} \kappa \left(u,_n^{(k)\,2} - u,_t^{(k)\,2} \right) dt \\ & - \frac{1}{2} \sum_{i=1}^{N} \left[\left| u,_t^{(k)} \, u,_n^{(k)} \right|\right] (B_i) - \sum_{i=1}^{N} \int_{\Gamma_i} (u,_{nt}^{(k)} + \kappa u,_t^{(k)}) u,_t^{(k)} \, dt \end{aligned} \tag{27}$$

[2] The reason for giving the last integral the form exhibited in Eq. (27) will become clear later.

and, by performing a second integration by parts,

$$\begin{aligned}\mathcal{G}(u^{(k)}) = & -\frac{1}{2}\sum_{i=1}^{N}\int_{\Gamma_i}\kappa(u,_n^{(k)\,2} - u,_t^{(k)\,2})\,dt - \frac{1}{2}\sum_{i=1}^{N}\left[\left| u,_t^{(k)}\, u,_n^{(k)} \right|\right](B_i) \\ & + \sum_{i=1}^{N}\left[\left| u,_{nt}^{(k)} + \kappa u,_t^{(k)} \right|\right] u^{(k)}(B_i) \\ & + \sum_{i=1}^{N}\int_{\Gamma_i}\left(u,_{nt}^{(k)} + \kappa u,_t^{(k)}\right),_t u^{(k)}\,dt.\end{aligned} \tag{28}$$

In the previous formulae the symbol $[|\cdot|]$ stands for the jump across the vertices B_i, evaluated according to the chosen orientation of $\partial\Omega$.

If $u^{(k)} \underset{H^2}{\longrightarrow} u$, one can pass to the limit in Eq. (28). Since $u,_n^{(k)} \underset{L^2(\Gamma_i)}{\longrightarrow} u,_n$ and $u,_t^{(k)} \underset{L^2(\Gamma_i)}{\longrightarrow} u,_t$ by Yakovlev's theorem and by the continuity of the embedding $H^{1/2}(\Gamma_i) \subset L^2(\Gamma_i)$, it follows that

$$\begin{aligned}\mathcal{G}(u) = & -\frac{1}{2}\sum_{i=1}^{N}\int_{\Gamma_i}\kappa\left(u,_n^2 - u,_t^2\right)dt - \lim_k\left\{\frac{1}{2}\sum_{i=1}^{N}\left[\left| u,_t^{(k)}\, u,_n^{(k)} \right|\right](B_i)\right. \\ & - \sum_{i=1}^{N}\left[\left| u,_{nt}^{(k)} + \kappa u,_t^{(k)} \right|\right] u^{(k)}(B_i) \\ & \left. - \sum_{i=1}^{N}\int_{\Gamma_i}\left(u,_{nt}^{(k)} + \kappa u,_t^{(k)}\right),_t u^{(k)}\,dt\right\}.\end{aligned} \tag{29}$$

Note that the existence of the limits of the first two terms in this equation implies that of the quantity within the braces and the independence of that limit from the approximation sequence. Indeed, this result can be localized at each single vertex by using a partition of unity.

The following theorem is a straightforward consequence of Eq. (29).

Theorem 4. *Let* $u \in H^2(\Omega)\cap H_0^1(\Omega)$. *Then*

$$\mathcal{G}(u) = -\frac{1}{2}\sum_{i=1}^{N}\int_{\Gamma_i}\kappa u,_n^2\,dt. \tag{30}$$

Proof. $H^2(\Omega)\cap H_0^1(\Omega)$ can be regarded as the completion of the smooth functions that vanish on $\partial\Omega$. If $\{u^{(k)}\}$ if any such approximation sequence for u, then $u^{(k)} \equiv 0$ on $\partial\Omega$, which implies that $\nabla u^{(k)}(B_i) = 0$. Thence, $u,_t^{(k)}$ and $u,_n^{(k)}$ vanish on the two sides of B_i. So, the term within the braces is zero for all k. □

For polygons, Theorem 4 confirms the conclusion obtained from the Gauss diagram. However, it also shows that, for curvilinear polygons or smooth domains, the total Gaussian curvature relative to functions in $H^2(\Omega)\cap H_0^1(\Omega)$ is generally different from zero. For instance, for the unit circle, where $\kappa = -1$, or more generally for convex domains, where $\kappa < 0$, we have $\mathcal{G}(u) > 0$ unless $\gamma u,_n = 0$. It is natural, then, that the space $H^2(\Omega)\cap H_0^1(\Omega)$ on the circle cannot be approximated by the analogous spaces defined on inscribed polygons, as is clear from Babuška's paradox. Regarded from another viewpoint, the presence of the boundary curvature in Eq. (30) emphasizes the fact that passing from the circle to the inscribed polygons is not a small perturbation as far as the basic properties of the function space $H^2(\Omega)\cap H_0^1(\Omega)$ are concerned.

Finally, it is worth mentioning a direct consequence of Theorem 4.

Theorem 5. *In $H^2(\Omega)\cap H_0^1(\Omega)$ the total Gaussian curvature is continuous with respect to the weak topology:*

$$\{u^{(k)}\},\ u \text{ in } H^2(\Omega)\cap H_0^1(\Omega),\ u^{(k)} \underset{H^2}{\rightharpoonup} u \quad \Rightarrow \quad \lim_k \mathcal{G}(u^{(k)}) = \mathcal{G}(u). \tag{31}$$

In fact, from Eq. (30), the total Gaussian curvature depends on the trace of $u,_n$ continuously in the norm of $L^2(\partial\Omega)$. Since it can be shown that $u^{(k)} \underset{H^2(\Omega)}{\rightharpoonup} u$ implies that $u,_n^{(k)} \underset{H^{1/2}(\partial\Omega)}{\rightharpoonup} u,_n$ and since the embedding of $H^{1/2}(\partial\Omega)$ into $L^2(\partial\Omega)$ is compact, the result follows.

Theorem 5 implies that the functional $\int_\Omega \det \nabla^2 u \, dx$ attains both its minimum and maximum in every weakly compact set. This property plays a part in the convergence of a non-classical discretization scheme for Kirchhoff-Love plates discussed by Davini [5].

4 Green's formula and related boundary operators

In the theory of plates the presence of Gaussian curvature in the strain energy functional

$$\frac{1}{2}\int_\Omega (|\Delta u|^2 - 2(1-\nu)\det \nabla^2 u)\, dx$$

is responsible for the special form of the boundary operators appearing in the equilibrium problem. The aim of this section is to analyze this aspect within the functional framework described above.

We calculate the Fréchet differential of the total Gaussian curvature at a point u

$$\delta\mathcal{G}(u)[v] = \int_\Omega \varepsilon_{\alpha\beta}(u,_{1\alpha}\ v,_{2\beta} + v,_{1\alpha}\ u,_{2\beta})\, dx.$$

If we restrict our attention to smooth functions for the moment, we can alternatively use formula (26) to calculate the differential. Omitting the upper labels for simplicity, we obtain

$$\delta\mathcal{G}(u)[v] = -\frac{1}{2}\int_{\partial\Omega}(2\kappa(u_{,n}\, v_{,n} + u_{,t}\, v_{,t}) + u_{,t}\, v_{,nt} + u_{,nt}\, v_{,t} - u_{,tt}\, v_{,n} - u_{,n}\, v_{,tt})\, dt. \quad (32)$$

Thence, after integrating by parts to eliminate the tangential derivatives of v, we get in succession

$$\delta\mathcal{G}(u)[v] = -\frac{1}{2}\sum_{i=1}^{N}[|u_{,n}\, v_{,t} - u_{,t}\, v_{,n}|](B_i) + \sum_{i=1}^{N}\int_{\Gamma_i}(u_{,tt} - \kappa u_{,n})v_{,n}\, dt - \sum_{i=1}^{N}\int_{\Gamma_i}(u_{,nt} + \kappa u_{,t})v_{,t}\, dt \quad (33)$$

and

$$\delta\mathcal{G}(u)[v] = -\frac{1}{2}\sum_{i=1}^{N}[|u_{,n}\, v_{,t} - u_{,t}\, v_{,n}|](B_i) + \sum_{i=1}^{N}\int_{\Gamma_i}(u_{,tt} - \kappa u_{,n})v_{,n}\, dt + \sum_{i=1}^{N}[|u_{,nt} + \kappa u_{,t}|]\, v(B_i) + \sum_{i=1}^{N}\int_{\Gamma_i}(u_{,nt} + \kappa u_{,t})_{,t}\, v\, dt. \quad (34)$$

Arguing by density enables us to extend Eq. (33), or equivalently Eq. (34), to all u and v in $H^2(\Omega)$.

Formally, for each u, $\delta\mathcal{G}(u)[\cdot]$ defines an element of the dual of the space of pairs (g_0, g_1) described by Yakovlev's theorem. It is possible, though, to extract more detailed information on the trace of the expressions appearing in the two integrals on the right-hand side of Eq. (34) by considering sections of that space.

Theorem 6. *Green's formula (34) uniquely defines the quantities* $[|u_{,nt} + \kappa u_{,t}|](B_i)$ *and two continuous operators* $u \mapsto \gamma(u_{,tt} - \kappa u_{,n})$ *and* $u \mapsto \gamma(u_{,nt} + \kappa u_{,t})_{,t}$ *from* $H^2(\Omega)$ *into* $\prod_{i=1}^{N}\widetilde{H}^{-1/2}(\Gamma_i)$ *and* $\prod_{i=1}^{N}\widetilde{H}^{-3/2}(\Gamma_i)$*, respectively, which are the extensions of the corresponding operators for smooth functions.*

Proof. Let v be smooth and vanish on $\partial\Omega$. Then $v = v_{,t} = v_{,n} = 0$ at the B_i. Thus, Eq. (34) reduces to the form $\sum_{i=1}^{N}\int_{\Gamma_i}(u_{,tt} - \kappa u_{,n})v_{,n}\, dt = \delta\mathcal{G}(u)[v]$ and, by continuity, it extends to all $v \in H^2(\Omega)\cap H_0^1(\Omega)$:

$$\sum_{i=1}^{N}\int_{\Gamma_i}(u_{,tt} - \kappa u_{,n})v_{,n}\, dt = \delta\mathcal{G}(u)[v] \quad \forall\, v \in H^2(\Omega)\cap H_0^1(\Omega) \quad (35)$$

Since $v \mapsto \gamma v_{,n}$ is a continuous map from $H^2(\Omega)\cap H_0^1(\Omega)$ onto $\prod_{i=1}^{N}\widetilde{H}^{1/2}(\Gamma_i)$ which admits a lifting and since $\delta\mathcal{G}(u)[\cdot]$ is a continuous functional on $H^2(\Omega)$, the previous equation uniquely defines $\gamma(u_{,tt} - \kappa u_{,n})$ as an element of $\prod_{i=1}^{N}\widetilde{H}^{-1/2}(\Gamma_i)$. With $\langle\cdot,\cdot\rangle_{\prod_{i=1}^{N}\widetilde{H}^{1/2}(\Gamma_i)}$ denoting the duality between the two spaces, a continuous extension of the inner product in L^2, Green's formula (35) can be written in the form

$$\langle u,_{tt} - \kappa u,_n , v,_n \rangle_{\prod_{i=1}^{N} \tilde{H}^{1/2}(\Gamma_i)} = \delta\mathcal{G}(u)[v] \quad \forall v \in H^2(\Omega) \cap H_0^1(\Omega) \tag{36}$$

Note that, by Eq. (35), $\delta\mathcal{G}(u)[v] = \delta\mathcal{G}(u)[w]\ \forall\, v, w \in H^2(\Omega)\cap H_0^1(\Omega)$ such that $\gamma v,_n = \gamma w,_n$. Therefore, the continuity of $\delta\mathcal{G}(\cdot)[\cdot]$ yields

$$|\langle u,_{tt} - \kappa u,_n , v,_n \rangle_{\prod_{i=1}^{N} \tilde{H}^{1/2}(\Gamma_i)}| \leq \inf_{w:\gamma w,_n = \gamma v,_n} \|w\|_{H^2(\Omega)} K \|u\|_{H^2(\Omega)}$$

and, from the existence of a lifting, it follows that

$$\|u,_{tt} - \kappa u,_n \|_{\prod_{i=1}^{N} \tilde{H}^{-1/2}(\Gamma_i)} := \sup_{v \in H^2(\Omega)\cap H_0^1(\Omega)} \frac{|\langle u,_{tt} - \kappa u,_n , v,_n \rangle_{\prod_{i=1}^{N} \tilde{H}^{1/2}(\Gamma_i)}|}{\|v,_n\|_{\prod_{i=1}^{N} \tilde{H}^{1/2}(\Gamma_i)}} \leq C\|u\|_{H^2(\Omega)}.$$

This implies that the operator $u \mapsto \gamma(u,_{tt} - \kappa u,_n)$, and thence the validity of formula (36), can be extended to all $u \in H^2(\Omega)$ by continuity.

Again, let v be smooth and such that $v,_n \equiv 0$ on $\partial\Omega$. It again follows that $v,_t = v,_n = 0$ at the B_i. Accordingly, the first two terms on the right-hand side of Eq. (33) vanish and the formula, extended by continuity to all $v \in H^2(\Omega)$ such that $\gamma v,_n = 0$, becomes:

$$-\sum_{i=1}^{N} \int_{\Gamma_i} (u,_{nt} + \kappa u,_t) v,_t \, dt = \delta\mathcal{G}(u)[v] \qquad \forall v \in H^2(\Omega) : \gamma v,_n = 0. \tag{37}$$

By Yakovlev's theorem there is a continuous lifting between $\{v \in H^2(\Omega) : \gamma v,_n = 0\}$ and the space V of functions

$$v \in \prod_{i=1}^{N} H^{3/2}(\Gamma_i) \cap H^1(\partial\Omega)$$

such that $v,_t \in \prod_{i=1}^{N} \tilde{H}^{1/2}(\Gamma_i)$ or, equivalently, the space

$$V_t = \prod_{i=1}^{N} \tilde{H}^{1/2}(\Gamma_i) \cap \left\{ \psi \in L^2(\Omega) : \int_{\partial\Omega} \psi \, dt = 0 \right\},$$

to which the $v,_t$ belong. In particular, Eq. (37) defines an element in the dual of the latter space and can be written as

$$-\sum_{i=1}^{N} \langle (u,_{nt} + \kappa u,_t), v,_t \rangle_{\tilde{H}^{1/2}(\Gamma_i)} = \delta\mathcal{G}(u)[v] \qquad \forall v \in H^2(\Omega) : \ \gamma v,_n = 0. \tag{38}$$

By iterating a previous argument this formula can be extended to all $u \in H^2(\Omega)$.

In particular, when $v,_t \in \widetilde{H}^{1/2}(\Gamma_i)$, Eq. (38) uniquely defines $(u,_{nt} + \kappa u,_t),_t \in \widetilde{H}^{-3/2}(\Gamma_i)$. It follows that $(u,_{nt} + \kappa u,_t) \in \widetilde{H}^{-1/2}(\Gamma_i)$ is defined up to a constant $c_i(u)$ in each arc. Accordingly, we can write

$$(u,_{nt} + \kappa u,_t) = (u,_{nt} + \kappa u,_t)^\circ + \sum_{i=1}^{N} c_i(u)\chi_{\Gamma_i}, \tag{39}$$

with χ_{Γ_i} the characteristic function of Γ_i and $(u,_{nt} + \kappa u,_t)^\circ \in \prod_{i=1}^{N} \widetilde{H}^{-1/2}(\Gamma_i)$ such that its distributional derivative does not contain atoms at the end points of the arcs Γ_i.

Equation (39) yields

$$(u,_{nt} + \kappa u,_t),_t = (u,_{nt} + \kappa u,_t)^\circ,_t + \sum_{i=1}^{N} [|c_i(u)|]\,\delta(B_i). \tag{40}$$

If we consider $v \in V$ in Eq. (38) and take account of Eq. (40), we get the formula

$$\begin{aligned} &\sum_{i=1}^{N} [|c_i(u)|]\, v(B_i) - \sum_{i=1}^{N} \langle (u,_{nt} + \kappa u,_t)^\circ, v,_t \rangle_{\widetilde{H}^{1/2}(\Gamma_i)} \\ &\quad = \sum_{i=1}^{N} [|c_i(u)|]\, v(B_i) + \langle (u,_{nt} + \kappa u,_t)^\circ,_t\,, v \rangle \\ &\quad = \delta\mathcal{G}(u)[v] \quad \forall\, v \in H^2(\Omega) : \gamma v,_n = 0. \end{aligned} \tag{41}$$

By the fact that of $v(B_i)$ is arbitrary, this equation uniquely determines the $[|c_i(u)|]$.

It is tempting to identify $[|c_i(u)|]$ with $[|(u,_{nt} + \kappa u,_t)|]\,(B_i)$. Notice, however, that these quantities need not be the difference of the left and right ordinary limits of $(u,_{nt} + \kappa u,_t)$ because this notion might be meaningless for general functions $u \in H^2(\Omega)$. $\square$

In a curvilinear system of coordinates $\{t, \zeta\}$ where t is the arc-length on $\partial\Omega$ and ζ the distance from $\partial\Omega$ measured along the rays orthogonal to it, torsional and bending moments at the boundary for the Kirchhoff-Love model are given by

$$\begin{aligned} M_{tn} &= (1-\nu)(u,_{nt} + \kappa u,_t), \\ M_{nn} &= u,_{nn} + \nu(u,_{tt} - \kappa u,_\nu). \end{aligned} \tag{42}$$

Furthermore, the Laplacian operator takes the form

$$\Delta u = u,_{tt} + u,_{nn} - \kappa u,_\nu\,.$$

It follows that

$$\Delta u - M_{nn} = (1-\nu)(u,_{tt} - \kappa u,_n). \tag{43}$$

Thus, to within a material factor, the terms analyzed in Theorem 6 are the essential ingredients of the boundary operators of the theory of plates. For instance, for the condition of simple support, Eq. (36) gives

$$u \in H^2(\Omega) \cap H_0^1(\Omega), \ \int_{\partial\Omega} \kappa u_{,n}\, , v_{,n}\, dt = -\delta\mathcal{G}(u)[v] \quad \forall\, v \in H^2(\Omega) \cap H_0^1(\Omega),$$

which is in accord with formula (30), and Eq. (43) provides the Neumann boundary operator of the corresponding equilibrium problem (cf. [10]):

$$M_{nn} := \Delta u + (1-\nu)\kappa u_{,n}\,. \tag{44}$$

5 Local character of the total Gaussian curvature and averages of the traces near the corners of the domain

As we have seen, because of the existence of a continuous functional such as $\mathcal{G}(u)$ which conveys the peculiarities of the boundary and of the traces of u and $u_{,n}$ inside Ω, the presence of corners substantially changes the scenario of the function space $H^2(\Omega)$. In this final section we focus on the role of the corners as localized sources of important effects on $\mathcal{G}(u)$. In particular, we study how averages of the traces of u on curves in the vicinity of a corner are affected. To do so, it is useful to abandon the notion of duality used so far and substitute it by other operations that allow us to localize the analysis more easily.

We shall restrict to a convex Ω. We introduce a sequence of sub-domains $\Omega_h \subset \Omega$ obtained by rounding the corners B_i by arcs tangent to $\partial\Omega$ and of radius ρ_h, with $\rho_h \underset{h\to\infty}{\to} 0$. Since the domains Ω_h are now of class $C^{1,1}$, the traces $\gamma u \in H^{3/2}(\partial\Omega_h)$ and $\gamma u_{,n} \in H^{1/2}(\partial\Omega_h)$ [11]. Then, by repeating the calculations that follow Eq. (25), we get

$$\mathcal{G}_{\Omega_h}(u) = -\frac{1}{2}\int_{\partial\Omega_h} \kappa u_{,n}^2\, dt - \frac{1}{2}\int_{\partial\Omega_h} \kappa u_{,t}^2\, dt - \langle u_{,nt}\, , u_{,t}\rangle_{H^{1/2}(\partial\Omega_h)}. \tag{45}$$

For given h, consider an interior neighborhood of $\partial\Omega_h$ and introduce the normal system of coordinates (t, ζ), with $\zeta \in (-\delta, 0]$ the distance (with sign) from $\partial\Omega_h$ measured along the outward normal, and t the arc-length along $\partial\Omega_h$ taken from some origin. Denote the curve at distance ζ from $\partial\Omega_h$ by $\partial\Omega_h(\zeta)$. Then, the following theorem holds.

Lemma 2. *For all $u \in H^2(\Omega)$ and for all h,*

$$\langle u_{,nt}\, , u_{,t}\rangle_{H^{1/2}(\partial\Omega_h)} = -\frac{1}{2}\int_{\partial\Omega_h} \kappa u_{,t}^2\, dt + \frac{1}{2}\left(\frac{d}{d\zeta}\int_{\partial\Omega_h(\zeta)} u_{,t'}^2\, dt'\right)_{\zeta=0}. \tag{46}$$

Here t' denotes the arc-length along $\partial\Omega_h(\zeta)$ and the differentiability of $\int_{\partial\Omega_h(\zeta)} u_{,t'}^2\, dt'$ at $\zeta = 0$ follows from the proof of the theorem.

Proof. The proof is by density. Let $\{u^{(k)}\}$ be a sequence with $u^{(k)} \in C^2(\overline{\Omega})$, so that the duality on the left-hand side of Eq. (45) reduces to an integral. Note that, on $\partial\Omega_h(\zeta)$, $u^{(k)}{}_{,n} = \dfrac{\partial u^{(k)}}{\partial \zeta}(t,\zeta)$ and $u^{(k)}{}_{,t'} = \dfrac{\partial u^{(k)}}{\partial t}(t,\zeta)\dfrac{dt}{dt'}(t,\zeta)$, where $t = \hat{t}(t',\zeta)$ is regarded as a function of t' and ζ and $\dfrac{dt}{dt'}$ denotes $\dfrac{\partial \hat{t}}{\partial t'}$. It follows that

$$\int_{\partial\Omega_h(\zeta)} u^{(k)}{}_{,t'}\, u^{(k)}{}_{,nt'}\, dt' = \int_{\partial\Omega_h(\zeta)} \frac{\partial u^{(k)}}{\partial t}\frac{\partial}{\partial t}\left(\frac{\partial u^{(k)}}{\partial \zeta}\right)\left(\frac{dt}{dt'}\right)^2 dt'. \tag{47}$$

By changing integration variable and taking into account the fact that the derivations with respect to ζ and t commute, the term on the right-hand side becomes

$$\frac{1}{2}\int_{\partial\Omega_h(\zeta)} \frac{\partial}{\partial \zeta}\left(\left(\frac{\partial u^{(k)}}{\partial t}\right)^2\right)\frac{dt}{dt'}\, dt.$$

We observe that the notation here is misleading because, due to the new parametrization, the integration domain is $\partial\Omega_h$ and is in fact independent of ζ.

Taking this into account we see that

$$\begin{aligned}&\frac{1}{2}\int_{\partial\Omega_h(\zeta)} \frac{\partial}{\partial \zeta}\left(\left(\frac{\partial u^{(k)}}{\partial t}\right)^2\right)\frac{dt}{dt'}\, dt\\ &= \frac{1}{2}\frac{d}{d\zeta}\int_{\partial\Omega_h(\zeta)}\left(\frac{\partial u^{(k)}}{\partial t}\right)^2\frac{dt}{dt'}\,dt - \frac{1}{2}\int_{\partial\Omega_h(\zeta)}\left(\frac{\partial u^{(k)}}{\partial t}\right)^2\frac{\partial}{\partial \zeta}\left(\frac{dt}{dt'}\right)dt.\end{aligned} \tag{48}$$

Thus, the term on the right-hand side of Eq. (47) can be written as

$$\begin{aligned}&\int_{\partial\Omega_h(\zeta)} \frac{\partial u^{(k)}}{\partial t}\frac{\partial}{\partial t}\left(\frac{\partial u^{(k)}}{\partial \zeta}\right)\left(\frac{dt}{dt'}\right)^2 dt'\\ &= \frac{1}{2}\frac{d}{d\zeta}\int_{\partial\Omega_h(\zeta)}\left(\frac{\partial u^{(k)}}{\partial t'}\right)^2 dt' - \frac{1}{2}\int_{\partial\Omega_h(\zeta)}\left(\frac{\partial u^{(k)}}{\partial t}\right)^2\frac{\partial}{\partial \zeta}\left(\frac{dt}{dt'}\right)dt,\end{aligned} \tag{49}$$

by changing the integration variable again.

The derivative $\dfrac{dt}{dt'}$ equals the ratio between the radii of curvature of $\partial\Omega_h$ and $\partial\Omega_h(\zeta)$. Denoting them by $\rho_h(0)$ and $\rho_h(\zeta)$, we get

$$\frac{dt}{dt'} = \frac{\rho_h(0)}{\rho_h(\zeta)} \tag{50}$$

where $\rho_h(\zeta) = \rho_h(0) + \zeta$. It follows that

$$\frac{\partial}{\partial \zeta}\left(\frac{dt}{dt'}\right) = -\frac{1}{\rho_h(\zeta)}\left(\frac{dt}{dt'}\right). \tag{51}$$

Then, on taking account of (49) and (51), Eq. (47) reads

$$\int_{\partial\Omega_h(\zeta)} u^{(k)}{}_{,t'}\, u^{(k)}{}_{,nt'}\, dt' = \frac{1}{2}\frac{d}{d\zeta}\int_{\partial\Omega_h(\zeta)} \left(\frac{\partial u^{(k)}}{\partial t'}\right)^2 dt' + \frac{1}{2}\int_{\partial\Omega_h(\zeta)} \left(\frac{\partial u^{(k)}}{\partial t'}\right)^2 \frac{1}{\rho_h(\zeta)}\, dt'. \quad (52)$$

Introduce the functions

$$g^k(\zeta) := \int_{\partial\Omega_h(\zeta)} u^{(k)}{}_{,t'}\, u^{(k)}{}_{,nt'}\, dt',$$

$$l^k(\zeta) := \frac{1}{2}\int_{\partial\Omega_h(\zeta)} \left(\frac{\partial u^{(k)}}{\partial t'}\right)^2 \frac{1}{\rho_h(\zeta)}\, dt',$$

$$f^k(\zeta) := \frac{1}{2}\int_{\partial\Omega_h(\zeta)} \left(\frac{\partial u^{(k)}}{\partial t'}\right)^2 dt',$$

and write Eq. (52) in the form

$$\frac{d}{d\zeta} f^k(\zeta) = g^k(\zeta) - l^k(\zeta). \quad (53)$$

The functions g^k, l^k, and f^k are continuous up to $\zeta = 0$, and converge, respectively, to

$$\begin{aligned} \lim_k g^k(\zeta) &= \langle u_{,nt'}\,, u_{,t'}\rangle_{H^{1/2}(\partial\Omega_h(\zeta))}, \\ \lim_k l^k(\zeta) &= \frac{1}{2}\int_{\partial\Omega_h(\zeta)} \frac{1}{\rho_h(\zeta)} u_{,t'}^2\, dt', \\ \lim_k f^k(\zeta) &= \frac{1}{2}\int_{\partial\Omega_h(\zeta)} u_{,t'}^2\, dt', \end{aligned} \quad (54)$$

when $u^{(k)} \to u$ in $H^2(\Omega)$, as is easily seen. Furthermore, the convergence of g^k and l^k is uniform in ζ. It follows that

$$\lim_k \frac{d}{d\zeta} f^k(\zeta) = \frac{d}{d\zeta}\lim_k f^k(\zeta) = \frac{1}{2}\frac{d}{d\zeta}\int_{\partial\Omega_h(\zeta)} u_{,t'}^2\, dt'.$$

Passing to the limit in Eq. (53) then yields the equation

$$\begin{aligned} \frac{1}{2}\frac{d}{d\zeta}\int_{\partial\Omega_h(\zeta)} u_{,t'}^2\, dt' &= \langle (u_{,n})_{,t'}\,, u_{,t'}\rangle_{H^{1/2}(\partial\Omega_h(\zeta))} \\ &\quad - \frac{1}{2}\int_{\partial\Omega_h(\zeta)} \frac{1}{\rho_h(\zeta)} u_{,t'}^2\, dt', \end{aligned} \quad (55)$$

which proves the result when this is calculated at $\zeta = 0$.

The uniform convergence of $l^k(\zeta)$ in Eq. $(54)_2$ is straightforward. We next consider that of $g^k(\zeta)$ in Eq. $(54)_1$. From the definition of $g^k(\zeta)$ it follows that

$$\begin{aligned}
&|g^k(\zeta) - \langle u_{,nt'}, u_{,t'}\rangle_{H^{1/2}(\partial\Omega_h(\zeta))}| \\
&\le \left|\langle u^{(k)}{}_{,nt'}, u^{(k)}{}_{,t'} - u_{,t'}\rangle_{H^{1/2}(\partial\Omega_h(\zeta))}\right| + \left|\langle u^{(k)}{}_{,nt'} - u_{,nt'}, u_{,t'}\rangle_{H^{1/2}(\partial\Omega_h(\zeta))}\right| \\
&\le \|u^{(k)}{}_{,t'} - u_{,t'}\|_{H^{1/2}(\partial\Omega_h(\zeta))} \|u^{(k)}{}_{,nt'}\|_{H^{-1/2}(\partial\Omega_h(\zeta))} \\
&\quad + \|u_{,t'}\|_{H^{1/2}(\partial\Omega_h(\zeta))} \|u^{(k)}{}_{,n} - u_{,n}{}_{,t'}\|_{H^{-1/2}(\partial\Omega_h(\zeta))}.
\end{aligned} \tag{56}$$

Therefore, since $v \in H^2(\Omega)$ is mapped continuously into the traces of $v_{,t}$ and $v_{,n}$ and the derivative maps $H^{1/2}(\partial\Omega_h(\zeta))$ continuously into $H^{-1/2}(\partial\Omega_h(\zeta))$, it follows that

$$|g^k(\zeta) - \langle (u_{,n})_{,t'}, u_{,t'}\rangle_{H^{1/2}(\partial\Omega_h(\zeta))}| \le c_h \|u\|_{H^2(\Omega)} \|u^{(k)} - u\|_{H^2(\Omega)} \tag{57}$$

where c_h is a constant independent of ζ (and h). □

From Lemma 2 we get an alternative representation formula for the total Gaussian curvature on Ω_h:

$$\mathcal{G}_{\Omega_h}(u) = -\frac{1}{2}\int_{\partial\Omega_h} \kappa u_{,n}^2\, dt - \frac{1}{2}\left(\frac{d}{d\zeta}\int_{\partial\Omega_h(\zeta)} u_{,t'}^2\, dt'\right)_{\zeta=0}. \tag{58}$$

Denote the circular arcs smoothing the corners B_k by $\gamma_h^{(k)}$ and by $\gamma_h^{(k)}(\zeta)$ the corresponding arcs on $\partial\Omega_h(\zeta)$.

Lemma 3.

$$\forall u \in H^2(\Omega)\cap H_0^1(\Omega) \qquad \left(\frac{d}{d\zeta}\int_{\partial\Omega_h(\zeta)\setminus\bigcup_k \gamma_h^{(k)}(\zeta)} u_{,t'}^2\, dt'\right)_{\zeta=0} = 0. \tag{59}$$

Proof. We focus on a connected component of $\partial\Omega_h(\zeta)\setminus\bigcup_k\gamma_h^{(k)}(\zeta)$ and call it $\Gamma_h^{(k)}(\zeta)$. Consider the integral

$$\int_{\Gamma_h^{(k)}(\zeta)} u_{,t'}^2\, dt' = \int_{\Gamma_h^{(k)}(\zeta)} u_{,t}^2 \frac{dt}{dt'}\, dt$$

and calculate it for $\zeta = \epsilon$. As $\frac{dt}{dt'}$ is continuous in ζ and tends to 1 for $\epsilon \to 0$ (cf. (50)), there is a constant C such that

$$\int_{\Gamma_h^{(k)}(\epsilon)} u_{,t'}^2\, dt' \le C\int_{\Gamma_h^{(k)}(\epsilon)} u_{,t}^2\, dt.$$

Note that $u_{,t} = 0$ on $\Gamma_h^{(k)}(0)$ because it is contained in $\partial\Omega$. Then, an argument of Nečas (see [11, Chap. 2, Theorem 4.4]) can be adapted to this case yielding

$$\int_{\Gamma_h^{(k)}(\epsilon)} u_{,t'}^2\, dt' \le C\epsilon\|u\|_{H^2(\Omega_h\setminus\Omega_h(\epsilon))},$$

where $\Omega_h(\epsilon)$ is the domain bounded by $\partial\Omega_h(\epsilon)$. Recalling that the integral over $\Gamma_h^{(k)}(\zeta = 0)$ vanishes, it follows that

$$\left(\frac{d}{d\zeta}\int_{\partial\Omega_h(\zeta)\setminus\cup_k\gamma_h^{(k)}(\zeta)} u_{,t'}^2\, dt'\right)_{\zeta=0} = \lim_{\epsilon\to 0}\frac{1}{\epsilon}\int_{\Gamma_h^{(k)}(\epsilon)} u_{,t'}^2\, dt' = 0.$$

□

Taking account of the lemma and splitting the first integral on the right-hand side of Eq. (58) we get

$$\begin{aligned}\mathcal{G}_{\Omega_h}(u) = &-\frac{1}{2}\int_{\partial\Omega_h\setminus(\cup_k\gamma_h^{(k)})} \kappa u_{,n}^2\, dt \\ &-\frac{1}{2}\sum_{k=1}^{n}\left[\int_{\gamma_h^{(k)}} \kappa u_{,n}^2\, dt + \left(\frac{d}{d\zeta}\int_{\gamma_h^{(k)}(\zeta)} u_{,t'}^2\, dt'\right)_{\zeta=0}\right] \\ &\forall u \in H^2(\Omega)\cap H_0^1(\Omega).\end{aligned} \tag{60}$$

Note that the lemma is true whenever $u = 0$ on the $\Gamma_h^{(k)}$. Thus, the remark applies to the total Gaussian curvature of plates partially supported at the boundary. This means that for a polygonal domain all is decided by the terms in the sum. Thus, two functions that coincide in a neighborhood of every vertex B_k have the same total Gaussian curvature. This is what is meant by the expression *local character of the total Gaussian curvature* in the title of this section.

More generally, if we pass to the limit with respect to h, by the continuity of $\mathcal{G}(\cdot)$ and by Theorem 4 we see that

$$\begin{aligned}&\lim_{h\to\infty}\sum_{k=1}^{n}\left[\frac{1}{\rho_h}\int_{\gamma_h^{(k)}} u_{,n}^2\, dt - \left(\frac{d}{d\zeta}\int_{\gamma_h^{(k)}(\zeta)} u_{,t'}^2\, dt'\right)_{\zeta=0}\right] = 0 \\ &\forall u \in H^2(\Omega)\cap H_0^1(\Omega),\end{aligned} \tag{61}$$

where we have taken into account the fact that $\kappa = -\frac{1}{\rho_h}$ on $\gamma_h^{(k)}$. One can in fact use a partition of unity to show that for each corner the limit is zero.

The following stronger result holds.

Theorem 7. *For all* $u \in H^2(\Omega)\cap H_0^1(\Omega)$,

$$\begin{aligned}&\lim_{h\to\infty}\frac{1}{\rho_h}\int_{\gamma_h^{(k)}} u_{,n}^2\, dt = 0, \\ &\lim_{h\to\infty}\frac{d}{d\zeta}\int_{\gamma_h^{(k)}(\zeta)} u_{,t'}^2\, dt'\Big|_{\zeta=0} = 0.\end{aligned} \tag{62}$$

Proof. Since

$$\lim_{h\to\infty}\left[\frac{1}{\rho_h}\int_{\gamma_h^{(k)}} u,_n^2\,dt - \left(\frac{d}{d\zeta}\int_{\gamma_h^{(k)}(\zeta)} u_{,t'}^2\,dt'\right)_{\zeta=0}\right] = 0,$$

it is enough to prove the first inequality.

Take a local system of coordinates with the x_2-axis containing the bisectrix of the angle of the circular sector relative to the arc $\gamma_h^{(k)}$ and call $2\varphi_k$ its aperture. Let $x_2 = a(x_1)$, $x_2 = a_h(x_1)$, with $|x_1| \le \delta_h = \rho_h \sin\varphi_k$, be the Cartesian representations of the arc $\gamma_h^{(k)}$ and the corresponding curve $\hat{\gamma}_h^{(k)}$ on $\partial\Omega$, respectively. From

$$\left|u_{,\alpha}(x_1, a_h(x_1)) - u_{,\alpha}(x_1, a(x_1))\right| \le \int_{a(x_1)}^{a_h(x_1)} \left|\frac{\partial u_{,\alpha}}{\partial x_2}\right| dx_2,$$

by Hölder's inequality and by the fact that $a(x_1)$ is a Lipschitz function, it follows that

$$\begin{aligned}\left|u_{,\alpha}(x_1, a_h(x_1)) - u_{,\alpha}(x_1, a(x_1))\right|^2 &\le |a_h(x_1) - a(x_1)| \int_{a(x_1)}^{a_h(x_1)} \left|\frac{\partial u_{,\alpha}}{\partial x_2}\right|^2 dx_2\\ &\le C\rho_h \int_{a(x_1)}^{a_h(x_1)} \left|\frac{\partial u_{,\alpha}}{\partial x_2}\right|^2 dx_2.\end{aligned}$$

By integrating both sides with respect to x_1 we obtain the inequality

$$\int_{-\delta_h}^{\delta_h} |u,_\alpha(x_1, a_h(x_1)) - u,_\alpha(x_1, a(x_1))|^2\,dx_1 \le C\rho_h \|u\|_{H^2(\Omega\setminus\Omega_h)}.$$

Hence,

$$\frac{1}{\rho_h}\int_{\gamma_h^{(k)}} |u,_\alpha|^2\,dt \le C\frac{1}{\rho_h}\int_{\hat{\gamma}_h^{(k)}} |u,_\alpha|^2\,dt + o(1), \tag{63}$$

where we have substituted the integration with respect to x_1 by that with respect to the arc-length and adjusted the constant C.

If the origin of the arc-length in the second integral is chosen to coincide with the vertex B_k and the length of $\hat{\gamma}_h^{(k)}$ is denoted by $|\hat{\gamma}_h^{(k)})|$, Eq. (63) gives

$$\frac{1}{\rho_h}\int_{\gamma_h^{(k)}} |u,_\alpha|^2\,dt \le C\frac{|\hat{\gamma}_h^{(k)}|}{\rho_h}\int_{\hat{\gamma}_h^{(k)}} \frac{|u,_\alpha|^2}{|t|}\,dt + o(1).$$

As $h \to \infty$ the ratio $\frac{|\hat{\gamma}_h^{(k)}|}{\rho_h}$ remains bounded and $\int_{\hat{\gamma}_h^{(k)}} \frac{|u,_\alpha|^2}{|t|}\,dt$ tends to zero by Gagliardo's theorem, because $u,_\alpha \in H^{1/2}(\partial\Omega)$. It follows that

$$\lim_h \frac{1}{\rho_h}\int_{\gamma_h^{(k)}} |u,_\alpha|^2\,dt = 0. \tag{64}$$

Then

$$\lim_h \frac{1}{\rho_h} \int_{\gamma_h^{(k)}} |u,_n|^2 \, dt \leq \lim_h \frac{1}{\rho_h} \int_{\gamma_h^{(k)}} (|u,_1|^2 + |u,_2|^2) \, dt = 0$$

and the result is proved.

Acknowledgements. I am indebted to various colleagues for many useful discussions. I wish to thank in particular G. Dal Maso and R. Paroni.

References

[1] Babuska, I. (1961): Stability of the domain of definition with respect to fundamental problems in the theory of partial differential equations especially in connection with the theory of elasticity. I, II. Czechoslovak Math. J. **11**, 76–105; 165–203 (Russian)

[2] Babuska, I., Pitkäranta, J. (1990): The plate paradox for hard and soft simple support. SIAM J. Math. Anal. **21**, 551–576

[3] Babuska, I., Scapolla, T. (1989): Benchmark computation and performance evaluation for a rhombic plate bending problem. Internat. J. Numer. Methods Engrg. **28**, 155–179

[4] Davini, C., Pitacco, I. (1998): Relaxed notions of curvature and a lumped strain method for elastic plates. SIAM J. Numer. Anal. **35**, 677–691

[5] Davini, C. (2002): Γ-convergence of external approximations in boundary value problems involving the bi-Laplacian. J. Comput. Appl. Math. To appear

[6] Gagliardo, E. (1957): Caratterizzazioni delle tracce sulla frontiera relative ad alcune classi di funzioni in n variabili. Rend. Sem. Mat. Univ. Padova, **27**, 284–305

[7] Grisvard, P. (1985): Elliptic problems in non-smooth domains. Pitman, London

[8] Grisvard, P. (1992): Singularities in boundary value problems. Springer, Berlin

[9] Lions, J.L., Magenes, E. (1972): Non-homogeneous boundary value problems and applications. Vol. I, Springer, Heidelberg

[10] Maz'ya, V.G., Nazarov, S.A. (1987): Paradoxes of limit passage in solutions of boundary value problems involving the approximation of smooth domains by polygonal domains. Math. USSR Izvestiya **29**, 511–533

[11] Nečas, J. (1967): Les méthodes directes en théorie des équations elliptiques. Masson, Paris

[12] Pogorelov, A.V. (1978): The Minkowski multidimensional problem. Winston, Washington, DC

[13] Suraci, S. (1997): Un nuovo metodo di approssimazione in analisi strutturale: applicazioni e confronti con il metodo degli elementi finiti. Thesis. Universitá degli Studi di Udine, Udine

[14] Yakovlev, G.N. (1961): Boundary properties of functions of class $W_p^{(l)}$ on regions with angular points. Soviet Math. Dokl. **2**, 1177–1180

Rank 1 Convexity for a Class of Incompressible Elastic Materials

J. Ernest Dunn, Roger Fosdick, Ying Zhang

Abstract. We consider the class of incompressible elastic solids for which the stored energy function $W(\cdot)$ depends on the deformation gradient $\mathbf{F}$ through $|\mathbf{F}|^2$, $W(\mathbf{F}) = \phi(\kappa)$, $\kappa = \sqrt{|\mathbf{F}|^2 - 3}$. We show that $W(\cdot)$ is rank 1 convex at a given $\mathbf{F} \Longleftrightarrow \phi(\cdot)$ is non-decreasing at $\sqrt{|\mathbf{F}|^2 - 3}$ and has $\sqrt{|\mathbf{F}|^2 - 3}$ as a point of convexity.

1 Introduction and setting

For incompressible elastic materials, the stored energy function $W(\cdot) : \mathcal{U} \to \mathbb{R}$ is defined for deformation gradient tensors $\mathbf{F} \in \mathcal{U}$, where

$$\mathcal{U} \equiv \{\mathbf{F} |\det \mathbf{F} = 1\}. \tag{1.1}$$

Considered as an 8-dimensional surface embedded in the 9-dimensional linear space $\mathbb{R}^3 \otimes \mathbb{R}^3$ of all second order tensors, it is easy to see that a *normal* to $\mathcal{U}$ at $\mathbf{F} \in \mathcal{U}$ is given by $\mathbf{F}^{-1T} \in \mathbb{R}^3 \otimes \mathbb{R}^3$ and that the linear *tangent space* $\mathcal{T}_{\mathcal{U}}$ to $\mathcal{U}$ at $\mathbf{F} \in \mathcal{U}$ is characterized by

$$\mathcal{T}_{\mathcal{U}}(\mathbf{F}) = \{\mathbf{A} \in \mathbb{R}^3 \otimes \mathbb{R}^3 \mid \mathbf{A} \cdot \mathbf{F}^{-1T} = 0\}. \tag{1.2}$$

Clearly, if $\mathbf{T} \in \mathbb{R}^3 \otimes \mathbb{R}^3$ is any second order tensor, then

$$\mathbf{A} \equiv \mathbf{T} - \frac{\mathbf{T} \cdot \mathbf{F}^{-1T}}{|\mathbf{F}^{-1T}|^2}\mathbf{F}^{-1T} \in \mathcal{T}_{\mathcal{U}}(\mathbf{F}), \tag{1.3}$$

and this is a general representation formula for the form of every member of the tangent space $\mathcal{T}_{\mathcal{U}}(\mathbf{F})$.

A property of $\mathcal{U}$ and the tangent space $\mathcal{T}_{\mathcal{U}}(\mathbf{F})$, for $\mathbf{F} \in \mathcal{U}$, that is particularly relevant to our work here is the following observation: *if* $\mathbf{F} \in \mathcal{U}$, *then* $\mathbf{F} + \mathbf{a} \otimes \mathbf{b} \in \mathcal{U}$ *if and only if* $\mathbf{a} \otimes \mathbf{b} \in \mathcal{T}_{\mathcal{U}}(\mathbf{F})$.

We shall assume that $W(\cdot) : \mathcal{U} \to \mathbb{R}$ is differentiable. Thus, for every $\mathbf{F}_0 \in \mathcal{U}$, there exists a unique $W_{\mathbf{F}}(\mathbf{F}_0) \in \mathbb{R}^3 \otimes \mathbb{R}^3$ such that, for any differentiable curve $\mathbf{F}(t) \in \mathcal{U}$, $t \in (-1, 1)$, with $\mathbf{F}(0) = \mathbf{F}_0$ and $\dot{\mathbf{F}}(0) = \dot{\mathbf{F}}_0 \in \mathcal{T}_{\mathcal{U}}(\mathbf{F}_0)$, we have

$$\dot{W}(\mathbf{F}(t))_{|t=0} = W_{\mathbf{F}}(\mathbf{F}_0) \cdot \dot{\mathbf{F}}_0. \tag{1.4}$$

By uniqueness, $W_{\mathbf{F}}(\mathbf{F}_0) \in \mathcal{T}_{\mathcal{U}}(\mathbf{F}_0)$.

The stored energy function $W(\cdot) : \mathcal{U} \to \mathbb{R}$ is rank 1 convex at $\tilde{\mathbf{F}} \in \mathcal{U}$ if

$$W(\mathbf{F}) \geq W(\tilde{\mathbf{F}}) + W_{\mathbf{F}}(\tilde{\mathbf{F}}) \cdot (\mathbf{F} - \tilde{\mathbf{F}}) \tag{1.5}$$

for all $\mathbf{F} \in \mathcal{U}$ with $\mathbf{F} - \tilde{\mathbf{F}} =$ rank 1. Since $W_{\mathbf{F}}(\tilde{\mathbf{F}}) \in \mathcal{T}_{\mathcal{U}}(\tilde{\mathbf{F}})$ and because of the observation above, it is clear that an alternative way of expressing the rank 1 convexity of $W(\cdot)$ at $\tilde{\mathbf{F}} \in \mathcal{U}$ is to require

$$W(\tilde{\mathbf{F}} + \mathbf{a} \otimes \mathbf{b}) \geq W(\tilde{\mathbf{F}}) + W_{\mathbf{F}}(\tilde{\mathbf{F}}) \cdot (\mathbf{a} \otimes \mathbf{b}) \tag{1.6}$$

for all $\mathbf{a} \otimes \mathbf{b} \in \mathcal{T}_{\mathcal{U}}(\tilde{\mathbf{F}})$.

In this paper, we consider those isotropic, incompressible elastic materials for which the stored energy function depends upon the deformation gradient only through its magnitude, i.e.,

$$W(\mathbf{F}) = \psi(\mathrm{I}), \tag{1.7}$$

for all $\mathbf{F} \in \mathcal{U}$, where $\mathrm{I} = \mathrm{I}(\mathbf{F}) \equiv |\mathbf{F}|^2 = \operatorname{tr} \mathbf{F}\mathbf{F}^T$. Note that, since $\mathbf{F} \in \mathcal{U}$, $\mathrm{I}(\mathbf{F}) \geq 3$. Thus, $\psi(\cdot)$, which we take to be differentiable, is defined on $[3, \infty)$. If we introduce the (generalized) *shear strain* $\kappa = \kappa(\mathbf{F}) \equiv \sqrt{|\mathbf{F}|^2 - 3}$, then $\kappa \in [0, \infty)$ and we see that we may also write that

$$W(\mathbf{F}) = \psi(\mathrm{I}) = \phi(\kappa), \tag{1.8}$$

where $\kappa = \sqrt{\mathrm{I} - 3}$, $\mathrm{I} = \mathrm{I}(\mathbf{F})$, and where $\phi(\cdot) : [0, \infty) \to \mathbb{R}$ and is differentiable. We seek necessary and sufficient conditions on $\psi(\cdot)$ (equivalently, on $\phi(\cdot)$) for $W(\cdot)$ to be rank 1 convex at a deformation gradient $\mathbf{F} \in \mathcal{U}$. In [1], we studied the analogous problem for a compressible elastic material. While, in that case, a constitutive assumption of the form (1.7) is a bit peculiar, we nevertheless were able to determine such conditions. In the next section, we develop necessary and sufficient conditions for the incompressible materials analyzed here. In the context of incompressibility, the constitutive assumption (1.7) is not so strange. Indeed, neo-Hookean materials are just the special case of (1.8) afforded by $\phi(\kappa) \propto \kappa^2$. As we shall see, the compressible and incompressible cases have different resolutions.

2 Convexity theorem

In this section we prove the following

Theorem 2.1. *Suppose* (1.8) *holds. Then* $W(\cdot) : \mathcal{U} \to \mathbb{R}$ *is rank* 1 *convex at* $\tilde{\mathbf{F}} \in \mathcal{U}$ *if and only if* $\phi(\cdot) : [0, \infty) \to \mathbb{R}$ *is non-decreasing at* $\tilde{\kappa}$,

$$\phi'(\tilde{\kappa}) \geq 0, \tag{2.1}$$

and has $\tilde{\kappa}$ *as a point of convexity,*

$$\phi(\kappa) \geq \phi(\tilde{\kappa}) + \phi'(\tilde{\kappa})(\kappa - \tilde{\kappa}) \tag{2.2}$$

for all $\kappa \in [0, \infty)$, *where* $\tilde{\kappa} \equiv \sqrt{\tilde{\mathrm{I}} - 3}$, $\tilde{\mathrm{I}} = \mathrm{I}(\tilde{\mathbf{F}}) \equiv |\tilde{\mathbf{F}}|^2$.

Proof. First, from (1.7) and the definition of $W_{\mathbf{F}}$ from (1.4), it is straightforward to show that

$$W_{\mathbf{F}}(\mathbf{F}) = 2\left(\mathbf{F} - \frac{3}{|\mathbf{F}^{-1T}|^2}\mathbf{F}^{-1T}\right)\psi'(\mathrm{I}). \tag{2.3}$$

Then, using (1.7) we see that the rank 1 convexity condition (1.6) has the equivalent form

$$\psi(\tilde{\mathrm{I}} + 2\mathbf{a}\cdot\tilde{\mathbf{F}}\mathbf{b} + |\mathbf{a}|^2|\mathbf{b}|^2) \geq \psi(\tilde{\mathrm{I}}) + 2\mathbf{a}\cdot\tilde{\mathbf{F}}\mathbf{b}\psi'(\tilde{\mathrm{I}})$$

for all $\mathbf{a}\otimes\mathbf{b} \in \mathcal{T_U}(\tilde{\mathbf{F}})$. We introduce

$$\mathbf{c} \equiv \tilde{\mathbf{F}}^{-1T}\mathbf{b},$$

so that $\mathbf{a}\otimes\mathbf{b} \in \mathcal{T_U}(\tilde{\mathbf{F}}) \Longleftrightarrow \mathbf{a}\cdot\mathbf{c} = 0$, and write this last inequality as

$$\psi(\tilde{\mathrm{I}} + 2\mathbf{a}\cdot\tilde{\mathbf{B}}\mathbf{c} + |\mathbf{a}|^2\mathbf{c}\cdot\tilde{\mathbf{B}}\mathbf{c}) \geq \psi(\tilde{\mathrm{I}}) + 2\mathbf{a}\cdot\tilde{\mathbf{B}}\mathbf{c}\psi'(\tilde{\mathrm{I}}) \tag{2.4}$$

for all $\mathbf{a}$ and $\mathbf{c}$ such that $\mathbf{a}\cdot\mathbf{c} = 0$, where $\tilde{\mathbf{B}} = \tilde{\mathbf{F}}\tilde{\mathbf{F}}^T$ so that $\tilde{\mathrm{I}} = \operatorname{tr}\tilde{\mathbf{B}}$. As a special case, in order to arrive at (2.1), suppose we take $\mathbf{c}$ parallel to an eigenvector of $\tilde{\mathbf{B}}$. Then $\tilde{\mathbf{B}}\mathbf{c} = \lambda\mathbf{c}$, and for all $\mathbf{a}$ such that $\mathbf{a}\cdot\mathbf{c} = 0$ we see that $\mathbf{a}\cdot\tilde{\mathbf{B}}\mathbf{c} = 0$ and the inequality becomes

$$\psi(\tilde{\mathrm{I}} + \lambda|\mathbf{a}|^2|\mathbf{c}|^2) \geq \psi(\tilde{\mathrm{I}})$$

or equivalently, since $\lambda > 0$,

$$\psi(\mathrm{I}) \geq \psi(\tilde{\mathrm{I}})$$

for all $\mathrm{I} \geq \tilde{\mathrm{I}}$. Thus,

$$\psi'(\tilde{\mathrm{I}}) \geq 0,$$

and since (1.8) implies that $\phi'(\kappa) = 2\kappa\psi'(\mathrm{I})_{|\mathrm{I}=\kappa^2+3}$ we arrive at the equivalent inequality

$$\phi'(\tilde{\kappa}) \geq 0$$

to complete the proof of (2.1). Now, we replace $\mathbf{a}$ in the fundamental inequality (2.4) by $\alpha\mathbf{a}$ where $\alpha \in \mathbb{R}$. Then, we may write

$$\psi(\mathrm{I}^\star(\alpha)) \geq \psi(\tilde{\mathrm{I}}) + 2\alpha\mathbf{a}\cdot\tilde{\mathbf{B}}\mathbf{c}\psi'(\tilde{\mathrm{I}}), \tag{2.5}$$

where

$$\mathrm{I}^\star(\alpha) \equiv \tilde{\mathrm{I}} + 2\alpha\mathbf{a}\cdot\tilde{\mathbf{B}}\mathbf{c} + \alpha^2|\mathbf{a}|^2\mathbf{c}\cdot\tilde{\mathbf{B}}\mathbf{c}, \tag{2.6}$$

which must hold for all $\mathbf{a}$ and $\mathbf{c}$ such that $\mathbf{a}\cdot\mathbf{c}=0$, and for all $\alpha\in\mathbb{R}$. Clearly, the parabola $\mathrm{I}^\star(\alpha)$ is convex and has its minimum $\mathrm{I}^\star_{\min}$ at $\alpha=\alpha_m$ where

$$\alpha_m \equiv -\frac{\mathbf{a}\cdot\tilde{\mathbf{B}}\mathbf{c}}{|\mathbf{a}|^2\mathbf{c}\cdot\tilde{\mathbf{B}}\mathbf{c}} \tag{2.7}$$

with

$$\mathrm{I}^\star_{\min} \equiv \mathrm{I}^\star(\alpha_m) = \tilde{\mathrm{I}} - \frac{(\mathbf{a}\cdot\tilde{\mathbf{B}}\mathbf{c})^2}{|\mathbf{a}|^2\mathbf{c}\cdot\tilde{\mathbf{B}}\mathbf{c}}. \tag{2.8}$$

Also, it is clear that, for any $\mathrm{I}\geq\mathrm{I}^\star_{\min}$, the equation

$$\mathrm{I}^\star(\alpha)=\mathrm{I}$$

has two corresponding real roots $\alpha^-\leq\alpha^+$ which may be written in the form

$$\alpha^\pm=\alpha_m\pm\frac{\sqrt{\mathrm{I}-\mathrm{I}^\star_{\min}}}{\sqrt{|\mathbf{a}|^2\mathbf{c}\cdot\tilde{\mathbf{B}}\mathbf{c}}}.$$

Thus, (2.5) and (2.6) yield the inequality

$$\psi(\mathrm{I})\geq\psi(\tilde{\mathrm{I}})+2\alpha^\pm\mathbf{a}\cdot\tilde{\mathbf{B}}\mathbf{c}\psi'(\tilde{\mathrm{I}})$$

for all $\mathrm{I}\geq\mathrm{I}^\star_{\min}$, which is optimal for the choice α^+ if $\mathbf{a}\cdot\tilde{\mathbf{B}}\mathbf{c}\geq 0$ and optimal for the choice α^- if $\mathbf{a}\cdot\tilde{\mathbf{B}}\mathbf{c}\leq 0$. For this, we need to apply $\psi'(\tilde{\mathrm{I}})\geq 0$ as was shown earlier. In either case, from the definition (2.7) of α_m we readily see that

$$\psi(\mathrm{I})\geq\psi(\tilde{\mathrm{I}})+2\left(|\mathbf{a}\cdot\tilde{\mathbf{B}}\mathbf{c}|\frac{\sqrt{\mathrm{I}-\mathrm{I}^\star_{\min}}}{\sqrt{|\mathbf{a}|^2\mathbf{c}\cdot\tilde{\mathbf{B}}\mathbf{c}}}-\frac{(\mathbf{a}\cdot\tilde{\mathbf{B}}\mathbf{c})^2}{|\mathbf{a}|^2\mathbf{c}\cdot\tilde{\mathbf{B}}\mathbf{c}}\right)\psi'(\tilde{\mathrm{I}}).$$

Moreover, a similar application of the definition (2.8) of $\mathrm{I}^\star_{\min}$ allows us to transform this last inequality into the equivalent form

$$\psi(\mathrm{I})\geq\psi(\tilde{\mathrm{I}})+2\frac{(\mathbf{a}\cdot\tilde{\mathbf{B}}\mathbf{c})^2}{|\mathbf{a}|^2\mathbf{c}\cdot\tilde{\mathbf{B}}\mathbf{c}}\left(\sqrt{1+(\mathrm{I}-\tilde{\mathrm{I}})/\frac{(\mathbf{a}\cdot\tilde{\mathbf{B}}\mathbf{c})^2}{|\mathbf{a}|^2\mathbf{c}\cdot\tilde{\mathbf{B}}\mathbf{c}}}-1\right)\psi'(\tilde{\mathrm{I}}), \tag{2.9}$$

which is supposed to hold for all $\mathbf{a}\neq 0$ and $\mathbf{c}\neq 0$ such that $\mathbf{a}\cdot\mathbf{c}=0$, and for all $\mathrm{I}\geq\tilde{\mathrm{I}}-\frac{(\mathbf{a}\cdot\tilde{\mathbf{B}}\mathbf{c})^2}{|\mathbf{a}|^2\mathbf{c}\cdot\tilde{\mathbf{B}}\mathbf{c}}$. Thus, we have shown that the fundamental inequality (2.4) implies (2.9). It is also true that (2.9) together with $\psi'(\tilde{\mathrm{I}})\geq 0$ (equivalently $\phi'(\tilde{\kappa})\geq 0$) implies the fundamental inequality (2.4) and so (2.4) and (2.9) with $\psi'(\tilde{\mathrm{I}})\geq 0$ are equivalent. To see this, suppose $\mathbf{a}$ and $\mathbf{c}$ are given such that $\mathbf{a}\neq 0$, $\mathbf{c}\neq 0$ and $\mathbf{a}\cdot\mathbf{c}=0$. Further, recall (2.8) and suppose that $\mathrm{I}\in\mathbb{R}$ is such that $\mathrm{I}\geq\tilde{\mathrm{I}}-\frac{(\mathbf{a}\cdot\tilde{\mathbf{B}}\mathbf{c})^2}{|\mathbf{a}|^2\mathbf{c}\cdot\tilde{\mathbf{B}}\mathbf{c}}=\mathrm{I}^\star_{\min}$. Then, recalling the definition of $I^\star(\alpha)$ in (2.6), we see that there is an $\alpha \in \mathbb{R}$ such that $I^\star(\alpha) = I$. In this case, (2.9) may be written in the form

$$\psi(I^\star(\alpha)) \geq \psi(\tilde{I}) + 2(\tilde{I} - I^\star_{\min}) \left(\sqrt{1 + \frac{I^\star(\alpha) - \tilde{I}}{\tilde{I} - I^\star_{\min}}} - 1 \right) \psi'(\tilde{I}), \tag{2.10}$$

where, again, we recall that $I^\star(\alpha)$ is defined in (2.6). Now, we consider the second term on the right-hand side of (2.10) and observe that

$$\begin{aligned}
&(\tilde{I} - I^\star_{\min}) \left(\sqrt{1 + \frac{I^\star(\alpha) - \tilde{I}}{\tilde{I} - I^\star_{\min}}} - 1 \right) \\
&= \sqrt{\tilde{I} - I^\star_{\min}} \left(\sqrt{I^\star(\alpha) - I^\star_{\min}} - \sqrt{\tilde{I} - I^\star_{\min}} \right) \\
&= \frac{|\mathbf{a} \cdot \tilde{\mathbf{B}}\mathbf{c}|}{\sqrt{|\mathbf{a}|^2 \mathbf{c} \cdot \tilde{\mathbf{B}}\mathbf{c}}} \left(\sqrt{2\alpha \mathbf{a} \cdot \tilde{\mathbf{B}}\mathbf{c} + \alpha^2 |\mathbf{a}|^2 \mathbf{c} \cdot \tilde{\mathbf{B}}\mathbf{c} + \frac{(\mathbf{a} \cdot \tilde{\mathbf{B}}\mathbf{c})^2}{|\mathbf{a}|^2 \mathbf{c} \cdot \tilde{\mathbf{B}}\mathbf{c}}} - \frac{|\mathbf{a} \cdot \tilde{\mathbf{B}}\mathbf{c}|}{\sqrt{|\mathbf{a}|^2 \mathbf{c} \cdot \tilde{\mathbf{B}}\mathbf{c}}} \right) \\
&= \frac{|\mathbf{a} \cdot \tilde{\mathbf{B}}\mathbf{c}|}{\sqrt{|\mathbf{a}|^2 \mathbf{c} \cdot \tilde{\mathbf{B}}\mathbf{c}}} \frac{\left(|\alpha |\mathbf{a}|^2 \mathbf{c} \cdot \tilde{\mathbf{B}}\mathbf{c} + \mathbf{a} \cdot \tilde{\mathbf{B}}\mathbf{c}| - |\mathbf{a} \cdot \tilde{\mathbf{B}}\mathbf{c}| \right)}{\sqrt{|\mathbf{a}|^2 \mathbf{c} \cdot \tilde{\mathbf{B}}\mathbf{c}}} \\
&= |\mathbf{a} \cdot \tilde{\mathbf{B}}\mathbf{c}| (|\alpha - \alpha_m| - |\alpha_m|),
\end{aligned}$$

where, recall, α_m is defined in (2.7). Clearly,

$$\alpha_m(\alpha_m - \alpha) \leq |\alpha_m||\alpha - \alpha_m|$$

so that

$$-\alpha\alpha_m \leq |\alpha_m|(|\alpha - \alpha_m| - |\alpha_m|).$$

Thus, with the aid of (2.7) we find that

$$\alpha \mathbf{a} \cdot \tilde{\mathbf{B}}\mathbf{c} \leq |\mathbf{a} \cdot \tilde{\mathbf{B}}\mathbf{c}|(|\alpha - \alpha_m| - |\alpha_m|)$$

so that

$$\alpha \mathbf{a} \cdot \tilde{\mathbf{B}}\mathbf{c} \leq (\tilde{I} - I^\star_{\min}) \left(\sqrt{1 + \frac{I^\star(\alpha) - \tilde{I}}{\tilde{I} - I^\star_{\min}}} - 1 \right).$$

Finally, because $\psi'(\tilde{I}) \geq 0$, it then follows from (2.10) that

$$\psi(I^\star(\alpha)) \geq \psi(\tilde{I}) + 2\alpha \mathbf{a} \cdot \tilde{\mathbf{B}}\mathbf{c}\psi'(\tilde{I}),$$

and this is the inequality (2.5) which, as we already know, with $\psi'(\tilde{I}) \geq 0$ is equivalent to the fundamental inequality (2.4). Thus, (2.9) *together with* $\psi'(\tilde{I}) \geq 0$ *is equivalent to* (2.4), as was claimed. We now assume that $\psi'(\tilde{\mathrm{I}}) \geq 0$ and consider optimizing the inequality (2.9) by selecting the "best" possible vectors $\mathbf{a} \neq 0$ and $\mathbf{c} \neq 0$ with $\mathbf{a} \cdot \mathbf{c} = 0$. To do this, we first observe, in (2.9), that $\mathbf{a}$ and $\mathbf{c}$ appear only in the combination

$$f(\mathbf{a}, \mathbf{c}) = \frac{(\mathbf{a} \cdot \tilde{\mathbf{B}}\mathbf{c})^2}{|\mathbf{a}|^2 \mathbf{c} \cdot \tilde{\mathbf{B}}\mathbf{c}} \equiv \xi, \tag{2.11}$$

and that the expression

$$\xi \left(\sqrt{1 + \frac{\mathrm{I} - \tilde{\mathrm{I}}}{\xi}} - 1 \right)$$

as well as the interval $[\tilde{\mathrm{I}} - \xi, \infty)$ in (2.9) are increasing functions of ξ. Thus, the "best" choice for $\mathbf{a}$ and $\mathbf{c}$ is a solution of the problem

$$\max_{\substack{\mathbf{a} \neq 0, \mathbf{c} \neq 0 \\ \mathbf{a} \cdot \mathbf{c} = 0}} f(\mathbf{a}, \mathbf{c}).$$

Since $\tilde{\mathbf{B}}$ is positive definite and symmetric, it is convenient to introduce $\tilde{\mathbf{V}} \equiv \sqrt{\tilde{\mathbf{B}}}$ and to cast this extremum problem in terms of the substitution variables

$$\mathbf{x} \equiv \frac{\mathbf{a}}{|\mathbf{a}|}, \quad \mathbf{y} \equiv \frac{\tilde{\mathbf{V}}\mathbf{c}}{\sqrt{\mathbf{c} \cdot \tilde{\mathbf{B}}\mathbf{c}}}. \tag{2.12}$$

Clearly, under this substitution we have $f(\mathbf{a}, \mathbf{c}) = (\mathbf{x} \cdot \tilde{\mathbf{V}}\mathbf{y})^2$, and in terms of $\mathbf{x}$ and $\mathbf{y}$ the problem now becomes

$$\max_{\substack{|\mathbf{x}|=1, |\mathbf{y}|=1 \\ \mathbf{x} \cdot \tilde{\mathbf{V}}^{-1}\mathbf{y}=0}} (\mathbf{x} \cdot \tilde{\mathbf{V}}\mathbf{y})^2,$$

and the appropriate Lagrangian for the Lagrange multiplier method is given by

$$L \equiv (\mathbf{x} \cdot \tilde{\mathbf{V}}\mathbf{y})^2 - \lambda(\mathbf{x} \cdot \mathbf{x} - 1) - \mu(\mathbf{y} \cdot \mathbf{y} - 1) - \tau \mathbf{x} \cdot \tilde{\mathbf{V}}^{-1}\mathbf{y}.$$

The Euler equations are then

$$\left.\begin{aligned} &2(\mathbf{x} \cdot \tilde{\mathbf{V}}\mathbf{y})\tilde{\mathbf{V}}\mathbf{y} - 2\lambda\mathbf{x} - \tau\tilde{\mathbf{V}}^{-1}\mathbf{y} = 0, \\ &2(\mathbf{x} \cdot \tilde{\mathbf{V}}\mathbf{y})\tilde{\mathbf{V}}\mathbf{x} - 2\mu\mathbf{y} - \tau\tilde{\mathbf{V}}^{-1}\mathbf{x} = 0, \\ &\mathbf{x} \cdot \mathbf{x} = 1\ ,\ \mathbf{y} \cdot \mathbf{y} = 1\ ,\ \mathbf{x} \cdot \tilde{\mathbf{V}}^{-1}\mathbf{y} = 0. \end{aligned}\right\} \tag{2.13}$$

Because of the symmetry of $\tilde{\mathbf{V}}$, it is straightforward to see that

$$\lambda = \mu = (\mathbf{x} \cdot \tilde{\mathbf{V}}\mathbf{y})^2 \tag{2.14}$$

so that the first two equations of (2.13) may be written as

$$2\sqrt{\lambda}\tilde{\mathbf{V}}\mathbf{y} - 2\lambda\mathbf{x} - \tau\tilde{\mathbf{V}}^{-1}\mathbf{y} = 0,$$
$$2\sqrt{\lambda}\tilde{\mathbf{V}}\mathbf{x} - 2\lambda\mathbf{y} - \tau\tilde{\mathbf{V}}^{-1}\mathbf{x} = 0.$$

Clearly, by first adding and then subtracting these equations we develop, respectively,

$$\left(2\sqrt{\lambda}\tilde{\mathbf{V}} - \tau\tilde{\mathbf{V}}^{-1}\right)(\mathbf{x}+\mathbf{y}) = 2\lambda(\mathbf{x}+\mathbf{y}),$$
$$\left(2\sqrt{\lambda}\tilde{\mathbf{V}} - \tau\tilde{\mathbf{V}}^{-1}\right)(\mathbf{x}-\mathbf{y}) = -2\lambda(\mathbf{x}-\mathbf{y}),$$

and it is significant to notice that $\mathbf{x}+\mathbf{y}$ is orthogonal to $\mathbf{x}-\mathbf{y}$ because of the third and fourth equations of (2.13). Thus, $\mathbf{x}+\mathbf{y}$ and $\mathbf{x}-\mathbf{y}$ must correspond to *different* (i.e., orthogonal) eigenvectors of $2\sqrt{\lambda}\tilde{\mathbf{V}} - \tau\tilde{\mathbf{V}}^{-1}$. Since a set of orthonormal eigenvectors for $2\sqrt{\lambda}\tilde{\mathbf{V}} - \tau\tilde{\mathbf{V}}^{-1}$ is equivalently a set of orthonormal eigenvectors for $\tilde{\mathbf{V}}$, it is helpful to introduce $(\tilde{\nu}_1, \tilde{\mathbf{e}}_1)$ and $(\tilde{\nu}_2, \tilde{\mathbf{e}}_2)$ as representing any two eigenvalue–orthonormal eigenvector pairs for $\tilde{\mathbf{V}}$. Then, we may write

$$\mathbf{x}+\mathbf{y} = \beta\tilde{\mathbf{e}}_1, \quad \mathbf{x}-\mathbf{y} = \gamma\tilde{\mathbf{e}}_2,$$

or, equivalently,

$$\mathbf{x} = \frac{1}{2}(\beta\tilde{\mathbf{e}}_1 + \gamma\tilde{\mathbf{e}}_2), \quad \mathbf{y} = \frac{1}{2}(\beta\tilde{\mathbf{e}}_1 - \gamma\tilde{\mathbf{e}}_2), \tag{2.15}$$

where the last equation of (2.13) requires

$$\frac{1}{4}(\beta^2 + \gamma^2) = 1, \quad \frac{1}{4}\left(\frac{\beta^2}{\tilde{\nu}_1} - \frac{\gamma^2}{\tilde{\nu}_2}\right) = 0.$$

This yields

$$\beta^2 = \frac{4\tilde{\nu}_1}{\tilde{\nu}_1 + \tilde{\nu}_2}, \quad \gamma^2 = \frac{4\tilde{\nu}_2}{\tilde{\nu}_1 + \tilde{\nu}_2}, \tag{2.16}$$

and (2.14), (2.15) and (2.16) readily provide

$$\lambda = \mu = (\mathbf{x}\cdot\tilde{\mathbf{V}}\mathbf{y})^2 = (\tilde{\nu}_1 - \tilde{\nu}_2)^2.$$

Although τ can be easily determined, it is not essential to do so because we wished only to determine the maximum of $(\mathbf{x}\cdot\tilde{\mathbf{V}}\mathbf{y})^2$ subject to the stated constraints. We have found

$$\max_{\substack{|\mathbf{x}|=1,\ |\mathbf{y}|=1 \\ \mathbf{x}\cdot\tilde{\mathbf{V}}^{-1}\mathbf{y}=0}} (\mathbf{x}\cdot\tilde{\mathbf{V}}\mathbf{y})^2 = (\tilde{\nu}_{\max} - \tilde{\nu}_{\min})^2,$$

where $\tilde{\nu}_{\max}$ and $\tilde{\nu}_{\min}$ are the maximum and minimum eigenvalues of $\tilde{\mathbf{V}}$, respectively. Thus, we have achieved our immediate goal of optimizing the inequality (2.9). So, assuming that $\psi'(\tilde{\mathrm{I}}) \geq 0$, then *equivalent to* (2.9) we may write

$$\psi(\mathrm{I}) \geq \psi(\tilde{\mathrm{I}}) + 2(\tilde{\nu}_{\max} - \tilde{\nu}_{\min})^2 \left(\sqrt{1 + \frac{\mathrm{I} - \tilde{\mathrm{I}}}{(\tilde{\nu}_{\max} - \tilde{\nu}_{\min})^2}} - 1 \right) \psi'(\tilde{\mathrm{I}}) \qquad (2.17)$$

for all $\mathrm{I} \geq \tilde{\mathrm{I}} - (\tilde{\nu}_{\max} - \tilde{\nu}_{\min})^2$. But, with $\tilde{\mathrm{I}}$ given and fixed and $\psi'(\tilde{\mathrm{I}}) \geq 0$, we know that the right-hand side of (2.17) is an increasing function of $(\tilde{\nu}_{\max} - \tilde{\nu}_{\min})^2$ and so for the optimal inequality which is both necessary and sufficient for (2.17) (provided $\psi'(\tilde{\mathrm{I}}) \geq 0$), we seek the maximum of $(\tilde{\nu}_{\max} - \tilde{\nu}_{\min})^2$ subject to the constraint that $\tilde{\mathrm{I}} \geq 3$ is given, where

$$\tilde{\mathrm{I}} = \mathrm{I}(\tilde{\mathbf{F}}) = |\tilde{\mathbf{F}}|^2 = |\tilde{\mathbf{V}}|^2 = \tilde{\nu}_{\max}^2 + \tilde{\nu}_{\min}^2 + \frac{1}{\tilde{\nu}_{\max}^2 \tilde{\nu}_{\min}^2}.$$

By differentiation, this maximum is readily seen to occur at $\tilde{\nu}_{\max} = \frac{1}{\tilde{\nu}_{\min}}$, and when this holds we have $\tilde{\nu}_{\max}\tilde{\nu}_{\min} = 1$ and so it follows that

$$(\tilde{\nu}_{\max} - \tilde{\nu}_{\min})^2_{|\tilde{\nu}_{\max} = \frac{1}{\tilde{\nu}_{\min}}} = \tilde{\mathrm{I}} - 3 = \tilde{\kappa}.$$

Thus, by taking account of (1.8) and assuming that $\psi'(\tilde{\mathrm{I}}) \geq 0$, we see that (2.17) may be written equivalently as

$$\phi(\kappa) \geq \phi(\tilde{\kappa}) + \phi'(\tilde{\kappa})(\kappa - \tilde{\kappa})$$

for all $\kappa \in [0, \infty)$, and we see that (2.2) holds which completes the proof of the theorem. □

Acknowledgements. The National Science Foundation, Grant No. DMS-9531925, and the National Natural Science Foundation of China are gratefully acknowledged for their support of this research. R.F. also acknowledges the gracious hospitality of Professor Salvatore Marzano of the Politecnico di Bari, Italy, during his tenure in Bari as a 1997 Visiting CNR Professor.

Reference

[1] Dunn, J.E., Fosdick, R. (1994): The Weierstrass condition for a special class of elastic materials. J. Elasticity **34**, 167–184

A Continuum Description of Diatomic Systems

Pasquale Giovine

Abstract. In previous works complete dynamical equations of balance for a diatomic crystalline system were obtained by partially linearizing the equations stated for a binary mixture of elastic bodies in absence of diffusion. The method of approximation, similar to one proposed by Signorini within the theory of elasticity, used the hypothesis that the only relative motion between phases was infinitesimal, the mean being finite. Here we describe the theory and then obtain the complete linear equations in the isotropic elastic case; further, we investigate the propagation of plane harmonic waves and compare our results with those of previous models.

1 Introduction

We imagine the material element of a diatomic crystalline system as a cell of a centered cubic crystal with the central atom belonging to a different species from those in the corners and free to move, within the cell, without participating exactly in the mean motion (see [1, 2]); hence we model the medium as superimposed elastic continua which may interact and undergo individual motions.

Also we want to avoid the phenomenon of diffusion between components because the relative displacement has to remain confined to less than the atomic spacing. Thus, in [3, 4], by introducing a "small" scalar parameter ϵ which measures the removal between constituents, we systematically used expansions in powers of ϵ in the mechanical equations of balance for a binary mixture of Truesdell [5] and in constitutive fields, respectively, without imposing restrictions on the mean motion.

The partially linearized equations of motion, obtained for the diatomic medium, fit a model of continuum with vectorial microstructure (see Sect. 19 of [6]) and generalize the proposals of previous continuous theories (see [7, 8] in which, for example, constitutive equations do not satisfy the principle of equipresence in Sect. 96 of [9]).

In this paper we give a full description of the theory and then we obtain the complete linear case for an isotropic elastic diatomic continuum. An application to the study of the phenomenon of propagation of plane harmonic waves shows the existence of two kinds of dispersive solutions: the acoustical branch, lower in frequency, approaches the classical elastic non-dispersive velocity from below as the wave number nears zero; the optical one, higher in frequency, is highly dispersive and has non-zero cut–off frequencies corresponding to some diatomic resonance.

At the end we observe that results, obtained in [1, 2] for diatomic media, are included in our model as particular cases.

2 Kinematics and conservation of mass in interacting continua

We briefly introduce essential definitions from [3]. We consider two elastic bodies $\mathcal{C}_1$ and $\mathcal{C}_2$ which simultaneously occupy the same open regions of space; we assume that the reference placement $\mathcal{B}_*$ is a homogeneous natural state for the two bodies so that the mass densities ρ_{1*} and ρ_{2*} of constituents are constant and the diatomic medium is free of stresses when in the reference placement. If x_{i*} is the place occupied by a material element of $\mathcal{C}_i$ in the reference placement $\mathcal{B}_*$, the motion of $\mathcal{C}_i$ is a "sufficiently" smooth invertible mapping

$$y = x_i(x_{i*}, \tau) \tag{1}$$

of $\mathcal{C}_i$ onto a time-sequence of placements $\mathcal{B}_\tau$ in space, with the time τ in the interval (τ_0, τ_1): "sufficiently" smooth to justify all the calculations which are made in the sequel.

By hypothesis, and in agreement with [5], places y in each placement $\mathcal{B}_\tau$ are taken simultaneously by one particle from each of the two elastic media, each particle being the image, respectively, of elements $x_{1*} \in \mathcal{C}_1$ and $x_{2*} \in \mathcal{C}_2$ in $\mathcal{B}_*$, usually distinct.

The velocity v_i, the acceleration a_i, the deformation gradient F_i and its determinant ι_i of the ith constituent are, respectively,

$$v_i := \frac{\partial x_i}{\partial \tau}(x_{i*}, \tau)_{|x_{i*}=\text{const.}}, \qquad a_i := \frac{\partial^2 x_i}{\partial \tau^2}(x_{i*}, \tau)_{|x_{i*}=\text{const.}}, \tag{2}$$

$$F_i(x_{i*}, \tau) := \frac{\partial x_i}{\partial x_{i*}}(x_{i*}, \tau), \qquad \iota_i(x_{i*}, \tau) := \det F_i(x_{i*}, \tau). \tag{3}$$

Each body $\mathcal{C}_i$ has its own mass and consequently its mass density ρ_i in $\mathcal{B}_\tau$. The total mass density ρ and the microstructural mass density σ are defined as:

$$\rho := \rho_1 + \rho_2 \quad \text{and} \quad \sigma := \nu_* \rho_2 - (1 - \nu_*)\rho_1, \tag{4}$$

where $\nu_* = \frac{\rho_{1*}}{\rho_*}$ and $\rho_* = \rho_{1*} + \rho_{2*}$ are the concentration of the first constituent and the total mass density of the body in the reference placement $\mathcal{B}_*$, respectively, both constant because of the homogeneity of $\mathcal{B}_*$.

Here we neglect phenomena involving creation of mass or vacancy; thus the mass of each body is conserved and the following equations of conservation of mass are valid:

$$(\rho_1 \iota_1)(x_{1*}, \tau) = \rho_{1*} \quad \text{and} \quad (\rho_2 \iota_2)(x_{2*}, \tau) = \rho_{2*}. \tag{5}$$

Now, by applying the inverse motion of $\mathcal{C}_i$ to y, we see that

$$x_{i*} = x_{i*}(y, \tau); \tag{6}$$

we can define the mean x_* and the removal functions s_* of places x_{1*} and x_{2*} of $\mathcal{B}_*$, beginning at y in $\mathcal{B}_\tau$, for all τ:

$$x_* := \nu_* x_{1*} + (1 - \nu_*) x_{2*} \quad \text{and} \quad s_* := x_{2*} - x_{1*}. \tag{7}$$

Because of the absence of diffusion we expect that the mean x_* is in $\mathcal{B}_*$; in fact places x_{1*} and x_{2*} should belong to the same cell of the centered cubic crystal in our model. From definitions (7), we have

$$x_{1*} = x_* - (1 - \nu_*)s_* \quad \text{and} \quad x_{2*} = x_* + \nu_* s_* \tag{8}$$

(see Fig. 1).

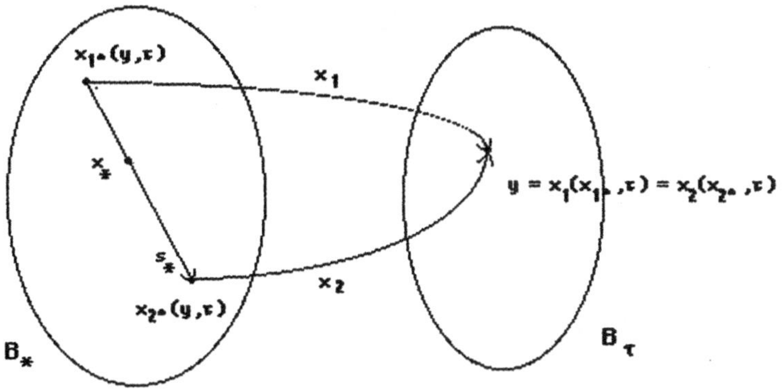

Fig. 1. Graph of means and removals

But the position of the mean x_* in $\mathcal{B}_*$ is itself occupied by elements of both bodies. Thus, by using the motion (1) of $\mathcal{C}_i$ to x_* reciprocally, we can introduce, similar to means and removals, the following formal definitions of average displacement x and relative displacement (or drift) d between atomic species calculated in x_* at time τ:

$$x(x_*, \tau) := \nu_* x_1(x_*, \tau) + (1 - \nu_*)x_2(x_*, \tau) \quad \text{and} \tag{9}$$
$$d(x_*, \tau) := x_2(x_*, \tau) - x_1(x_*, \tau); \tag{10}$$

hence, we also have

$$x_1(x_*, \tau) = [x - (1 - \nu_*)d](x_*, \tau) \quad \text{and} \quad x_2(x_*, \tau) = (x + \nu_* d)(x_*, \tau) \tag{11}$$

(see Fig. 2).

Remark. The relative displacements $d(x_{1*}(y, \tau), \tau)$ and $d(x_{2*}(y, \tau), \tau)$ are exactly those vectors $r_{21}(y, \tau)$ and $-r_{12}(y, \tau)$, respectively, which appear in Eqs. (2.13) of [10] in a more general context.

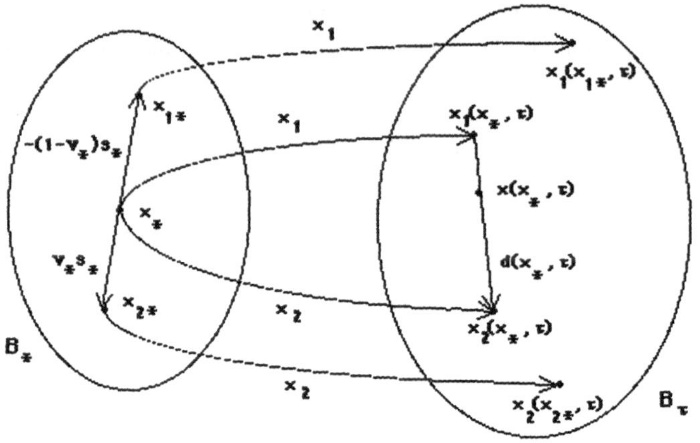

Fig. 2. Graph of average displacements and drifts

3 First partial linearization

We now specify the condition, mentioned in the introduction, that the relative displacement must remain confined to less than the atomic spacing. We require that the removal s_* be small and calculate Taylor formulae beginning from x_*, of first order in s_* and pertinent to motions 1, for $i = 1, 2$,

$$
\begin{aligned}
y = x_1(x_{1*}, \tau) &= x_1(x_*, \tau) - (1-\nu_*)[F_1(x_*, \tau)]s_* + o(s_*) \qquad (12)\\
&= x - (1-\nu_*)d - (1-\nu_*)\{\text{Grad}[x - (1-\nu_*)d]\}s_* + o(s_*),\\
y = x_2(x_{2*}, \tau) &= x_2(x_*, \tau) + \nu_*[F_2(x_*, \tau)]s_* + o(s_*) = \qquad (13)\\
&= x + \nu_* d + \nu_*[\text{Grad}(x + \nu_* d)]s_* + o(s_*),
\end{aligned}
$$

where Grad denotes differentiation with respect to x_*, while $o(s_*)$ is an infinitesimal vector, for $s_* \to 0$, of higher order than the Euclidean norm $\|s_*\|$ of s_*; the second equalities in formulas (12) and (13) take into account relations $(3)_1$ and (8), and the third relations (11).

Instead of carrying our analysis further in terms of orders of approximation compared to powers of s_*, we prefer to consider a special class of motions: each is characterized by the choice of a parameter ϵ, the dependence on which is "sufficiently" smooth and such that, when $\epsilon = 0$, it corresponds to a motion with null drift. In the specification of a particular problem, the parameter ϵ may arise naturally as a quantity which can be increased from zero; here we suppose that s_* is of the form

$$s_* = \epsilon s \qquad (14)$$

and display terms of lower order in ϵ in all quantities related to the motion that depend on ϵ in this way. Hence we assume that

$$\begin{aligned} x(x_*, \tau, \epsilon) &= (x^0 + \epsilon x^1)(x_*, \tau) + o(\epsilon), \\ d(x_*, \tau, \epsilon) &= \epsilon d^1(x_*, \tau) + o(\epsilon), \end{aligned} \tag{15}$$

where, here and in the sequel, $o(\epsilon)$ represents any scalar, vectorial or tensorial field such that, for the related norm, $\|o(\epsilon)\| < \delta|\epsilon|^2$ with $\delta > 0$.

When $\epsilon = 0$, we have a motion with null drift, that is, an admissible motion as desired; therefore, if we define tensors $F^i := \operatorname{Grad} x^i$, for $i = 0, 1$, we have $\iota^0 := \det F^0 > 0$. By inserting expressions (14) and (15) in relations (2), (8), (12) and (13) , we obtain

$$y = (x^0 + \epsilon x^1)(x_*, \tau) + o(\epsilon), \quad d(x_*, \tau) = -\epsilon(F^0 s)(x_*, \tau) + o(\epsilon), \tag{16}$$

$$\begin{aligned} &a_1(x_{1*}, \tau) \\ &= \left\{\frac{\partial^2 x^0}{\partial \tau^2} + \epsilon\left[\frac{\partial^2 x^1}{\partial \tau^2} + (1 - \nu_*)\left(F^0 \frac{\partial^2 s}{\partial \tau^2} + 2\frac{\partial F^0}{\partial \tau}\frac{\partial s}{\partial \tau}\right)\right]\right\}(x_*, \tau) + o(\epsilon), \\ &a_2(x_{2*}, \tau) \\ &= \left\{\frac{\partial^2 x^0}{\partial \tau^2} + \epsilon\left[\frac{\partial^2 x^1}{\partial \tau^2} - \nu_*\left(F^0 \frac{\partial^2 s}{\partial \tau^2} + 2\frac{\partial F^0}{\partial \tau}\frac{\partial s}{\partial \tau}\right)\right]\right\}(x_*, \tau) + o(\epsilon), \end{aligned} \tag{17}$$

$$\begin{aligned} F_1(x_{1*}, \tau) &= \{F^0 + \epsilon[F^1 + (1 - \nu_*)F^0 \operatorname{Grad} s]\}(x_*, \tau) + o(\epsilon), \\ F_2(x_{2*}, \tau) &= \{F^0 + \epsilon[F^1 - \nu_* F^0 \operatorname{Grad} s]\}(x_*, \tau) + o(\epsilon), \end{aligned} \tag{18}$$

$$\begin{aligned} \iota_1(x_{1*}, \tau) &= \iota^0(x_*, \tau)\{1 + \epsilon\{\operatorname{tr}[(F^0)^{-1}F^1] + (1 - \nu_*) \operatorname{Div} s\}(x_*, \tau)\} + o(\epsilon), \\ \iota_2(x_{2*}, \tau) &= \iota^0(x_*, \tau)\{1 + \epsilon\{\operatorname{tr}[(F^0)^{-1}F^1] - \nu_* \operatorname{Div} s\}(x_*, \tau)\} + o(\epsilon). \end{aligned} \tag{19}$$

It is worth remarking that the motion y and the average displacement x are equal up to the first order in ϵ, while the first two terms in expansions of accelerations, deformation gradients and their determinants coincide; so, for the diatomic medium, we shall be interested principally in studying the equations of motion in terms of average and relative quantities instead of peculiar ones.

Similar expansions for peculiar mass densities ρ_1 and ρ_2 ensue from the respective Taylor formulae and continuity equations (5) by considering, for ρ and σ defined in (4), evaluations as in (15) and by inserting them in (5) with relations (19); for $\epsilon = 0$ we deduce that

$$\rho^0\iota^0 = \rho_* \quad \text{and} \quad \sigma^0 = 0, \tag{20}$$

which is what we expect for the admissible motion of a single body, while, for first order in ϵ,

$$\rho^1 + \rho^0 \operatorname{tr}[(F^0)^{-1} F^1] = 0 \quad \text{and} \quad \sigma^1 = \nu_*(1 - \nu_*)[\rho^0 \operatorname{Div} s - (\operatorname{Grad} \rho^0) \cdot s]. \tag{21}$$

From (20) it follows that

$$\rho_{1*} = \nu_* \rho^0 \iota^0 \quad \text{and} \quad \rho_{2*} = (1 - \nu_*) \rho^0 \iota^0, \tag{22}$$

while, finally, ρ_1 and ρ_2 are expressed, up to first order in ϵ, by

$$\rho_1(x_{1*}, \tau) = \nu_* \{\rho^0 \{1 - \epsilon \{\operatorname{tr}[(F^0)^{-1} F^1] + (1 - \nu_*) \operatorname{Div} s\}\}(x_*, \tau),$$
$$\rho_2(x_{2*}, \tau) = (1 - \nu_*) \{\rho^0 \{1 - \epsilon \{\operatorname{tr}[(F^0)^{-1} F^1] - \nu_* \operatorname{Div} s\}\}(x_*, \tau).$$

4 Dynamics of interacting continua and constitutive equations for diatomic systems

The equations of balance for linear momentum and moment of momentum for a binary mixture in local form were stated in [5], when current fields were viewed as depending on $(y, \tau) \in \mathcal{B}_\tau \times (\tau_0, \tau_1)$. Here we are interested in a mixture of elastic bodies, in the absence of chemical reactions, which move, excluding phenomena of diffusion; hence we prefer to formulate those equations in material form by applying the inverse motion (6) of $\mathcal{C}_i$, for $i = 1, 2$, in order to exploit the consequences. They are:

$$\rho_{1*} a_1(x_{1*}, \tau) = (\operatorname{Div} P_1(x_*, \tau))_{|x_* = x_{1*}} + \rho_{1*} f_1(x_{1*}, \tau) + g_1(x_{1*}, \tau), \tag{23}$$
$$\rho_{2*} a_2(x_{2*}, \tau) = (\operatorname{Div} P_2(x_*, \tau))_{|x_* = x_{2*}} + \rho_{2*} f_2(x_{2*}, \tau) + g_2(x_{2*}, \tau), \tag{24}$$
$$\operatorname{skw}\left[\left(\iota_1^{-1} P_1 F_1^T\right)(x_{1*}, \tau) + \left(\iota_2^{-1} P_2 F_2^T\right)(x_{2*}, \tau)\right] = 0, \tag{25}$$
$$\left(\iota_1^{-1} g_1\right)(x_{1*}, \tau) + \left(\iota_2^{-1} g_2\right)(x_{2*}, \tau) = 0, \tag{26}$$

where $\operatorname{skw} D := \frac{1}{2}(D - D^T)$ represents the antisymmetric part of a second order tensor D. In these equations, for each constituent $\mathcal{C}_i$, P_i is the Piola–Kirchhoff stress tensor, f_i the density per unit mass of body forces and g_i the referential density per unit volume of internal actions; also, the material variable was specified in x_{1*} or x_{2*} in order to precise the use, to obtain each equation, of the application (6) with $i = 1$ or 2, respectively.

The balance of angular momentum (25) represents the condition of symmetry of the Cauchy stress tensor for the mixture as a whole, while the restriction (26) applies because the sum of internal forces in local form has to be null.

In [4] we proposed constitutive equations of fields P_i and g_i for isotropic elastic bodies $\mathcal{C}_i$ satisfying the principles of equipresence and material indifference. The chosen set of characteristics of deformation, endowed with the necessary invariance properties with regard to rigid displacements from the reference placement $\mathcal{B}_*$, were

$$E := \frac{1}{2}(F_1^T F_1 - I), \quad A := \frac{1}{2}(F_1^T F_2 - I) \quad \text{and} \quad s_*, \tag{27}$$

in which the first two choices are standard in elastic mixtures, while the last made was because we allow only strictly local interactions between constituents in the theory, that is, interactions between atoms of different species belonging to the same cubic material element as described in the introduction (see [2]) also; a model for the weakly non-local case was proposed in [**?**]. E is a symmetric tensor, while A is the sum of its skew part W and symmetric part $N := \operatorname{sym} A = \frac{1}{2}(A + A^T)$; moreover, all the geometric variables vanish simultaneously if and only if the displacement of the body is rigid.

By inserting the expansions (18) for the deformation gradients F_1 and F_2 in definitions $(27)_{1,2}$, we obtain

$$E = E^0 + \epsilon E^1 + o(\epsilon), \; N = N^0 + \epsilon N^1 + o(\epsilon), \; W = \epsilon W^1 + o(\epsilon), \tag{28}$$

where

$$E^0 = N^0 = \frac{1}{2}[(F^0)^T F^0 - I], \tag{29}$$

$$E^1 = \operatorname{sym}[(F^0)^T F^1 + (1 - \nu_*)(F^0)^T F^0 \operatorname{Grad} s], \tag{30}$$

$$N^1 = \operatorname{sym}\left[(F^0)^T F^1 + \left(\frac{1}{2} - \nu_*\right)(F^0)^T F^0 \operatorname{Grad} s\right], \tag{31}$$

$$W^1 = -\frac{1}{2}\operatorname{skw}[(F^0)^T F^0 \operatorname{Grad} s] \tag{32}$$

(I is the identity tensor).

Explicit representations of the Piola–Kirchhoff stress tensor P_i and of the internal actions vector g_i for isotropic elastic bodies were obtained in Sect. 5 of [4] with the use of the representation theorems of [11], the systematic expansion of the fields with respect to the parameter ϵ, the insertion of relations (28) and the fact that the balance equations (25) and (26) are satisfied identically.

But here, as we observed after the formulae (19), we are interested in the mean and relative motions of the diatomic system; hence we prefer to give the general constitutive expressions for the total Piola–Kirchhoff stress tensor P, for the microstructural tensor K and for the internal microstructural actions g. They are defined as follows: $P(x_*, \tau) := (P_1 + P_2)(x_*, \tau)$, $K(x_*, \tau) := [\nu_* P_2 - (1 - \nu_*)P_1](x_*, \tau)$ and $g(x_*, \tau) := g_1(x_*, \tau)\,[= -g_2(x_*, \tau) + o(\epsilon)]$, while their evaluations, similar to those assumed in (15) for the fields x and d, are, respectively:

$$P = (P^0 + \epsilon P^1)(x_*, \tau) + o(\epsilon), \quad K = \epsilon K^1(x_*, \tau) + o(\epsilon),$$
$$\text{and} \quad g = \epsilon g^1(x_*, \tau) + o(\epsilon) \tag{33}$$

with

$$P^0 = F^0\left[\alpha_0 I + \alpha_1 E^0 + \alpha_2 (E^0)^2\right], \tag{34}$$

$$\begin{aligned} P^1 &= F^1\left[\alpha_0 I + \alpha_1 E^0 + \alpha_2 (E^0)^2\right] \\ &\quad + F^0\Big\{\xi_0 I + \xi_1 E^0 + \xi_2 (E^0)^2 + \xi_3 E^1 + \xi_4 N^1 \\ &\quad + \operatorname{sym}\left[E^0(\xi_5 E^1 + \xi_6 N^1 + \xi_7 W^1) + (E^0)^2(\xi_8 E^1 + \xi_9 N^1 + \xi_{10} W^1)\right]\Big\}, \end{aligned} \tag{35}$$

$$\begin{aligned} K^1 &= \nu_*(1-\nu_*)F^0 \operatorname{Grad} s\left[\alpha_0 I + \alpha_1 E^0 + \alpha_2 (E^0)^2\right] - F^0\Big\{\chi_0 I + \chi_1 E^0 \\ &\quad + \chi_2 (E^0)^2 + \chi_3 E^1 + \chi_4 N^1 + \operatorname{sym}\Big[E^0(\chi_5 E^1 + \chi_6 N^1 + \chi_7 W^1) \\ &\quad + (E^0)^2(\chi_8 E^1 + \chi_9 N^1 + \chi_{10} W^1)\Big] + \chi_{11} W^1 \\ &\quad + \operatorname{skw}\Big[E^0(\chi_{12} E^1 + \chi_{13} N^1 + \chi_{14} W^1) \\ &\quad + (E^0)^2(\chi_{15} E^1 + \chi_{16} N^1) + E^0(\chi_{17} E^1 + \chi_{18} N^1)(E^0)^2\Big]\Big\}, \end{aligned} \tag{36}$$

$$g^1 = -\,F^0\left[\eta_0 I + \eta_1 E^0 + \eta_2 (E^0)^2\right] s; \tag{37}$$

the coefficients α_m and η_m, for $m = 0, 1, 2$, ξ_p, for $p = 3, \ldots, 10$, and χ_q, for $q = 3, \ldots, 18$, are scalar functions of the principal invariants of E^0, i.e., of $\omega_1 := \operatorname{tr} E^0$, $\omega_2 := \frac{1}{2}\left\{(\operatorname{tr} E^0)^2 - \operatorname{tr}\left[(E^0)^2\right]\right\}$ and $\omega_3 := \det E^0$, while ξ_p and χ_p, for $p = 0, 1, 2$, also depend on the invariants $\omega_4 := \operatorname{tr} E^1$, $\omega_5 := E^0 \cdot E^1$, $\omega_6 := (E^0)^2 \cdot E^1$, $\omega_7 := \operatorname{tr} N^1$, $\omega_8 := E^0 \cdot N^1$ and $\omega_9 := (E^0)^2 \cdot N^1$. Further, by requiring that the reference placement $\mathcal{B}_*$ is a natural state for the body for which the residual stresses P_* and K_* vanish and the residual internal actions g_* remain constant, i.e., $E^0 = E^1 = N^1 = W^1 = 0$, when $F_1 = F_2 = I$ (and $F^0 = I$, $F^1 = \operatorname{Grad} s = 0$), we have

$$\begin{aligned} &\alpha_0(0,0,0) = 0, \quad \xi_0(\omega_j = 0; j = 1, \ldots, 9) = 0 \\ &\text{and} \qquad\qquad\quad \chi_0(\omega_j = 0; j = 1, \ldots, 9) = 0. \end{aligned} \tag{38}$$

When $\epsilon = 0$ the two continua $\mathcal{C}_1$ and $\mathcal{C}_2$ have the same motion and all microstructural fields vanish, while the Piola–Kirchhoff tensor P reduces to the classical non-linear form $P = F^0\left[\alpha_0 I + \alpha_1 E^0 + \alpha_2 (E^0)^2\right]$ for a single isotropic elastic material with E^0 the material strain tensor of the total body.

5 Equations of motion for the diatomic medium

Now we are able to obtain the equations for the average and the relative motion of the diatomic system. By summing the equations of balance for linear momentum (23) and (24) and by substituting the expansions (17) for accelerations and (33) for constitutive fields, in addition to expressions (22) for the referential mass densities, we obtain the following equation of motion up to first order in ϵ for the elastic mixture considered, depending on $(x_*, \tau) \in \mathcal{B}_* \times (\tau_0, \tau_1)$:

$$\rho_* \left(\frac{\partial^2 x^0}{\partial \tau^2} + \epsilon \frac{\partial^2 x^1}{\partial \tau^2} \right) = \mathrm{Div}(P^0 + \epsilon P^1) + \rho_*(f^0 + \epsilon f^1); \tag{39}$$

instead, by subtracting Eq. (23) times $(1 - \nu_*)$ from Eq. (24) times ν_*, we get the equation which governs the relative motion between the different atomic species

$$\begin{aligned} &\epsilon \rho_* \nu_* (1 - \nu_*) \frac{\partial^2 d^1}{\partial \tau^2} \\ &\quad = \epsilon \left\{ \mathrm{Div}\, K^1 + \nu_*(1 - \nu_*) \left[\rho_* b^1 + (\mathrm{Grad}\, P^0)(\mathrm{Grad}\, s)^T \right] - g^1 \right\}, \end{aligned} \tag{40}$$

where we have used the following expansions for the assigned external actions, in agreement with the other fields:

$$\begin{aligned} f &:= [\nu_* f_1 + (1 - \nu_*) f_2] = (f^0 + \epsilon f^1)(x_*, \tau) + o(\epsilon), \\ b &:= (f_2 - f_1) = \epsilon\, b^1(x_*, \tau) + o(\epsilon). \end{aligned}$$

The systematic method of approximation used to obtain these equations is very similar to the procedure followed by Signorini and other authors (see Sects. 63–65 of [9]) within the theory of finite elastic deformations. Also, equating like powers of ϵ in (39) and (40) yields three systems of differential equations that, with associated boundary conditions other than the constitutive relations (34)–(37), govern the motion of an elastic isotropic diatomic medium in our approximation, when related boundary–value problems are resolved.

We observe that our equations of motion fit those which appear in the theory of continua with vectorial microstructure [6], when we define as densities of kinetic coenergy $\chi := \frac{1}{2} \epsilon\, \nu_*(1 - \nu_*) \left(\frac{\partial d^1}{\partial \tau} \right)^2$ and of internal microstructural actions $\vartheta := \epsilon \left[g^1 - \nu_*(1 - \nu_*)(\mathrm{Grad}\, P^0)(\mathrm{Grad}\, s)^T \right]$; in any case, we should also require that relative displacements of constituents with respect to average displacement x are divergence free (see Eq. (2.5) of [7]) to obtain a complete equivalence: in fact, in this case, we can demonstrate that the mass coefficient σ^1 in $(21)_2$ vanishes and hence, by considering relations $(20)_1$ and $(21)_1$, we are left, up to first order in ϵ, with a single equation of conservation of mass for our body as in [6].

6 The complete linear theory in the isotropic elastic case

We next study the propagation of elastic waves in our model of diatomic media and to compare our results with the continuum model of NaCl-type lattices proposed in [1] and with the one-dimesional pseudo-continuum model for ordered structures of particles of two components in [2].

Let u be the average displacement field of the mixture from the reference placement $\mathcal{B}_*$,

$$u(x_*, \tau, \epsilon) := x(x_*, \tau, \epsilon) - x_*; \tag{41}$$

the complete infinitesimal theory models physical situations in which the displacement gradient Grad u is also, in some sense, sufficiently small that its square can be neglected, as is the case, for example, when that square is itself $o(\epsilon)$ (even if, in general, we could assume that there were two different orders of approximation, not necessarily related: the relative order, restricted by the atomic spacing, and the average order).

Hence, if we suppose that $x^0(x_*, \tau) \equiv x_*$ during the motion of the diatomic continuum, from $(15)_1$ we have

$$F^0 = I \quad \text{and} \quad \mathrm{Grad}\, u = \epsilon F^1 + o(\epsilon); \tag{42}$$

by substituting in the characteristics of deformation (29)–(32) we obtain the following representations for them:

$$E^0 = N^0 = 0, \qquad E^1 = \mathrm{sym}\left[F^1 + (1 - \nu_*)\,\mathrm{Grad}\, s\right], \tag{43}$$

$$W^1 = -\frac{1}{2}\mathrm{skw}\,(\mathrm{Grad}\, s). \quad N^1 = \mathrm{sym}\left[F^1 + \left(\frac{1}{2} - \nu_*\right)\mathrm{Grad}\, s\right], \tag{44}$$

The linear expressions in the average displacement gradient for the constitutive fields are then obtained by inserting relations (38), (43) and (44) in Eqs. (34)–(37):

$$P^0 = 0,\; P^1 = \mathrm{tr}(\lambda_1 F^1 + \lambda_2\,\mathrm{Grad}\, s)I + 2\mathrm{sym}(\mu_1 F^1 + \mu_2\,\mathrm{Grad}\, s), \tag{45}$$

$$\begin{aligned} K^1 = & -\mathrm{tr}(\lambda_3 F^1 + \lambda_4\,\mathrm{Grad}\, s)I - 2\mathrm{sym}(\mu_3 F^1 + \mu_4\,\mathrm{Grad}\, s) \\ & - 2\mu_5\,\mathrm{skw}(\mathrm{Grad}\, s) \quad \text{and} \quad g^1 = -\lambda_5\, s, \end{aligned} \tag{46}$$

where the real coefficients λ_m and μ_m, for $m = 1, \ldots, 5$, are all constants (and appropriately related to the constitutive representations of the scalar coefficients in (34)–(37), when (43) and (44) are inserted).

We observe that, in the complete linear theory, the total Piola–Kirchhoff stress tensor P is then symmetric, while the microstructural tensor K also maintains a skew part.

By inserting these constitutive relations in the balance equations (39) and (40), we find the general system of partial differential equations which governs the motion of a linear diatomic medium in the isotropic elastic case, that is, in components:

$$\begin{aligned} \rho_* x^1_{j,\tau\tau} &= (\lambda_1 + \mu_1)x^1_{i,ij} + \mu_1 x^1_{j,ii} + (\lambda_2 + \mu_2)s_{i,ij} \\ &\quad + \mu_2 s_{j,ii} + \rho_* f^1, \\ \rho_*\nu_*(1 - \nu_*)s_{j,\tau\tau} &= (\lambda_3 + \mu_3)x^1_{i,ij} + \mu_3 x^1_{j,ii} + (\lambda_4 + \mu_4 + \mu_5) \\ &\quad s_{i,ij} + (\mu_4 - \mu_5)s_{j,ii} - \lambda_5 s_j - \rho_*\nu_*(1 - \nu_*)b^1, \end{aligned} \tag{47}$$

where we have used the relation $d^1 = -s$ obtained from (16), observed that $f^0 = b^0 = 0$ and eliminated the scalar parameter ϵ which multiplies both sides of the equations; letters after commas represent partial derivations with respect to the time τ or to the specified material coordinate.

7 Application to plane harmonic waves

To investigate the propagation of plane waves in the elastic diatomic system, characterized by the system (47), we seek a solution of the form

$$(x^1, s)(x_*, \tau) = (a, b)e^{i(\phi n\cdot x_* - \omega\tau)}, \tag{48}$$

where a and b are the vectorial amplitudes, ϕ the wave number, n the unit vector denoting the direction of wave propagation and ω the angular frequency.

By substituting (48) in the system of linear equations (47) in the absence of external body forces, i.e., $f^1 = b^1 = 0$, and by considering the vector decomposition $c = (n\cdot c)n + (n\times c)\times n$ for a and b, we are led to the following system of four equations, the first two in the direction of propagation n and the last two in the plane normal to n itself:

$$\begin{aligned} &\left[(\lambda_1 + 2\mu_1)\phi^2 - \rho_*\omega^2\right] n\cdot a + (\lambda_2 + 2\mu_2)\phi^2\, n\cdot b = 0,\\ &(\lambda_3 + 2\mu_3)\phi^2\, n\cdot a + \left[\lambda_5 + (\lambda_4 + 2\mu_4)\phi^2 - \rho_*\nu_*(1-\nu_*)\omega^2\right] n\cdot b = 0, \end{aligned} \tag{49}$$

$$\begin{aligned} &(\mu_1\phi^2 - \rho_*\omega^2)\,(n\times a)\times n + \mu_2\phi^2\,(n\times b)\times n = 0,\\ &\mu_3\phi^2\,(n\times a)\times n + \Big[\lambda_5 + (\mu_4 - \mu_5)\phi^2\\ &\qquad\qquad - \rho_*\nu_*(1-\nu_*)\omega^2\Big]\,(n\times b)\times n = 0. \end{aligned} \tag{50}$$

The systems (49) and (50) are constituted from two linear homogeneous algebraic equations in the scalars $n\cdot a$ and $n\cdot b$ in the vectors $(n\times a)\times n$ and $(n\times b)\times n$, respectively, which have a non-trivial solution if the secular determinant of the matrix of coefficients vanishes, i.e., if

$$\begin{aligned} &\rho_*^2\nu_*(1-\nu_*)\omega^4 - \rho_*\{\lambda_5 + [\lambda_4 + 2\mu_4 + \nu_*(1-\nu_*)(\lambda_1 + 2\mu_1)]\phi^2\}\omega^2\\ &\quad + \{(\lambda_1 + 2\mu_1)[\lambda_5 + (\lambda_4 + 2\mu_4)\phi^2] - (\lambda_2 + 2\mu_2)(\lambda_3 + 2\mu_3)\phi^2\}\phi^2 = 0 \end{aligned} \tag{51}$$

and

$$\rho_*^2 \nu_*(1-\nu_*)\omega^4 - \rho_*\{\lambda_5 + [\mu_4 - \mu_5 + \nu_*(1-\nu_*)\mu_1]\phi^2\}\omega^2 \\ + \mu_1\lambda_5\phi^2 + [\mu_1(\mu_4-\mu_5) - \mu_2\mu_3]\phi^4 = 0, \quad (52)$$

respectively.

The dispersion relations (51) and (52) govern the propagation of purely longitudinal and transverse waves, respectively, in the isotropic elastic diatomic continuum: for a given wave number ϕ, there are two real ω^2-roots, if the discriminant is positive, i.e., if $(\lambda_2 + 2\mu_2)(\lambda_3 + 2\mu_3) > 0$ and $\mu_2\mu_3 > 0$, respectively.

By solving Eq. (51) for ω^2 and expanding for small ϕ, we obtain two longitudinal solutions

$$\omega_a^2 = \frac{\lambda_1 + 2\mu_1}{\rho_*}\phi^2 + O(\phi^4) \quad \text{and} \\ \omega_o^2 = \frac{\lambda_5}{\rho_*\nu_*(1-\nu_*)} + \frac{\lambda_4 + 2\mu_4}{\rho_*\nu_*(1-\nu_*)}\phi^2 + O(\phi^4), \quad (53)$$

corresponding to the acoustical and the optical mode, respectively. If $\lambda_1 + 2\mu_1 > 0$, $\lambda_5 > 0$ and $\lambda_4 + 2\mu_4 > 0$, both ω_a and ω_o are real for real ϕ, and we have two branches on a diagram of ω versus ϕ: the acoustical which starts from the origin with positive slope $\sqrt{\frac{\lambda_1+2\mu_1}{\rho_*}}$ and the optical which emanates from $\left(0, \sqrt{\frac{\lambda_5}{\rho_*\nu_*(1-\nu_*)}}\right)$ with null slope and positive curvature $\frac{\lambda_4+2\mu_4}{\sqrt{\rho_*\nu_*(1-\nu_*)\lambda_5}}$. Hence, we can observe that the group velocity $\upsilon := \frac{d\omega}{d\phi}$ has the usual longitudinal elastic velocity $\sqrt{\frac{\lambda_1+2\mu_1}{\rho_*}}$ in the acoustic mode as long wave limit for $\phi \to 0$, while it vanishes in the optical mode.

By now solving Eq. (52), we obtain instead the transverse solutions

$$\omega_a^2 = \frac{\mu_1}{\rho_*}\phi^2 + O(\phi^4) \quad \text{and} \\ \omega_o^2 = \frac{\lambda_5}{\rho_*\nu_*(1-\nu_*)} + \frac{\mu_4 - \mu_5}{\rho_*\nu_*(1-\nu_*)}\phi^2 + O(\phi^4). \quad (54)$$

If $\mu_1 > 0$, $\lambda_5 > 0$ and $\mu_4 > \mu_5$, here we also have two branches on the diagram which start from the same places as in the longitudinal case, but now the acoustical branch has positive slope $\sqrt{\frac{\mu_1}{\rho_*}}$, while the optical one still has null slope but positive curvature $\frac{\mu_4-\mu_5}{\sqrt{\rho_*\nu_*(1-\nu_*)\lambda_5}}$. Thus, in the long wave limit, the group velocity υ has the common transversal velocity of sound $\sqrt{\frac{\mu_1}{\rho_*}}$ in the acoustic mode, while it again vanishes in the optical mode.

The cut–off frequencies $\omega_a = 0$ and $\omega_o = \sqrt{\frac{\lambda_5}{\rho_*\nu_*(1-\nu_*)}}$, obtained from equations (53) and (54) by setting $\phi = 0$ and valid for both longitudinal and transverse waves, coincide with one of those given by Brillouin [12] for NaCl-lattices; hence, in our elastic model, the constant λ_5 depicts the spring between Na and Cl atoms in a

NaCl crystal and corresponds to the constant $2(\alpha + 8\gamma)$ of the continuum model of Mindlin [1].

In general, we can easily recognize results in [1] for NaCl-lattices from our previous developments by appropriately identifying terms in the dispersion relation (28) of [1] with ours in (51) and (52). In the one-dimensional case, instead our system (47) reduces to Eqs. (2.29) and (2.30) proposed in [2] for ordered structures of particles of two components by simply imposing the equalities $\lambda_1 + 2\mu_1 = 2B$, $\lambda_4 + 2\mu_4 = -2\nu_*(1 - \nu_*)B$, $\lambda_5 = 2C$ and $\lambda_2 + 2\mu_2 = \lambda_3 + 2\mu_3 = (2\nu_* - 1)B$; thus, we can also obtain results in [2] for longitudinal modes as a particular case of our model.

Acknowledgements. This research was supported by MURST through the project "Modelli Matematici per la Scienza dei Materiali".

Reference

[1] Mindlin, R. D. (1968): Theories of elastic continua and crystal lattice theories. In: Kröner, E. (ed.) Mechanics of generalized continua. Springer, Berlin, pp. 312–320

[2] Fischer-Hjalmars, I. (1982): Micropolar phenomena in ordered structures. In: Brulin, O., Msieh,R.K.T. (eds.) Mechanics of micropolar media. World Scientific, Singapore, pp. 1–34

[3] Giovine, P. (1993): Continui biatomici. I. Equazioni di bilancio. Boll. Un. Mat. Ital. B **7**, 23–43

[4] Giovine, P. (1999): On the dynamics of elastic diatomic continua. Atti del XIV Congresso Nazionale dell'AIMETA '99, Como, Meccanica Generale II. Associazione Italiana di Meccanica Teorica e Applicata, pp. 1–10

[5] Truesdell, C. (1969): Thermodynamics of diffusion. Lecture 5. In: Truesdell, C., Rational thermodynamics. McGraw-Hill, New York

[6] Capriz, G. (1989): Continua with microstructure. (Springer Tracts in Natural Philosophy vol. **35**). Springer, New York.

[7] Tiersten, H.F., Jahanmir, M. (1977): A theory of composites modeled as interpenetrating solid continua. Arch. Rational Mech. Anal. **65**, 153–192

[8] Demiray, H. (1973): A continuum theory of elastic solids with diatomic structure. Int. J. Engrg. Sci. **11**, 1237–1246

[9] Truesdell, C., Noll, W. (1965): The non-linear field theories of mechanics. (Handbuch der Physik, Bd. III/3) Springer, Berlin

[10] Pop, J.J., Bowen, R.M. (1978): A Theory of mixtures with a long range spatial interactions. Acta Mech. **29**, 21–34

[11] Zheng, Q.S. (1994): Theory of representations for tensor functions – A unified invariant approach to constitutive equations. Appl. Mech. Rev. **47**, 545–587

[12] Brillouin, L. (1953): Wave propagation in periodic structures. 2nd ed. Dover, New York

Induced Anisotropy, Particle Spin, and Effective Moduli in Granular Materials

James T. Jenkins, Luigi La Ragione

Abstract. We consider the incremental response of a deformed aggregate of spherical particles that interact through elastic, frictional contacts. We show that when the principal axes of the incremental strain do not coincide with the principal axes of the existing strain, the average rotation of the particles is not the same as the average rotation of the aggregate. We incorporate this difference when calculating the effective moduli of the deformed aggregate.

1 Introduction

We consider a random aggregate of identical spheres that interact through elastic, frictional contacts and calculate the effective elastic moduli for a strained aggregate. We adopt the assumption that the relative displacement of the centers of a pair of contacting spheres is determined by the average strain and rotation of the aggregate and the average rotation of the spheres about their centers. We assume that the packing is initially isotropic and remains so. However, because both the normal and tangential contact stiffness depend upon the normal component of the contact force, the stiffnesses vary with the orientation of the contact with respect to the prinicpal axes of strain. This we refer to as induced anisotropy. We assume that the deviatoric strain is small relative to the volume strain. Then, when induced anisotropy is present and the aggregate is subjected to an increment in strain with principal axes that are not parallel to the existing strain, the symmetry of the stress requires that the average rotation of the spheres must differ from the average rotation of the aggregate. This results in an additional contribution to the moduli. We calculate the difference in the average rotations and the resulting effective moduli.

2 Contact force

We focus our attention on a pair of contacting spheres, label them A and B, and denote the vector from the center of A to the center of B by $\mathbf{d}^{(BA)}$. We write the increment $\dot{\mathbf{F}}^{(BA)}$in the contact force exerted by particle B on particle A in terms of the increment $\dot{\mathbf{u}}^{(BA)}$in the relative displacement of the points of contact:

$$\dot{F}_i^{(BA)} = K_{ij}^{(BA)}\dot{u}_j^{(BA)},$$

where $\mathbf{K}^{(BA)}$ is the contact stiffness.

In this paper, we assume that the contact stiffness is given in terms of the unit vector $\widehat{\mathbf{d}}^{(BA)}$ in the direction of $\mathbf{d}^{(BA)}$ by

$$K_{ij}^{(BA)} = K_N^{(BA)} \widehat{d}_i^{(BA)} \widehat{d}_j^{(BA)} + K_T^{(BA)} (\delta_{ij} - \widehat{d}_i^{(BA)} \widehat{d}_j^{(BA)}). \qquad (1)$$

Here $K_N^{(BA)}$ and $K_T^{(BA)}$ are the normal and tangential contact stiffness, assumed to be given here in terms of the normal component δ of the compressive displacement of the centers of the particles, the diameter d of the spheres, and their material properties by

$$K_N^{(BA)} = \frac{\mu d^{1/2}}{(1-\nu)} \delta^{1/2},$$

where μ and ν are, respectively, the shear modulus and Poisson ratio of the material of the spheres, and

$$K_T^{(BA)} = \frac{2\mu d^{1/2}}{(2-\nu)} \delta^{1/2}. \qquad (2)$$

The relations (1) and (2) are simpler than those employed, for example, by Thornton and Randall [1] in that the stiffness matrix is diagonal, only the elastic contribution the tangential stiffness is included, and frictional sliding is ignored. Frictional sliding could be incorporated; in this case, the sliding contacts must be identified and their stiffness set equal to zero (e.g., [2]). However, in a calculation of effective moduli associated, for example, with the transmission of waves with wavelengths that are long compared to the diameter of the particles, it may be appropriate to focus on the elastic response of the contact.

3 Contact displacement

The increment $\dot{\mathbf{u}}^{(BA)}$in contact displacement may be written in terms of the increments $\dot{\mathbf{c}}^{(B)}$ and $\dot{\mathbf{c}}^{(A)}$ in the translations of the centers of the two spheres and the increments $\dot{\omega}^{(B)}$and $\dot{\omega}^{(A)}$ in their rotations about their centers by

$$\dot{u}_i^{(BA)} = \dot{c}_i^{(B)} - \dot{c}_i^{(A)} - \frac{1}{2}\varepsilon_{ijk} \left(\dot{\omega}_j^{(B)} + \dot{\omega}_j^{(A)} \right) d_k^{(BA)}.$$

Here we assume that the relative displacement of the centers is equal to the sum of the increments in average strain $\dot{\mathbf{E}}$ and average rotation $\dot{\mathbf{W}}$ of the aggregate based upon the average position of the particle centers:

$$\dot{c}_i^{(B)} - \dot{c}_i^{(A)} = \left(\dot{E}_{ij} + \dot{W}_{ij} \right) d_j^{(BA)},$$

and that the increments in the rotation of the spheres about their centers is equal to the average increment $\dot{\Omega}^{\times}$ of this rotation:

$$\dot{\omega}_j^{(A)} = \dot{\omega}_j^{(B)} = \dot{\Omega}_j^{\times}.$$

Then

$$\dot{u}_i^{(BA)} = \left(\dot{E}_{ij} + \dot{W}_{ij} - \dot{\Omega}_{ij}\right) d_j^{(BA)},$$

where

$$\dot{\Omega}_{ik} = \varepsilon_{ijk}\dot{\Omega}_j^{\times}.$$

It is necessary to distinguish between the average rotation based upon the displacements of the particle centers and the average spin about the centers because as anisotropies develop in the state of the material, these need not be equal [3–5]. Their difference is then determined by the requirement that the stress be symmetric.

When the incremental displacement of the centers is given in terms of the averages of the increments, the normal component δ of the compressive displacement is given in terms of the average strain by

$$\delta = -\widehat{d}_i^{(BA)} E_{ij} d_j^{(BA)}.$$

This is how the stiffness depends upon the existing average strain.

4 Incremental stress

For a given $\dot{\mathbf{F}}^{(BA)}$, the incremental stress $\dot{\mathbf{T}}$ may be written as the average over all N particles in a region of homogeneous distortion that is identified with the continuum point as

$$\dot{T}_{ij} = \left\langle \frac{1}{V^{(A)}} \sum_{n=1}^{N^{(A)}} \dot{F}_i^{(nA)} d_j^{(nA)} \right\rangle \equiv \frac{1}{2N} \sum_{A=1}^{N} \frac{1}{V^{(A)}} \sum_{n=1}^{N^{(A)}} \dot{F}_i^{(nA)} d_j^{(nA)}, \tag{3}$$

where $N^{(A)}$ is the number of particles in contact with particle A and $V^{(A)}$ is the volume occupied by particle A and its nearest neighbors, including the space between the particles.

An analytical expression for the stress increment may be obtained by employing the continuous analog of Eq. (3). This is phrased in terms of the number of particles per unit volume n and a contact distribution function $f(\hat{\mathbf{d}})$ defined so that $f(\hat{\mathbf{d}})d\boldsymbol{\alpha}$ is the number of contacts in the element of solid angle $d\boldsymbol{\alpha}$ centered at $\hat{\mathbf{d}}$. For an isotropic distribution of contacts, $f = k/4\pi$, where k is the coordination number. Then

$$\dot{T}_{ij} = \frac{n}{2} \iint f(\hat{\mathbf{d}}) K_{ik} d_j d_l d\alpha \dot{L}_{kl},$$

where the integration is over all solid angles and

$$\dot{L}_{kl} \equiv \dot{E}_{kl} + \dot{W}_{kl} - \dot{\Omega}_{kl}.$$

Using such a formula, Digby [6] and Walton [7] consider isotropic aggregates and calculate the effective shear and bulk moduli. Their expression for the effective shear modulus $\overline{\mu}$ is, for example,

$$\overline{\mu} = \frac{1}{15} nkd^3 \frac{\mu}{(1-\nu)} \frac{(5-4\nu)}{(2-\nu)} \left[\frac{9}{8} \frac{(1-\nu)}{\mu} \frac{p}{nkd^3} \right]^{1/3},$$

where p is the confining pressure.

Taking $k = 5.36$ as determined in their numerical simulations, Cundall et al. [8, 9] find the predicted shear modulus to be three times that measured in their experiments and numerical simulations. On the other hand, Norris and Johnson [10] adopt a value of $k = 9$ that may be appropriate to their much higher confining pressures, and find the predicted values of wave speeds based on these to be in reasonable agreement with those measured in experiments by Domenico [11]. These differing results highlight the importance of the coordination number in the mechanics of these materials and indicate its range of variation when friction is present. Makse et al. [12] use numerical simulations to make clear how the coordination number increases with increasing confining pressure for frictionless and frictional contacts.

We next use the mean field approximation to show that, when anisotropy is induced in the distribution of contact stiffness by the average strain and the principal axes of the increment are not parallel to those of the existing strain, the symmetry of the stress requires that the average rotation of the aggregate must differ from the average rotation of the particles.

5 Average spin

Using the continuous form of the discrete averages and supposing that the initial state is isotropic, we find that the expression for the incremental stress is

$$\dot{T}_{ij} = \frac{n}{2} \frac{k}{4\pi} d^2 \left(\iint K_{im}^{(BA)} \widehat{d}_n^{(BA)} \widehat{d}_j^{(BA)} d\alpha \right) \dot{L}_{mn}, \tag{4}$$

and the requirement that the incremental in stress be symmetric,

$$\varepsilon_{pij} \dot{T}_{ij} \equiv 0,$$

may be written as

$$0 \equiv \varepsilon_{mpj} \left(\iint K_T^{(BA)} \widehat{d}_q^{(BA)} \widehat{d}_j^{(BA)} d\alpha \right) \dot{L}_{pq}. \tag{5}$$

In this paper, we assume that an initially isotropic aggregate is first compressed isotropically and then subjected to a deviatoric strain. To reflect this, we write the strain as the sum of its isotropic and deviatoric parts,

$$E_{ij} = \frac{e}{3} \delta_{ij} + \widehat{E}_{ij}.$$

Furthermore, we assume that the deviatoric part of the strain is small compared to the isotropic part. This is reasonable for at least a range of loading prior to localization and failure. Finally, we expand the term in the stiffness that involves the strain and retain only terms that are linear in the ratio of the deviatoric to the isotropic strain:

$$\left(-\widehat{d}_i^{(BA)} E_{ij} \widehat{d}_j^{(BA)}\right)^{1/2} \doteq \left(-\frac{e}{3}\right)^{1/2} - \frac{1}{2}\left(-\frac{e}{3}\right)^{-1/2} \widehat{d}_i^{(BA)} \widehat{E}_{ij} \widehat{d}_j^{(BA)}.$$

So, for example,

$$K_T^{(BA)} \doteq \frac{2\mu d}{(2-\nu)}\left(-\frac{e}{3}\right)^{1/2}\left(1 + \frac{3}{2} e^{-1} \widehat{d}_k^{(BA)} \widehat{d}_l^{(BA)} \widehat{E}_{kl}\right).$$

We employ this in the expression (5) for the symmetry of the stress increment:

$$0 \equiv \varepsilon_{mpj}\left[\iint \left(1 + \frac{3}{2} e^{-1} \widehat{d}_k^{(BA)} \widehat{d}_l^{(BA)} \widehat{E}_{kl}\right) \widehat{d}_q^{(BA)} \widehat{d}_j^{(BA)} d\alpha\right] \dot{L}_{pq}.$$

The integrations may be carried out with the assistance of the identities obtained by applying the divergence theorem to the tensor products on the unit sphere,

$$\iint \widehat{d}_q^{(BA)} \widehat{d}_j^{(BA)} d\alpha = \iiint r_{q},_{j}\, dv = \frac{4\pi}{3}\delta_{qj}$$

and

$$\iint \widehat{d}_k^{(BA)} \widehat{d}_l^{(BA)} \widehat{d}_q^{(BA)} \widehat{d}_j^{(BA)} d\alpha = \iiint \left(r_k r_l r_q\right),_{j}\, dv = \frac{4\pi}{15} X_{klqj},$$

where

$$X_{klqj} \equiv \delta_{kj}\delta_{lq} + \delta_{lj}\delta_{kq} + \delta_{qj}\delta_{kl}.$$

The result is

$$0 \equiv \varepsilon_{mpj}\left(\delta_{qj} + \frac{3}{10} X_{klqj} e^{-1} \widehat{E}_{kl}\right) \dot{L}_{pq},$$

or

$$0 \equiv \varepsilon_{mpj}\left(\delta_{jq} + \frac{3}{5} e^{-1} \widehat{E}_{jq}\right)\left(\dot{W}_{pq} - \dot{\Omega}_{pq}\right) + \frac{3}{5} e^{-1} \varepsilon_{mpj} \widehat{E}_{jq} \dot{E}_{pq}. \tag{6}$$

The second term is that due to the anisotropy induced in the distribution of stiffness by the strain. We write (6) in terms of the axial vectors of the antisymmetric tensors:

$$0 \equiv \varepsilon_{mpj}\varepsilon_{plq}\left(\delta_{jq} + \frac{3}{5} e^{-1} \widehat{E}_{jq}\right)\left(\dot{W}_l^{\times} - \dot{\Omega}_l^{\times}\right) + \frac{3}{5} e^{-1} \varepsilon_{mpj} \widehat{E}_{jq} \dot{E}_{pq},$$

or

$$0 \equiv \left(\delta_{lm} - \frac{3}{10} e^{-1} \widehat{E}_{lm}\right)\left(\dot{W}_l^{\times} - \dot{\Omega}_l^{\times}\right) - \frac{3}{10} e^{-1} \varepsilon_{mpj} \widehat{E}_{jq} \dot{E}_{pq}.$$

The latter is the equation that we wish to solve for the difference of the two average rotations. The approximate inversion of the matrix of coefficients in this equation is

$$\left(\delta_{lm} - \frac{3}{10}e^{-1}\widehat{E}_{lm}\right)^{-1} = \delta_{lm} + \frac{3}{10}e^{-1}\widehat{E}_{lm}.$$

Consequently,

$$\dot{W}_l^{\times} - \dot{\Omega}_l^{\times} \equiv \frac{3}{10}e^{-1}\left(\delta_{lm} + \frac{3}{10}e^{-1}\widehat{E}_{lm}\right)\varepsilon_{mpj}\widehat{E}_{jq}\dot{E}_{pq},$$

or, to the order of the approximation,

$$\dot{W}_l^{\times} - \dot{\Omega}_l^{\times} \equiv \frac{3}{10}e^{-1}\varepsilon_{lpj}\widehat{E}_{jq}\dot{E}_{pq}.$$

The form of this solution most convenient in the calculation of the stress increment is

$$\dot{W}_{mn} - \dot{\Omega}_{mn} = \frac{3}{10}e^{-1}\varepsilon_{mln}\varepsilon_{lpj}\widehat{E}_{jq}\dot{E}_{pq}.$$

We note that ,when the eigenvectors of the strain increment are parallel to the eigenvectors of the deviatoric strain, the increments of the average rotations are equal.

We next calculate the stress increment and show that it is influenced by the induced anisotropy, both explicitly and through the difference in the average rotations.

6 Effective moduli

The effective moduli are the coefficients of $\dot{\mathbf{E}}$ in the expression for the incremental stress. In order to calculate the effective moduli, we adopt the expression for the incremental stress, made symmetric by incorporating the difference in the average rotations:

$$\dot{T}_{ij} = \frac{n}{2}\frac{k}{4\pi}d^2\left(\iint K_{im}^{(BA)}\widehat{d}_n^{(BA)}\widehat{d}_j^{(BA)}\,d\alpha\right)$$
$$\left(\dot{E}_{mn} + \frac{3}{10}e^{-1}\varepsilon_{mqn}\varepsilon_{qpl}\widehat{E}_{lk}\dot{E}_{pk}\right).$$

In this, the approximation to the complete stiffness matrix which corresponds to that for K_T is

$$K_{ij}^{(BA)} = \frac{\mu d}{(1-\nu)}\left(-\frac{e}{3}\right)^{1/2}\left(1+\frac{3}{2}e^{-1}\widehat{d}_k^{(BA)}\widehat{d}_l^{(BA)}\widehat{E}_{kl}\right)\widehat{d}_i^{(BA)}\widehat{d}_j^{(BA)}$$
$$+\frac{2\mu d}{(2-\nu)}\left(-\frac{e}{3}\right)^{1/2}\left(1+\frac{3}{2}e^{-1}\widehat{d}_k^{(BA)}\widehat{d}_l^{(BA)}\widehat{E}_{kl}\right)$$
$$\times\left(\delta_{ij} - \widehat{d}_i^{(BA)}\widehat{d}_j^{(BA)}\right).$$

With this,

$$\int K_{im}^{(BA)}\widehat{d}_n^{(BA)}\widehat{d}_j^{(BA)}d\alpha$$

$$= \frac{\mu d}{(1-\nu)}\left(-\frac{e}{3}\right)^{1/2}\left[\iint \widehat{d}_i^{(BA)}\widehat{d}_m^{(BA)}\widehat{d}_n^{(BA)}\widehat{d}_j^{(BA)}d\alpha + \frac{3}{2}e^{-1}\widehat{E}_{kl}\iint \widehat{d}_i^{(BA)}\widehat{d}_m^{(BA)}\widehat{d}_k^{(BA)}\widehat{d}_l^{(BA)}\widehat{d}_n^{(BA)}\widehat{d}_j^{(BA)}d\alpha\right]$$

$$+ \frac{2\mu d}{(2-\nu)}\left(-\frac{e}{3}\right)^{1/2}\left[\iint \left(\delta_{im} - \widehat{d}_i^{(BA)}\widehat{d}_m^{(BA)}\right)\widehat{d}_n^{(BA)}\widehat{d}_j^{(BA)}d\alpha + \frac{3}{2}e^{-1}\widehat{E}_{kl}\iint \left(\delta_{im} - \widehat{d}_i^{(BA)}\widehat{d}_m^{(BA)}\right)\widehat{d}_k^{(BA)}\widehat{d}_l^{(BA)}\widehat{d}_n^{(BA)}\widehat{d}_j^{(BA)}d\alpha\right].$$

We carry out the integrations with the assistance of the additional identity

$$\iint \widehat{d}_i^{(nA)}\widehat{d}_j^{(nA)}\widehat{d}_k^{(nA)}\widehat{d}_l^{(nA)}\widehat{d}_p^{(nA)}\widehat{d}_q^{(nA)}d\alpha = \iiint \left(r_i r_j r_k r_l r_p\right)_{,q} dv = \frac{4\pi}{105}Y_{ijklpq},$$

where

$$Y_{ijklpq} \equiv \delta_{iq}X_{jklp} + \delta_{jq}X_{klpi} + \delta_{kq}X_{lpij} + \delta_{lq}X_{pijk} + \delta_{pq}X_{ijkl}.$$

To the order of the approximation, the result is

$$\dot{T}_{ij} = \frac{nk}{30}\frac{\mu d^3}{(1-\nu)}\left(-\frac{e}{3}\right)^{1/2}\left[\left(X_{imnj} + \frac{3}{14}e^{-1}\widehat{E}_{kl}Y_{imklnj}\right)\delta_{mp}\delta_{nk} + \frac{3}{10}e^{-1}X_{imnj}\varepsilon_{mqn}\varepsilon_{qpl}\widehat{E}_{lk}\right]\dot{E}_{pk}$$

$$+ \frac{nk}{3}\frac{\mu d^3}{(2-\nu)}\left(-\frac{e}{3}\right)^{1/2}\left[\left(\delta_{im}\delta_{nj} - \frac{1}{5}X_{imnj}\right)\left(\delta_{pm}\delta_{kn} + \frac{3}{10}e^{-1}\varepsilon_{mqn}\varepsilon_{qpl}\widehat{E}_{lk}\right) + \frac{3}{10}e^{-1}\widehat{E}_{kl}\left(\delta_{im}X_{klnj} - \frac{1}{7}Y_{imklnj}\right)\delta_{pm}\delta_{kn}\right]E_{pk}.$$

The contributions to the effective moduli due to the induced anisotropy are linear in the ratio of the average deviatoric strain to the average volume strain, and there is an additional contribution when the eigenvectors of the incremental strain are not parallel to those of the deviatoric part of the average strain.

7 Conclusion

We have used a relatively simple, approximate theory to show that anisotropy in the distribution of the contact stiffnesses induced by straining may result in an average rotation of the particles that is different from the average rotation of the aggregate. This, in turn, can influence the effective elastic moduli. These are new effects that are associated with the additional internal degrees of freedom of the granular aggregate.

References

[1] Thornton, C., Randall, C.W. (1988): Application of theoretical contact mechanics to solid particle system simulation. In: Satake, M., Jenkins, J.T. (eds.) Micromechanics of granular materials. Elsevier, Amsterdam, pp. 133–142

[2] Jenkins, J.T., Strack, O.D.L. (1993): Mean-field inelastic behavior of random arrays of identical spheres. Mech. Materials **16**, 25–33

[3] Koenders, M.A. (1990): Localized deformation using higher-order stress-strain theory. J. Energy Resources Tech. **112**, 51–53

[4] Jenkins, J.T. (1991): Anisotropic elasticity for random arrays of identical spheres. In: Wu, J.J. et al. (eds.) Modern theory of anisotropic elasticity and its applications. SIAM, Philadelphia, pp. 368–377

[5] Jenkins, J.T., La Ragione, L. (2001): Particle spin in anisotropic granular materials. Internat. J. Solids Structures **38**, 1063–1069

[6] Digby, P.J (1981): The effective elastic moduli of porous granular rocks. J. Appl. Mech. **48**, 803–808

[7] Walton, K. (1987): The effective elastic moduli of a random packing of spheres. J. Mech. Phys. Solids **35**, 213–216

[8] Cundall, P.A., Jenkins, J.T., Ishibashi, I. (1989): Evolution of elastic moduli in a deforming granular assembly. In: Biarez, J., Gourvès, R. (eds.) Powders and grains. Balkema, Rotterdam, pp. 319–322

[9] Jenkins, J.T., Cundall, P.A., Ishibashi, I. (1989): Micromechanical modeling of granular materials with the assistance of experiments and numerical simulations. In: Biarez, J., Gourvès, R. (eds.) Powders and grains. Balkema, Rotterdam, pp. 257–264

[10] Norris, A.N., Johnson, D.L. (1997): Nonlinear elasticity of granular media. J. Appl. Mech. **64**, 39–49

[11] Domenico, S.N. (1977): Elastic properties of unconsolidated porous sand reservoirs. Geophys. **42**, 1339–1368

[12] Makse, H.A., Gland, N., Johnson, D.L., Schwartz, L.M. (1999): Why effective medium theory fails in granular materials. Phys. Rev. Lett. **83**, 5070–5073

[13] Jenkins, J.T. (1988): Volume change in small strain axisymmetric deformations of a granular material. In: Satake, M., Jenkins, J.T. (eds.) Micromechanics of granular materials. Elsevier, Amsterdam, pp. 245–252

[14] Chen, Y.C., Ishibashi, I., Jenkins, J.T. (1988): Dynamic shear modulus and fabric. I. Depositional and induced anisotropy. Géotechnique **38**, 25–32

On Uniqueness in Nonlinear Homogeneous Elasticity

Robin J. Knops

Abstract. In line with the procedure developed by Knops and Stuart, conservation laws are combined with general notions of convexity to derive some new uniqueness results for simple (affine) boundary value and initial boundary problems in nonlinear homogeneous elasticity.

1 Introduction

The author and co-workers [19–21] have considered uniqueness of smooth solutions to simple boundary value problems in homogeneous nonlinear elastostatics on a region that is star-shaped. Either the strain energy function is, for example, rank-one convex and strictly quasi-convex at a constant deformation gradient, or the complementary strain energy function possesses corresponding properties. Here, we extend these arguments and discuss static and dynamic problems for a nonlinear elastic body subject to zero body force and specified boundary deformation, and, for the most part, retain a quasi-convex and rank-one convex strain energy function. Uniqueness is proved in elastostatics for an affine deformation boundary value problem and for a star-shaped body containing regions in different phases, or for certain unbounded bodies when only strict rank-one convexity is required.

In the initial boundary value problem, global existence of a smooth solution is generally impossible (cp. Dafermos [6]). Nevertheless, Dafermos and Hrusa [7], and Hughes et al. [17] have proved local existence and uniqueness of a smooth solution under the assumption of a strongly-elliptic strain energy function. (Wheeler [32] and Dafermos [6] have given further uniqueness proofs to the same problem.) Subject to the same conditions, we demonstrate continuous dependence of a smooth solution on initial data and consequently provide yet another proof of uniqueness. We then seek to replace strong-ellipticity by (strict) quasi-convexity and (strict) rank-one convexity of the strain energy function at a particular displacement gradient. For simplicity, discussion is confined to the special problem of a three-dimensional star-shaped region with null initial and displacement boundary data.

Section 2 describes the static and dynamic problems and associated properties including those of the energy-momentum tensor. Section 3 introduces, for subsequent employment, conservation laws in n spatial dimensions and the generalised notions of convexity. Uniqueness is established in Sect. 4 for a star-shaped bounded elastic body that contains regions in different phases in equilibrium under zero body force and an affine deformation prescribed on the external surface. The strain energy function is assumed strictly quasi-convex and rank-one convex at a given constant deformation gradient. Section 4 also considers certain unbounded regions occupied by a homogeneous elastic material again in equilibrium under zero body force and affine deformation boundary conditions but for a strain energy function that is only strictly rank-one convex at a specified deformation gradient. Finally, Sect. 5 treats uniqueness in nonlinear homogeneous elastodynamics. Continuous dependence on initial data, and hence uniqueness, is first established for smooth solutions subject to general data and a strongly-elliptic strain energy function. We then study uniqueness of the null solution in the class of smooth solutions for the three-dimensional initial boundary value problem subject to null data and on a star-shaped region. We show that strong-ellipticity may be replaced by (strict) quasi-convexity and (strict) rank-one convexity at the null element.

Existence of a smooth solution is always assumed. Indicial notation is adopted throughout, although for brevity, and without confusion, a direct notation is occasionally introduced. Latin suffixes range over $1, 2, 3, \ldots, n$, the summation convention is employed and the comma notation indicates partial differentiation with respect to a spatial variable.

2 Basic equations

A nonlinear homogeneous elastic material in its reference configuration occupies the n-dimensional region $\Omega \subseteq \mathbb{R}^n$ whose boundary $\partial\Omega$ is piecewise continuously differentiable. A further restriction is imposed later on $\partial\Omega$. Points of Ω are given by $x_i, i = 1, \ldots n$, with respect to a fixed Cartesian coordinate system, and the unit outward normal on $\partial\Omega$ is denoted by the vector N with components N_i.

We are concerned with both static and dynamic problems. In the static case, the deformation is given by the smooth vector function $u(x)$ and the body is deformed and maintained in equilibrium under zero body-force and specified boundary values of the deformation. The deformation gradient, with components $u_{i,j}$, is in the class M^+ where

$$M^+ = \{A \in M^{n\times n} : \det A > 0\}, \tag{2.1}$$

and $M^{n\times n}$ represents the set of $n \times n$ matrices. The abbreviated notation ∇u is sometimes used to denote the deformation gradient.

The elastic body possesses a strain energy function $W : M^+ \to \mathbb{R}$ per unit volume defined on the deformation gradients in M^+. For $W \in C^1(M^+, \mathbb{R})$, the Piola- Kirchhoff stress tensor is given by the relation

$$\sigma_{ij} = \partial W/\partial u_{i,j}. \tag{2.2}$$

The deformation is a *smooth equilibrium solution* when $u \in C^2(\Omega, \mathbb{R}^n) \cup C^1(\bar{\Omega}, \mathbb{R}^n)$, $W \in C^1(M^+, \mathbb{R})$ and the equilibrium equations

$$\sigma_{ij,j} = 0, \quad x \in \Omega, \tag{2.3}$$

are satisfied, while on the boundary we have

$$u_i(x) = g_i(x), \quad x \in \partial\Omega, \tag{2.4}$$

where g is a specified vector function.

Subsequent results involve the energy-momentum, or Eshelby, tensor, which, in terms of the strain energy function, deformation gradient and Piola–Kirchhoff stress tensor, is defined to be

$$b_{ij} = W\delta_{ij} - \sigma_{kj}u_{k,i}, \quad x \in \Omega, \tag{2.5}$$

where δ_{ij} is the Kronecker delta. The tensor, introduced by Eshelby [9] and discussed by authors such as Chadwick [3] and Hill [15], has been employed in the study of inhomogeneous elastic materials (see, e.g., Maugin [23], Kienzler and Herrmann [18] and the references cited there), and occurs, for instance, in the variation of total strain energy

$$V(u) = \int_{\Omega} W(\nabla u(x))dx \tag{2.6}$$

when the reference configuration Ω is allowed to alter (see Gelfand and Fomin [12] and Olver [27]). For homogeneous elastic materials the tensor b_{ij} satisfies the equations

$$b_{ij,j} = 0, \quad x \in \Omega, \tag{2.7}$$

which are analogous to the equilibrium equation (2.3) satisfied by the Piola–Kirchhoff stress tensor.

It follows immediately from (2.3) and (2.7) and the divergence theorem that

$$\int_{\partial\Omega} \sigma_{ij}N_j dS = \int_{\partial\Omega} b_{ij}N_j dS = 0. \tag{2.8}$$

Now let Ω consist of a region Ω_1 embedded in the region Ω_2 so that $\Omega = \Omega_1 \cup \Omega_2 \cup \Gamma$ where Γ is the interfacial surface between Ω_1 and Ω_2 and $\partial\Omega = \partial\Omega_2 \backslash \Gamma$. (A greater number of subdivisions could be considered but this generality is omitted for simplicity). We suppose that the material occupying Ω remains in equilibrium under zero body-force and the specified boundary condition (2.4) on $\partial\Omega$, but that in Ω_1 and Ω_2 the material experiences different phases. We assume that across the interface Γ there is continuity of the deformation and the Piola–Kirchhoff stress vector, whereas the previous smoothness conditions on the deformation and strain energy function are assumed to hold separately in Ω_1 and Ω_2. Consequently, the deformation is piece-wise smooth in Ω. When the equilibrium configuration of Ω satisfying (2.4) is regarded as a minimiser of the total strain energy (2.6), it may be shown (Gelfand and Fomin [12]) that the Weierstrass–Erdmann corner conditions are satisfied across Γ. This is equivalent to the continuity across Γ of the energy-momentum vector

$$B_i = b_{ij}N_j. \tag{2.9}$$

(The statement is no longer true when Ω_1 and Ω_2 comprise different regions of an elastic composite (see, e.g., Hill [15]). We conclude that the balances (2.8) continue to hold for the region $\Omega = \Omega_1 \cup \Omega_2 \cup \Gamma$.

The discontinuity in the deformation gradient across Γ is affine. This fact follows from the well-known general lemma due originally to Hadamard:

Lemma 2.1. *Consider a region $\Omega \subset \mathbb{R}^n$ and let $u_i, v_i \in C^1(\bar{\Omega}, \mathbb{R}^n)$, with $u_i(x) = v_i(x)$, $x \in \partial\Omega$, where $\partial\Omega$ is the Lipschitz continuous surface of Ω. Then*

$$u_{i,j} - v_{i,j} = \lambda_i N_j \tag{2.10}$$

where $\lambda_i = \partial(u_i - v_i)/\partial N$ is the normal derivative of $u_i - v_i$ on $\partial\Omega$ and N_j is the unit outward normal on $\partial\Omega$.

The initial boundary value problem of elastodynamics that is discussed here requires the local existence on some finite time interval $[0, T]$ of a smooth deformation $u_i(x, t)$ such that $u \in C^2(\Omega, \mathbb{R}^n) \cap C^1(\bar{\Omega}, \mathbb{R}^n) \cap C^2([0, T], \mathbb{R}^n)$ and (2.1) is satisfied. The strain energy function $W \in C^1(M^+, \mathbb{R})$ and Piola–Kirchhoff stress tensor are introduced as in the equilibrium boundary value problem and body-force is again assumed to be zero. The equilibrium equations (2.3) are replaced by

$$\sigma_{ij,j} = \rho \ddot{u}_i, \quad (x, t) \in \Omega \times [0, T], \tag{2.11}$$

where t denotes the time variable, a superposed dot indicates partial differentiation with respect to time, and ρ denotes the mass density which is assumed positive and constant.

The initial and boundary conditions are respectively specified by

$$u_i(x, 0) = u_i^0(x), \quad \dot{u}_i(x, 0) = \dot{u}_i^0(x), \quad x \in \Omega, \tag{2.12}$$

and

$$u_i(x, t) = f_i(x, t), \quad x \in \partial\Omega \times [0, T], \tag{2.13}$$

where the vector functions u^0, $\dot{u}^0$, and f are given and assumed compatible on $\partial\Omega \times \{0\}$.

As mentioned in the introductory section, one objective is to establish local uniqueness to the null initial boundary value problem (2.11)–(2.13) in the absence of strong-ellipticity. Uniqueness of the solution to the affine deformation boundary value problem on an unbounded region is also sought. Conditions on the strain energy function together with certain conservation laws required to achieve these aims are discussed in the next section.

3 Conservation laws. Constitutive restrictions

The first part of this section derives conservation laws required in the subsequent proofs of uniqueness. Two such laws are obtained in the equilibrium problem from properties of the energy-momentum tensor (2.5) (cp. Hill [15]; see also Chadwick [3], Günther [14], Knowles and Sternberg [22] and Green [13]). They include the counterpart of the Pohožaev identity [28] and its generalisation by Pucci and Serrin [29] to elliptic systems of partial differential equations. (See also van der Vorst [31].)

The second part of the section recalls the general notions of convexity. A complete description of these and related concepts now standard in discussions of existence have been given by Ball [1], Dacorogna [5], and Müller [26] and the further references cited in these papers.

3.1 Conservations laws

While the generality achieved in [29, 31] may be useful in extending the results of this paper, for present purposes it suffices to confine attention to the following two special cases which can be derived *ab initio*. We first consider a bounded region Ω and then discuss an unbounded region.

The first conservation law is represented by $(2.8)_2$ which by (2.5) becomes

$$\int_{\partial\Omega} \left[N_i W - N_j \sigma_{kj} u_{k,i}\right] dS = 0. \tag{3.1}$$

The second conservation law, obtained on multiplying (2.7) by x_i and integrating over Ω, is given by

$$n \int_{\Omega} W(\nabla u(x))dx = \int_{\partial\Omega} \left[N_k x_k W + N_j \sigma_{ij} \left(u_i - x_k u_{i,k}\right)\right] dS. \tag{3.2}$$

It has been used by Knops and Stuart [19] to prove uniqueness in the affine displacement boundary value problem for a star-shaped region.

We remark that, when $\Omega = \Omega_1 \cup \Omega_2 \cup \Gamma$ (see Sect. 2 for notation), and the elastic material in different phases occupies Ω_1 and Ω_2, then, by virtue of the continuity across the interface Γ of the deformation, traction and energy-momentum vector, the conservation laws (3.1) and (3.2) continue to hold but with $\partial\Omega = \partial\Omega_2 \backslash \Gamma$.

For an unbounded region, the identity corresponding to (3.1) is obtained by integrating (2.7) over the region $\Omega \cap B(R)$, where $B(R)$ is a sphere of radius R and centre at the origin (cp. Esteban and Lions [10]). We also assume that the strain energy function satisfies $W \in L^1(\Omega)$ and the deformation vector u_i belongs to the Sobolev space $W^{1,2}(\Omega)$.

It follows that

$$\begin{aligned} \int_{\partial\Omega \cap B(R)} N_j b_{ij} dS &= - \int_{\Omega \cap \partial B(R)} N_j b_{ij} dS \\ &\le \int_{\Omega \cap \partial B(R)} \left\{ |W| + \left(\sigma_{ij}\sigma_{ij}\right)^{\frac{1}{2}} \left(u_{k,l} u_{k,l}\right)^{\frac{1}{2}} \right\} dS. \end{aligned} \tag{3.3}$$

By hypothesis the right-hand side of (3.3) tends to zero as $R \to \infty$ and hence the desired conservation law is given by

$$\lim_{R \to \infty} \int_{\partial\Omega \cap B(R)} \left[N_i W - N_j \sigma_{kj} u_{k,i}\right] dS = 0. \tag{3.4}$$

A similar argument may be used to extend the conservation law (3.2) to an unbounded region, but the general expression is not required in the sequel.

To derive conservation laws in the dynamic problems we consider the space and time dependent factor

$$a(x,t)\dot{u}_i + b_j(x,t) u_{i,j} + c(x,t) u_i, \tag{3.5}$$

where $a(x,t), b_j(x,t), c(x,t)$ are differentiable arbitrary functions to be specifically chosen later. In particular, (3.5) specialises to the (infinitesimal) generators of symmetry groups for Noether's theorem in the variational calculus (cp. Gelfand and Fomin [12] and Olver [27]; see also Fletcher [11] and Delph [8]). An antecedent is Friedrich's "a–b–c" method, and its development by, for instance, Morawetz [24]. Multiplication of (2.11) by (3.5) and integration over both Ω and the time interval $(t_1, t_2) \subseteq [0, T]$ after rearrangement leads to

$$\int_{t_1}^{t_2}\int_{\Omega(\eta)} P_1(x,\eta)\,dx\,d\eta = \int_{t_1}^{t_2}\int_{\partial\Omega(\eta)} N_j P_{2j}(x,t)\,dS\,d\eta + \int_{\Omega(\eta)} P_3(x,\eta)\,dx\Big|_{t_1}^{t_2}, \tag{3.6}$$

where here and subsequently $\Omega(t)$ denotes integration over Ω at time t, and

$$\begin{aligned} P_1(x,t) &= \dot{a}E(x,t) + b_{j,j}L(x,t) + b_{k,j}\sigma_{ij,k}u_i \\ &\quad + c\left(\rho\dot{u}_i\dot{u}_i - \sigma_{ij}u_{i,j}\right) + \sigma_{ij}u_i\left(b_{k,kj} - c_{,j}\right) \\ &\quad - \tfrac{1}{2}\ddot{c}\rho u_i u_i - a_{,j}\sigma_{ij}\dot{u}_i + \dot{b}_k\rho\dot{u}_i\dot{u}_{i,k}, \end{aligned} \tag{3.7}$$

$$P_{2j}(x,t) = b_i B_{ij} + \left(b_{j,k} - c\delta_{kj}\right)\sigma_{ik}u_i - a\sigma_{ij}\dot{u}_i, \tag{3.8}$$

$$P_3(x,t) = aE(x,t) + b_k\rho u_{i,k}\dot{u}_i + c\rho u_i\dot{u}_i - \tfrac{1}{2}\dot{c}\rho\dot{u}_i\dot{u}_i, \tag{3.9}$$

and

$$E(x,t) = W(\nabla u(x,t)) + \tfrac{1}{2}\rho\dot{u}_i(x,t)\dot{u}_i(x,t), \tag{3.10}$$

$$B_{ij}(x,t) = L(x,t)\delta_{ij} - \sigma_{kj}u_{k,i}, \tag{3.11}$$

$$L(x,t) = W(\nabla u(x,t)) - \tfrac{1}{2}\rho\dot{u}_i\dot{u}_i. \tag{3.12}$$

Obviously, $E(x,t)$ represents the local total energy function, whereas $L(x,t)$ is the Lagrangian.

On setting $a = 1, b_i = c = 0$, we deduce from (3.6) the identity

$$\int_{\Omega(t_2)} E(x,t_2)\,dx = \int_{\Omega(t_1)} E(x,t_1)\,dx + \int_{t_1}^{t_2}\int_{\partial\Omega(\eta)} N_j\sigma_{ij}\dot{u}_i\,dSd\,\eta, \tag{3.13}$$

which gives the time evolution of total energy.

We next select $a = t, b_i = x_i, c = 1$ and, after a straightforward calculation using (2.11), obtain from (3.6) the conservation law

$$\int_{t_1}^{t_2}\int_{\Omega(\eta)}\Big[(n+1)W(\nabla u(x,\eta)) - \tfrac{1}{2}(n-3)\rho\dot{u}_i(x,\eta)\dot{u}_i(x,\eta) + 2\rho u_i(x,\eta)\ddot{u}_i(x,\eta)\Big]\,dx\,d\eta$$

$$= \int_{t_1}^{t_2} \int_{\partial\Omega(\eta)} [N_i x_i W(\nabla u(x,\eta) + N_j \sigma_{ij}(\nabla u(x,\eta) \left\{u_i(x,\eta) - x_k u_{i,k}(x,\eta)\right\}$$
$$- \tfrac{1}{2} N_k x_k \rho \dot{u}_i(x,\eta)\dot{u}_i(x,\eta) - \eta N_j \sigma_{ij}(\nabla u(x,\eta))\dot{u}_i(x,\eta)]\, dS\, d\eta$$
$$+ \int_{\Omega(\eta)} [\eta E(x,\eta) + \rho x_k u_{i,k}(x,\eta)\dot{u}_i(x,\eta) + \rho u_i(x,\eta)\dot{u}_i(x,\eta)]\, dx \Big|_{t_1}^{t_2}. \tag{3.14}$$

The identity (3.14) is employed, with arguments of the functions suppressed, in the uniqueness proof of the null initial boundary value problem of nonlinear elastodynamics.

3.2 Constitutive restrictions for the strain energy function

We recall general notions of convexity and some related concepts required in Sects. 4 and 5.

Definition 3.1. The function $W \in C(M^+, \mathbb{R})$ is quasi-convex at $A \in M^+$ if and only if

$$\int_D W(A + \nabla\xi(x))dx \geq \int_D W(A)dx = W(A)|D|, \tag{3.15}$$

for all non-empty open bounded subsets $D \subseteq \mathbb{R}^n$, and all functions $\xi \in W_0^{1,\infty}(D, \mathbb{R}^n)$ which vanish on ∂D such that $A + \nabla\xi \in M^+$, $x \in D$. The volume of D is denoted by $|D|$. (See, e.g., [26]).

Note that by means of a covering argument it may be shown that the definition of quasi-convexity is independent of the choice of region D (see, e.g., [26]).

Definition 3.2. The function W is strictly quasi-convex at $A \in M^+$ if and only if W is quasi-convex at A and equality holds in (3.15) only when $\xi \equiv 0$.

Definition 3.3. The function W is (strictly) quasi-convex if and only if W is (strictly) quasi-convex at A for all (constant) $A \in M^+$.

Definition 3.4. The function $W \in C(M^+, \mathbb{R})$ is rank-one convex at $A \in M^+$ if and only if

$$W(A + sa\otimes b) \leq sW(A + a\otimes b) + (1 - s)W(A), \tag{3.16}$$

for every $a \in \mathbb{R}^n$, $b \in \mathbb{R}^n$ and all $s \in [0, 1]$ such that $A + a\otimes b \in M^+$. Here, $a\otimes b$ is the tensor product of the vectors a, b, and has Cartesian components $a_i b_j$.

Definition 3.5. The function W is strictly rank-one convex at $A \in M^+$ if and only if W is rank-one convex at A and equality holds in (3.16) only when $a\otimes b = 0$ or $s(1 - s) = 0$.

Definition 3.6. The function W is (strictly) rank-one convex if and only if W is (strictly) rank-one convex at A for all (constant) $A \in M^+$.

When $W \in C^1(M^+, \mathbb{R})$, an equivalent expression of rank-one convexity at A is obtained on letting $s \to 0$ in (3.16) and using the definition of a derivative. We have for $A + a \otimes b \in M^+$, $A \in M^+$,

$$W(A + a \otimes b) \geq W(A) + \frac{\partial W(A)}{\partial A_{ij}} a_i b_j, \tag{3.17}$$

where components are introduced on the right to emphasize summation.

Definition 3.7. The function $W \in C^2(M^+, \mathbb{R})$ is strongly-elliptic at $A \in M^+$ if and only if

$$\frac{\partial^2 W(A)}{\partial A_{ij} \partial A_{kl}} a_i a_k b_j b_l > 0, \tag{3.18}$$

for all $a \in \mathbb{R}^n$, $b \in \mathbb{R}^n$ that are not identically zero.

The notion of strong-ellipticity has been employed in uniqueness studies for the displacement boundary value problem in linear elasticity (e.g., van Hove [16]) and for the initial deformation boundary value problem in nonlinear elasticity (Wheeler [32]). Its role in other respects is described by, e.g., Ciarlet [4].

The above definitions are not independent but are related by the following implications (see, e.g., the discussion by Ball [2] and Müller [26]):

(a) Quasi-convexity of $W \Rightarrow$ rank-one convexity of W (Morrey [25], Ball [1]).
(b) Strong-ellipticity of $W \Rightarrow$ strict rank-one convexity of $W \in C^2(M^+, \mathbb{R})$.
(c) Rank-one convexity of $W \in C^2(M^+, \mathbb{R})$ is equivalent to the *Legendre–Hadamard* condition

$$\frac{\partial^2 W(A)}{\partial A_{ij} \partial A_{kl}} a_i a_k b_j b_l \geq 0 \tag{3.19}$$

for all $A \in M^+$ and $a \in \mathbb{R}^n$, $b \in \mathbb{R}^n$.

We remark that we do not require the concept of poly-convexity (Ball [1]) which is also related to the above definitions and is used in existence proofs.

Reverse implications between the definitions do not always hold. A counterexample due to Šverák [30] demonstrates failure of the reverse implication in (a), at least in $n \geq 3$ dimensions, while in (b) the reverse implication is clearly false. This failure in (b) motivates the search for alternative conditions for uniqueness in elastodynamics that are explored in Sect. 5.

4 Uniqueness in nonlinear elastostatics

This section discusses two extensions of the result derived in [19]. The first is to the problem in which phase transitions may occur in the star-shaped region Ω which now contains subregions over whose interfaces the deformation gradient on adjacent sides is rank-one connected. The second extension considers certain unbounded regions and enables uniqueness to be proved under rank-one convexity alone.

Proposition 4.1. *Let* $\Omega = \Omega_1 \cup \Omega_2 \cup \Gamma$ *be an open star- shaped bounded region with Lipschitz smooth interface* Γ *between* Ω_1 *and* Ω_2 *and with* $\partial\Omega = \partial\Omega_2 \backslash \Gamma_1$. *Let* Ω *be occupied by a homogeneous elastic material of strain energy function* $W \in C^2(M^+, \mathbb{R})$ *and let there be an equilibrium solution*

$$u_i \in C(\Omega, \mathbb{R}^n) \cap C^2(\Omega_1, \mathbb{R}^n) \cap C^2(\Omega_2, \mathbb{R}^n).$$

Let $A \in M^+$ *be a constant matrix and let* W *be strictly quasi-convex and rank-one convex at* A. *Let* $d \in \mathbb{R}^n$ *be a constant vector and the deformation satisfy the boundary condition*

$$u_i(x) = A_{ij}x_j + d_i, \quad x \in \partial\Omega. \tag{4.1}$$

Then $u_i(x) = A_{ij}x_j + d_i$ *for all* $x_i \in \bar{\Omega}$.

Proof. Let $v_i = A_{ij}x_j + d_i$ for $x \in \bar{\Omega}$, and assume that $u_i \not\equiv v_i, x \in \Omega$. Then v_i is a smooth equilibrium solution satisfying the boundary condition (4.1). Furthermore, both $u_i(x)$ and $v_i(x)$ satisfy the generalised conservation law (3.2) which on subtraction yields

$$\begin{aligned} &n[V(u) - V(v)] \\ &= \int_{\partial\Omega} \left[N_k x_k \{W(\nabla u) - W(\nabla v)\} + N_j \sigma_{ij}(\nabla u) x_k \{v_{i,k} - u_{i,k}\}\right] dS \end{aligned} \tag{4.2}$$

where V is defined by (2.6) and $(2.8)_1$ has been employed for the composite region Ω. The assumed star-shapedness of Ω together with rank-one convexity of W at A implies that

$$V(u) \leq V(v), \tag{4.3}$$

which contradicts the strict quasi-convexity of W at A. (Observe that $u_i - v_i \in W_0^{1,\infty}(\Omega)$ and that $u_{i,j} - v_{i,j}$ on adjacent sides of Γ is rank-one connected.) Thus $u_i(x) \equiv v_i(x)$ for all $x \in \bar{\Omega}$.

The result of [19], and also Proposition 4.1, may be extended to an unbounded region upon using the appropriate generalisation of the conservation law (3.2). Details are omitted and instead we consider an alternative result valid for the special class of unbounded regions for which a constant vector H_i exists such that

$$H_i N_i \geq 0, \quad x \in \partial\Omega, \tag{4.4}$$

and $H_i N_i \not\equiv 0, x \in \partial\Omega$. Regions of this type, which include the half-space, cone and (semi-) infinite strip, have been discussed by Esteban and Lions [10] who applied a similar argument to the following in the context of non-existence of non-trivial solutions to a semi-elliptic equation subject to null Dirichlet boundary data. □

Proposition 4.2. *Suppose Ω is an unbounded open connected region whose boundary satisfies condition (4.4). Let $u_i(x)$ be a smooth equilibrium solution which, for constant $A \in M^+$ and $d \in \mathbb{R}^N$, satisfies the boundary condition*

$$u_i(x) = A_{ij}x_j + d_i, \quad x \in \partial\Omega, \tag{4.5}$$

and

$$\lim_{R\to\infty} \int_{\Omega\cap\partial B(R)} (u_{i,j} - A_{ij})(u_{i,j} - A_{ij})^{\frac{1}{2}} = 0, \tag{4.6}$$

where $B(R)$ is the sphere of radius R and centre at the origin of coordinates. Let $W \in C^2(M^+, \mathbb{R})$ be strictly rank-one convex at A and satisfy the asymptotic behaviour

$$\lim_{R\to\infty} \int_{\Omega\cap\partial B(R)} |W(\nabla u) - W(A)| dS = 0. \tag{4.7}$$

Then $u_i(x) = A_{ij}x_j + d_i$ for all $x \in \bar{\Omega}$.

Remark 4.1. The strain energy function W, determined only to within an arbitrary additive constant, may be replaced, without loss, by the function $\bar{W}(\nabla u) = W(\nabla u) - W(A)$. Consequently, condition (4.7) is satisfied whenever $\bar{W}(\nabla u(x)) \in L^1(\Omega)$.

Remark 4.2. When $A = 0$, condition (4.6) may be replaced by $u_i \in W^{1,2}(\Omega)$.

Proof. Let $v_i(x) = A_{ij}x_j + d_i$, $x \in \bar{\Omega}$, and assume that $u_i(x) \not\equiv v_i(x)$, $x \in \Omega$. The vector function v is a smooth equilibrium solution and by repeating the argument leading to (3.3) we obtain:

$$\begin{aligned}\int_{\partial\Omega\cap B(R)} &\left[N_i\{W(\nabla u) - W(A)\} - N_j\sigma_{kj}(\nabla u)(u_{k,i} - v_{k,i})\right] dS \\ &\leq \int_{\Omega\cap\partial B(R)} \Big[|W(\nabla u) - W(A)| + \{\sigma_{ij}(\nabla u)\sigma_{ij}(\nabla u)\}^{\frac{1}{2}} \\ &\qquad \{(u_{k,l} - A_{kl})(u_{k,l} - A_{kl}\}^{\frac{1}{2}}\Big] dS. \end{aligned} \tag{4.8}$$

On letting $R \to \infty$, we conclude that

$$\lim_{R\to\infty} \int_{\partial\Omega\cap B(R)} \left[N_i\{W(\nabla u) - W(A)\} - N_j\sigma_{kj}(\nabla u)(u_{k,i} - v_{k,i})\right] dS = 0, \tag{4.9}$$

where (4.6), (4.7) and the differentiability of W are used. But $u_i(x) = v_i(x)$ for $x \in \partial\Omega$ and therefore Lemma 2.1 with condition (4.4) enables (4.9) to be rewritten as

$$\lim_{R\to\infty} \int_{\partial\Omega\cap B(R)} H_i N_i \left[W(\nabla u) - W(A) - \sigma_{kj}(\nabla u)\lambda_k N_j\right] dS = 0. \tag{4.10}$$

Insertion into (4.10) of the assumption that W is strictly rank-one convex at A leads to a contradiction in view of (3.17). We conclude that $u_i(x) \equiv v_i(x)$, $x \in \bar{\Omega}$. □

5 Uniqueness in nonlinear elastodynamics

This section discusses some aspects of uniqueness of the smooth solution to the initial boundary value problem of nonlinear homogeneous elasticity. As already remarked in Sect. 1, in general smooth solutions do not exist globally in time so that our conclusions hold only on some maximal interval of finite duration.

In what follows we seek to replace strong-ellipticity by the weaker assumptions of strict quasi-convexity and strict rank-one convexity of the strain energy function. We establish in a three-dimensional star-shaped region the uniqueness of smooth solutions subject to prescribed initial and Dirichlet boundary data. It is beyond the present scope to discuss the proof other than for null data, although, by way of introduction, we apply elements of the approach to obtain an alternative derivation to that of Wheeler [32] for uniqueness that, however, retains non-zero data and the assumption of strong-ellipticity. In fact, we are able to establish continuous dependence on the initial data. Dafermos [6, p. 66] obtains a similar result in nonlinear thermoelasticity but for a broader solution class and convex entropy.

5.1 Uniqueness and continuous initial data dependence

We establish the following proposition.

Proposition 5.1. *The smooth solutions to the equations of motion* (2.11) *subject to prescribed boundary conditions* (2.13) *depend continuously on the initial data* (2.12) *in a sufficiently small time interval* $[0, T]$ *provided the strain energy function* W *satisfies the strong-ellipticity condition* (3.18) *uniformly in* $\Omega \times [0, T]$. *The smooth solutions are assumed to be in the class given by*

$$\sup_{\Omega\times[0,T]} u_{i,j}u_{i,j} \leq N^2 \tag{5.1}$$

for prescribed constant N.

Proof. Let $w_i(x,t) = u_i(x,t) - v_i(x,t)$, where u_i, v_i are smooth solutions to the equations of motion (2.11) satisfying the prescribed boundary conditons (2.13). The vector function w inherits the smoothness of u and v and satisfies

$$\sigma_{ij,j}(\nabla u) - \sigma_{ij,j}(\nabla v) = \rho \ddot{w}_i, \quad (x,t) \in \Omega \times [0, T] \tag{5.2}$$

$$w_i(x,t) = 0. \quad (x,t) \in \partial\Omega \times [0, T]. \tag{5.3}$$

Consider the function

$$F(t) = \int_0^t \int_{\Omega(\eta)} \rho w_i w_i dx d\eta + (T-t)\int_{\Omega(0)} \rho w_i w_i dx + Q, \quad 0 \leq t \leq T, \tag{5.4}$$

where Q is a positive constant to be determined. After differentiation, we have

$$\dot{F}(t) = \int_{\Omega(t)} \rho w_i w_i \, dx - \int_{\Omega(0)} \rho w_i w_i \, dx = 2\int_0^t \int_{\Omega(\eta)} \rho w_i w_{i,\eta} \, dx \, d\eta, \tag{5.5}$$

$$\ddot{F}(t) = 2\int_{\Omega(t)} \rho w_i \dot{w}_i \, dx \tag{5.6}$$

$$= 2\int_0^t \int_{\Omega(\eta)} \left(\rho w_{i,\eta} w_{i,\eta} + \rho w_i w_{i,\eta\eta}\right) dx\, d\eta + 2\int_{\Omega(0)} \rho w_i \dot{w}_i \, dx \tag{5.7}$$

$$= 2\int_0^t \int_{\Omega(\eta)} \left[\rho w_{i,\eta} w_{i,\eta} - \left\{\sigma_{ij}(\nabla u) - \sigma_{ij}(\nabla v)\right\} w_{i,j}\right] dx\, d\eta$$
$$+ 2\int_{\Omega(0)} \rho w_i \dot{w}_i dx, \tag{5.8}$$

where (5.2) and (5.3) and the divergence theorem have been used to obtain (5.8)

By means of a Taylor's series expansion and the assumption $W \in C^2(M^+, \mathbb{R})$, Wheeler [32] has proved that

$$\left(U_{ij} U_{ij}\right)^{\frac{1}{2}} \le k_1 w_{i,j} w_{ij}, \tag{5.9}$$

where k_1 is a positive constant and

$$U_{ij} = \sigma_{ij}(\nabla u) - \sigma_{ij}(\nabla v) - c_{ijkl} w_{k,l}, \tag{5.10}$$

with the coefficients $c_{ijkl}(x, t)$ defined by

$$c_{ijkl} = c_{klij} = \frac{\partial \sigma_{ij}(\nabla u)}{\partial u_{k,l}} = \frac{\partial^2 W(\nabla u)}{\partial u_{i,j} \partial u_{k,l}}. \tag{5.11}$$

It follows easily from a further proof in the same paper that there exists a positive constant k_2 such that, for $t \in [0, T]$,

$$\int_{\Omega(t)} \left(\rho \dot{w}_i \dot{w}_i + c_{ijkl} w_{i,j} w_{k,l}\right) dx \le k_2 \int_0^t \int_{\Omega(\eta)} w_{i,j} w_{i,j} \, dx\, d\eta + \int_{\Omega(0)} \rho \dot{w}_i \dot{w}_i \, dx. \tag{5.12}$$

Furthermore, the paper also demonstrates that uniform strong-ellipticity of the strain energy function implies a modified Gårding's inequality which in the present notation is expressed by

$$\int_{\Omega(t)} c_{ijkl} w_{i,j} w_{k,l} \, dx \ge k_3 \int_{\Omega(t)} w_{i,j} w_{i,j} \, dx + k_4 \int_{\Omega(t)} w_i w_i \, dx, \quad t \in [0, T], \tag{5.13}$$

for positive constants k_3, k_4.

Upon introducing U_{ij} from (5.10) into (5.8) followed by Schwarz's inequality and (5.9) and (5.12), we arrive at

$$\ddot{F}(t) \leq 4(1-\alpha)\int_0^t \int_{\Omega(\eta)} \rho w_{i,\eta} w_{i,\eta}\, dx\, d\eta - 4\alpha \int_0^t \int_{\Omega(\eta)} c_{ijkl} w_{i,j} w_{k,l}\, dx\, d\eta$$
$$+ 2(2\alpha - 1)k_2 \int_0^t \int_0^\eta \int_{\Omega(\tau)} w_{i,j} w_{i,j}\, dx\, d\tau\, d\eta$$
$$+ 2k_5 \int_0^t \int_{\Omega(\eta)} w_{i,j} w_{i,j}\, dx\, d\eta + D, \tag{5.14}$$

where $\alpha > 1$ is a constant to be chosen later and

$$k_5 = k_1 \sup_{\Omega\times[0,T]} (w_{i,j} w_{i,j})^{\frac{1}{2}}, \tag{5.15}$$

$$D = 2(2\alpha - 1)\int_{\Omega(0)} \rho \dot{w}_i \dot{w}_i\, dx + 2\int_{\Omega(0)} \rho w_i \dot{w}_i\, dx. \tag{5.16}$$

Note that k_5 is bounded by assumption (5.1). An application of (5.13) next yields

$$\ddot{F}(t) \leq 4(1-\alpha)\int_0^t \int_{\Omega(\eta)} \rho w_{i,\eta} w_{i,\eta}\, dx\, d\eta + [-4\alpha k_3 + 2(2\alpha - 1)Tk_2 + 2k_5]$$
$$\int_0^t \int_{\Omega(\eta)} w_{i,j} w_{i,j}\, dx\, d\eta + D, \tag{5.17}$$

and, on taking T sufficiently small and setting

$$\alpha > \max\left\{1, \frac{k_5 - Tk_2}{2(k_3 - Tk_2)}\right\}, \tag{5.18}$$

we obtain

$$\ddot{F}(t) \leq 4(1-\alpha)\int_0^t \int_{\Omega(\eta)} \rho w_{i,\eta} w_{i,\eta} dx d\eta + D. \tag{5.19}$$

Consequently, we have

$$F(t)\ddot{F}(t) + (\alpha - 1)\dot{F}^2 \leq 4(1-\alpha)S^2 + DF, \tag{5.20}$$

where by Schwarz's inequality

$$S^2 = \int_0^t \int_{\Omega(\eta)} \rho w_i w_i dx d\eta \int_0^t \int_{\Omega(\eta)} \rho w_{i,\eta} w_{i,\eta} dx d\eta$$
$$- \left(\int_0^t \int_{\Omega(\eta)} \rho w_i w_{i,\eta} dx dx\right)^2 \geq 0. \tag{5.21}$$

Finally, we choose $|D| = Q$ and conclude from (5.20) that

$$(F^\alpha)^{\cdot\cdot} \leq \alpha F^\alpha, \tag{5.22}$$

which on integration and recalling (5.4) and (5.5) yields

$$\int_0^t \int_{\Omega(\eta)} \rho w_i w_i dx d\eta \leq \left[T \int_{\Omega(0)} \rho w_i w_i dx + Q \right] (\cosh \sqrt{\alpha} t)^{\frac{1}{\alpha}}, \quad t \in [0, T] \tag{5.23}$$

and continuous dependence on the initial data is established. Uniqueness of the solution follows immediately and may be extended by iteration to the whole interval of existence. □

5.2 Uniqueness in the null initial boundary value problem

In this remaining subsection we seek to dispense with strong-ellipticity. It is convenient to assume that the vector function $u_i(x, t)$ represents the *displacement* and not the *deformation* as previously. The strain energy function and Piola–Kirchhoff stress tensor become functions of the displacement gradient, the form of the equations of motion (2.11) and of the respective conservation laws in Sect. 3.1 remain unaltered, while the data (2.12)–(2.13) must be interpreted appropriately.

The strain energy function is assumed to satisfy conditions of (strict) quasi-convexity and (strict) rank-one convexity at the null element, and we confine ourselves to the simplest three-dimensional initial boundary value problem in which the initial and boundary data are prescribed to be zero. The region Ω is assumed to be star-shaped. The uniqueness proof depends in part on the conservation laws of Sect. 3.1 and considers smooth displacements u_i that on the maximal interval of existence $[0, T]$ are bounded in the explicit sense that

$$\sup_{\Omega \times [0,T]} (u_i u_i + u_{i,j} u_{i,j} + u_{i,jk} u_{i,jk}) \leq M^2, \tag{5.24}$$

for prescribed constant M. As just stated, the displacement also satisfies Eqs. (2.11), the boundary values

$$u_i(x, t) = 0, \quad (x, t) \in \partial\Omega \times [0, T], \tag{5.25}$$

and the initial values

$$u_i(x, 0) = \dot{u}_i(x, 0) = 0, \quad x \in \Omega. \tag{5.26}$$

We assume that in some non-empty subinterval $(t_1, t_2) \subseteq [0, T]$ the displacement $u_i(x, t)$ is not identically zero. Consequently, and in view of (5.24), there exists a positive constant C such that

$$\int_{\Omega(t)} u_{i,j} u_{i,j} \, dx \leq C \int_{\Omega(t)} u_i u_i \, dx, \quad t \in (t_1, t_2). \tag{5.27}$$

Specific values of the constants M and C are not required.

We establish three preliminary lemmas preparatory to the uniqueness proof.

Lemma 5.1. *The total energy* $E(x,t)$ *given by* (3.10) *satisfies*

$$\int_{\Omega(t)} E(x,t)\,dx = |\Omega| W(\phi), \quad t \in [t_1, t_2], \tag{5.28}$$

where ϕ *represents the null element, and* $|\Omega|$ *is the volume of* Ω.

The proof is immediate from (3.13) on noting (5.25) and (5.26).

Without risk of confusion with definition (5.4), let the function $F(t)$ be defined by

$$F(t) = \int_{t_1}^{t} \int_{\Omega(\eta)} \rho u_i u_i \,dx\,d\eta, \tag{5.29}$$

which by differentiation and use of (5.26) yields

$$\dot{F}(t) = \int_{\Omega(t)} \rho u_i u_i \,dx = 2\int_{t_1}^{t} \int_{\Omega(\eta)} \rho u_i u_{i,\eta} \,dx\,d\eta, \tag{5.30}$$

$$\ddot{F}(t) = 2\int_{\Omega(t)} \rho u_i \dot{u}_i \,dx = 2\int_{t_1}^{t} \int_{\Omega(\eta)} \left(\rho u_{i,\eta} u_{i,\eta} + \rho u_i u_{i,\eta\eta}\right) dx\,d\eta. \tag{5.31}$$

Lemma 5.2. *Let the region* Ω *be star-shaped, and let the strain energy function* $W \in C^2(\Omega, \mathbb{R})$ *be strictly quasi-convex and strictly rank-one convex at the null element* ϕ. *Then, for the three-dimensional* ($n = 3$) *initial boundary value problem* (2.11), (5.25) *and* (5.26), *the smooth displacement* $u_i(x,t)$ *satisfies the inequality*

$$\begin{aligned}\ddot{F}(t) < {} & 4\int_{t_1}^{t} \int_{\Omega(\eta)} \rho u_{i,\eta} u_{i,\eta} \,dx\,d\eta \\ & - 8\int_{t_1}^{t} \int_{\Omega(\eta)} W(\nabla u(x,\eta))\,dx\,d\eta - 2\rho^{\frac{1}{2}} R \int_{\Omega(t)} W(\nabla u(x,t))\,dx \\ & + \left(cR\rho^{-\frac{1}{2}}\right)\dot{F}(t) + 2[4(t-t_1) + \rho^{\frac{1}{2}} R]|\Omega|(W(\phi)|, \quad t \in (t_1, t_2),\end{aligned} \tag{5.32}$$

where $R = \sup\limits_{\Omega}(x_k x_k)^{\frac{1}{2}}$.

Proof. On setting $n = 3$, we obtain from (3.14), (5.28), (5.25) and (5.26), the identity

$$\begin{aligned}2\int_{t_1}^{t} \int_{\Omega(\eta)} \rho u_i u_{i,\eta\eta} \,dx\,d\eta = {} & -4\int_{t_1}^{t} \int_{\Omega(\eta)} W(\nabla u(x,\eta))\,dx\,d\eta \\ & + \int_{\Omega(t)} \rho u_i \dot{u}_i \,dx + (t-t_1)|\Omega| W(\phi) + I_1(t) + I_2(t),\end{aligned} \tag{5.33}$$

where

$$I_1(t) = \int_{t_1}^{t} \int_{\partial\Omega(\eta)} [N_i x_i W(\nabla u(x,\eta)) - N_j \sigma_{ij}(\nabla u(x,\eta)) x_k u_{i,k}]\, dx\, d\eta, \tag{5.34}$$

$$I_2(t) = \int_{\Omega(t)} \rho x_k u_{i,k} \dot{u}_i \, dx. \tag{5.35}$$

Estimates are next derived for $I_1(t)$, $I_2(t)$. As Ω is star-shaped we may suppose, by translation of the origin if necessary, that

$$N_i x_i \geq 0, \quad x \in \partial\Omega. \tag{5.36}$$

Furthermore, the vanishing of $u_i(x,t)$ on $\partial\Omega$ implies in the notation of Lemma 2.1 that $u_{i,j} = \lambda_i N_j$, $x \in \partial\Omega$, and it follows by the strict rank-one convexity of W at ϕ that

$$W(\phi) = W(\nabla u - \nabla u) > W(\nabla u) - \sigma_{ij}(\nabla u)\lambda_i N_j, \quad t \in (t_1, t_2). \tag{5.37}$$

Consequently, from (5.37) and (5.36) we have

$$I_1(t) < \int_{t_1}^{t} \int_{\partial\Omega(\eta)} N_i x_i W(\phi)\, dx\, d\eta = 3(t - t_1)|\Omega| W(\phi). \tag{5.38}$$

The Schwarz and arithmetic-geometric mean inequalities applied to (5.35) lead to

$$\begin{aligned} I_2(t) &\leq \int_{\Omega(t)} \rho (x_k x_k)^{\frac{1}{2}} (\dot{u}_i \dot{u}_i)^{\frac{1}{2}} (u_{i,j} u_{i,j})^{\frac{1}{2}}\, dx, \quad t \in (t_1, t_2), \\ &\leq \rho^{\frac{1}{2}} R \left[\int_{\Omega(t)} E(x,t)\, dx + \tfrac{1}{2} C \int_{\Omega(t)} u_i u_i \, dx \right. \\ &\qquad \left. - \int_{\Omega(t)} W(\nabla u)\, dx \right], \quad t \in (t_1, t_2), \\ &\leq \rho^{\frac{1}{2}} R |\Omega| W(\phi) + \tfrac{1}{2} C R \rho^{-\frac{1}{2}} \dot{F}(t) - \rho^{\frac{1}{2}} R \int_{\Omega(t)} W(\nabla u)\, dx, \quad t \in (t_1, t_2), \end{aligned} \tag{5.39}$$

where (5.28), (5.27) and (5.30) have been employed.

The insertion of the inequalities (5.38)–(5.39) into (5.33) and use of $(5.31)_1$ leads to the stated inequality (5.32). □

Lemma 5.3. *Subject to the conditions of Lemma 5.2, the function $F(t)$ satisfies the inequality*

$$F(t)\ddot{F}(t) + (\alpha - 1)\dot{F}^2(t) < (C R \rho^{-\frac{1}{2}}) F(t) \dot{F}(t), \quad t \in (t_1, t_2), \tag{5.40}$$

where $\alpha > 1$ is a positive constant.

Proof. On combining (5.29), $(5.30)_2$ with (5.32) we obtain

$$F\ddot{F} + (\alpha - 1)\dot{F}^2 < 4(1-\alpha)S^2 + 4\alpha F \int_{t_1}^{t} \int_{\Omega(\eta)} \rho u_{i,\eta} u_{i,\eta}\, dx\, d\eta$$
$$- 8F \int_{t_1}^{t} \int_{\Omega(\eta)} W(\nabla u)\, dx\, d\eta + CR\rho^{-\frac{1}{2}} F\dot{F}$$
$$- 2\rho^{\frac{1}{2}} RF \int_{\Omega(t)} W(\nabla u)\, dx + 2\left[4(t-t_1) + \rho^{\frac{1}{2}} R\right] F|\Omega| W(\phi), \qquad (5.41)$$

where

$$S^2 \equiv \int_{t_1}^{t} \int_{\Omega(\eta)} \rho u_i u_i\, dx\, d\eta \int_{t_1}^{t} \int_{\Omega(\eta)} \rho u_{i,\eta} u_{i,\eta}\, dx\, d\eta$$
$$- \left(\int_{t_1}^{t} \int_{\Omega(\eta)} \rho u_i u_{i,\eta}\, dx\, d\eta \right)^2 \geq 0. \qquad (5.42)$$

The strict quasi-convexity of W at ϕ gives

$$\int_{\Omega(t)} W(\nabla u(x,t))\, dx > |\Omega| W(\phi), \quad t \in (t_1, t_2) \qquad (5.43)$$

which together with (5.28) and $\alpha > 1$ reduces (5.41) to (5.40).

The preliminary lemmas are now assembled and we can establish the main result of this subsection. □

Proposition 5.2. *Let the conditions of Lemma 5.2 be satisfied. Then the null displacement is the only smooth solution on its interval of existence to the three-dimensional nonlinear homogeneous elastic initial boundary value problem under null data.*

Proof. We first rewrite (5.40) as

$$(F^\alpha)^{\cdot\cdot} < (CR\rho^{-\frac{1}{2}})(F^\alpha)^{\cdot}, \quad t \in (t_1, t_2), \qquad (5.44)$$

which, since $\dot{F}(t) \not\equiv 0$ on (t_1, t_2), after integration yields

$$F^\alpha(t) < F^\alpha(t_1) + \left(\alpha\rho^{\frac{1}{2}} C^{-1} R^{-1}\right) F^{\alpha-1}(t_1)\dot{F}(t_1) \left[\exp\left\{CR\rho^{-\frac{1}{2}}(t-t_1)\right\}\right],$$
$$t \in [t_1, t_2]. \qquad (5.45)$$

But by (5.29)–(5.30), we have $F(t_1) = \dot{F}(t_1) = 0$. We conclude from (5.45) that $F(t) < 0$ for $t \in [t_1, t_2]$ which contradicts the non-vanishing of the displacement on (t_1, t_2). Consequently, $u_i(x,t) \equiv 0$, $t \in [0, T]$ and the proposition follows. □

Remark 5.1. The sign of $W(\phi)$ is not specified in the proof of the proposition, although when $W(\phi) = 0$ the condition of quasi-convexity at ϕ of the strain energy yields

$$\int_\Omega W(\nabla u) dx \geq 0, \quad t \in [0, T],$$

and uniqueness of the null solution under null data is immediate from the conservation of energy (5.28).

Remark 5.2. The strict convexity conditions on the strain energy may be relaxed to quasi-convexity and rank-one convexity at the null element without affecting the result.

References

[1] Ball, J.M. (1977): Convexity conditions and existence theorems in nonlinear elasticity. Arch. Rational Mech. Anal. **63**, 337–403
[2] Ball, J.M. (1980): Strict convexity, strong ellipticity, and regularity in the calculus of variations. Math. Proc. Cambridge. Philos. Soc. **87**, 501–513
[3] Chadwick, P. (1975): Applications of an energy-momentum tensor in non-linear elastostatics. J. Elasticity **5**, 249–258
[4] Ciarlet, P.G. (1988): Mathematical elasticity. Vol. I. Three-dimensional elasticity. North-Holland, Amsterdam
[5] Dacorogna, B. (1988): Direct methods in the calculus of variations. Springer, Berlin
[6] Dafermos, C.M. (2000): Hyperbolic conservation laws in continuum physics. Springer, Berlin
[7] Dafermos, C.M., Hrusa, W.J. (1985): Energy methods for quasilinear hyperbolic initial-boundary value problems. Applications to elastodynamics. Arch. Rational Mech. Anal. **87**, 267–292
[8] Delph, T.J. (1982): Conservation laws in linear elasticity based upon divergence transformations. J. Elasticity **12**, 385–393
[9] Eshelby, J.D. (1975): The elastic energy-momentum tensor. J. Elasticity **5**, 321–335
[10] Esteban, M.J., Lions, P.L. (1982): Existence and non existence results for semi-linear elliptic problems in unbounded domains. Proc. Roy. Soc. Edinburgh Ser. A, **93**, 1–14
[11] Fletcher, D.C. (1976): Conservation laws in linear elastodynamics. Arch. Rational Mech. Anal. **60**, 329–353
[12] Gelfand, I.M., Fomin, S.V. (1963): Calculus of variations. Prentice-Hall, Englewood Cliffs, NJ
[13] Green, A.E. (1973): On some general formulae in finite elastostatics. Arch. Rational Mech. Anal. **50**, 73–80
[14] Günther, W. (1962): Über einige Randintegrale der Elastomechanik. Abh. Braunschweig. Wiss. Ges. **14**, 54–72
[15] Hill, R. (1986): Energy-momentum tensors in elastostatics: some reflections on the general theory. J. Mech. Phys. Solids **34**, 305–317
[16] van Hove, L. (1947): Sur l'extension de la condition de Legendre du calcul des variations aux integrales multiples à plusieurs functions inconnues. Nederl. Akad. Wetensch. Proc. Ser. A, **50**, 18–23
[17] Hughes, T.J.R., Kato, T. Marsden, J.E. (1976/77): Well-posed quasi-linear second-order hyperbolic systems with applications to nonlinear elastodynamics and general relativity. Arch. Rational Mech. Anal. **63**, 273–294
[18] Kienzler, R., Herrmann, G. (2000): Mechanics in material space with applications to defect and fracture mechanics. Springer, Berlin
[19] Knops, R.J., Stuart, C.A. (1984): Quasiconvexity and uniqueness of equilibrium solutions in nonlinear elasticity. Arch. Rational Mech. Anal. **86**, 233–249

[20] Knops, R.J., Williams, H.T. (2000): On uniqueness of the null-traction boundary value problem in nonlinear elastostatics. In: Ioos, G., Gues, O., Nouri, A. (eds.) Trends in the applications of mathematics to mechanics. Chapman and Hall/CRC Press, London, pp. 26–32

[21] Knops, R.J., Trimarco, C., Williams, H.T. (2001): Uniqueness and complementary energy in nonlinear elasticity. Report. Department of Mathematics, Heriot-Watt University, Edinburgh

[22] Knowles, J.K., Sternberg, E. (1972): On a class of conservation laws in linearized and finite elastostatics. Arch. Rational Mech. Anal. **44**, 187–211

[23] Maugin, G. (1993): Material inhomogeneities in elasticity. Chapman and Hall, London

[24] Morawetz, C.S. (1961): The decay of solutions of the exterior initial-boundary value problem for the wave equation. Comm. Pure Appl. Math. **14**, 561–568

[25] Morrey, C.B., Jr. (1996): Multiple integrals in the calculus of variations. Springer, New York

[26] Müller, S. (1996): Variational methods for microstructure and phase transitions. In: Hildebrand, S., Stuwe, M. (eds.) Calculus of variations and geometric evolution problems (Lecture Notes in Mathematics, vol. 1713). Springer, Berlin, pp. 85–210

[27] Olver, P.J. (1986): Applications of Lie groups to differential equations. Springer, New York

[28] Pohožaev, S.I. (1965): Eigenfunctions of the equation $\Delta u + \lambda f(u) = 0$. Soviet Math. Dokl. **6**, 1408–1411 (English transl. of Dokl. Akad. Nauk. SSSR **165**, 36–39 (1965))

[29] Pucci, P., Serrin, J. (1986): A general variational identity. Indiana Univ. Math. J. **35**, 681–703

[30] Šverák, V. (1992): Rank-one convexity does not imply quasiconvexity. Proc. Roy. Soc. Edinburgh Ser. A, **120**, 185–189

[31] van der Vorst, R.C.A.M. (1991): Variational identities and applications to differential systems. Arch. Rational Mech. Anal. **116**, 375–398

[32] Wheeler, L. (1976/77): A uniqueness theorem for the displacement problem in finite elastodynamics. Arch. Rational Mech. Anal. **63**, 183–189

Twin Balance Laws for Bodies Undergoing Structured Motions

David R. Owen

Abstract. Non-classical, structured motions provide additive decompositions of velocity into a part due to disarrangements and a part without disarrangements. An analogous decomposition of the stress in the context of structured motions leads to a decomposition of the power of the same type. In this article, a postulate of invariance of the power under superposed rigid motions, both with and without disarrangements, is used to derive two equations for balance of forces and two equations for balance of moments. These "twin" balance laws and the resulting, reduced expression for the power will provide a starting point for field theories of bodies undergoing disarrangements.

1 Introduction

The purpose of this article is the derivation of two pairs of balance laws for a continuous body undergoing structured motions; each pair will be called a pair of "twin balance laws" for reasons that will become apparent in what follows. Specifically, twin balance laws for forces and twin balance laws for moments will be derived by adapting postulates that require or imply invariance of the power [1, 2] to the non-classical setting of structured motions. The key features provided by structured motions in this analysis are (i) a natural and unambiguous additive decomposition of the power that reflects two different types of submacroscopic changes and (ii) a broadening of the collection of classical motions that permits a strengthening of the consequences implied by invariance of the power.

Structured motions [3] provide a setting in which the influences on macroscopic deformations and velocities of both smooth and non-smooth sub-macroscopic geometrical changes can be identified in a simple, rigorous and direct manner. The velocity and deformation due to *non-smooth* sub-macroscopic changes are called the velocity and deformation *due to disarrangements* (literally, *due to the upsetting of an arrangement*), and both the macroscopic velocity and the macroscopic deformation gradient can be decomposed into a sum of a term due to disarrangements and a term without disarrangements. In Sect. 2, decompositions and identification relations obtained in an earlier study of structured motions [3] are recorded and discussed. These relations justify the attributives "without disarrangements" and "due to disarrangements" and permit the introduction in Sect. 2 of both "rigid motions due to disarrangments" and "rigid motions without disarrangements". The former represent macroscopically rigid motions that arise from time-like jumps in a sequence of approximating piecewise rigid motions, while the latter represent classical rigid motions. The manner in which either kind of rigid motion can be superposed on a given structured motion also is recorded in Sect. 2, and the effects of these superpositions on the velocities and deformations with and without disarrangements are noted.

Earlier studies of structured deformations and refinements of classical balance laws have established additive decompositions and corresponding identification relations not only for kinematical quantitites as discussed above but also for measures of contact forces [3–6]. These decompositions for contact forces resolve the limit of volume-averages of contact forces, or the "volume density of contact forces," into the *volume density of contact forces due to disarrangements*, a term due to the contact forces on the geometrical sites where approximating piecewise smooth motions experience space-like jumps, plus the *volume density of contact forces without disarrangements*, a term due to the contact forces on the boundary of a preassigned part of a body – even when that boundary is broken into pieces by the approximating piecewise smooth motions. This decomposition is recorded in Sect. 3, where it becomes clear that the volume density of contact forces without disarrangements always can be written as the divergence of the *stress without disarrangements*, a tensor field that represents the stress in a virgin configuration differing from the given reference configuration by a purely microscopic deformation ([3, Part Two, Sect. 1], [6]). Although the volume density of contact forces due to disarrangements generally does not take the form of the divergence of a tensor field, the analysis in Sect. 3 shows that there is a natural way of writing the volume density of contact forces due to disarrangements as the divergence of a tensor field, the *stress due to disarrangements*, plus a body force term, the *body force due to disarrangements* that reflects the macroscopic effects of spatial variations in voids appearing through approximating piecewise-smooth motions.

These considerations lead in Sect. 3 to a decomposition of the total body force (including inertial forces) in the virgin configuration into a sum of the total body force without disarrangements (including inertial forces without disarrangements) plus the total body force due to disarrangements (including inertial forces due to disarrangments). In this paper, the specific forms of the inertial forces with and without disarrangements are not derived, because they rest on notions of "acceleration with and without disarrangements" that require the notion of a "second-order structured deformation" [7] and go well beyond the scope of the present analysis. It is worth noting that the classical analysis of Noll [1] also leaves the form of the inertial forces unspecified. At the end of Sect. 3 transformation laws under the superposed rigid body motions considered in Sect. 2 are obtained for the stresses with and without disarrangements and for the total body forces with and without disarrangements.

In Sect. 4 the power expended in a structured motion is defined in terms of the kinematical and dynamical quantities discussed in Sects. 2 and 3. The power expended is assumed to be the sum of the "power without disarrangements" and the "power due to disarrangements", each term having an explicit expression in terms of quantities introduced in Sects. 2 and 3. With this extension to structured motions of the power expended, it is natural to extend the postulate of invariance of the power – imposed directly in [1], while following from invariance of certain terms in the equation of energy balance in [2] – to include both the cases of superposed rigid motions without disarrangements and superposed rigid motions due to disarrangements. It should be noted that in the classical arguments that motivated the present work

Noll [1] employed "changes of observers", whereas Green and Rivlin [2] employed superposed rigid motions.

In Sect. 5, the classical derivations of balance of forces and moments are carried out with a new outcome: invariance of the power under superposed rigid motions without disarrangements yields both a law of balance of forces without disarrangements $\text{BF}_\backslash$ and a law of balance of moments without disarrangements $\text{BM}_\backslash$, while invariance under superposed rigid motions due to disarrangements yields both a law of balance of forces due to disarrangements BF_d and a law of balance of moments due to disarrangements BM_d. The typographical expressions of the two laws of balance of forces, $\text{BF}_\backslash$ and BF_d, are identical except for interchange of the symbols $\backslash$ and d, as is also the case for the equations expressing the two laws of balance of moments, $\text{BM}_\backslash$ and BM_d. Accordingly, the pair of balance laws ($\text{BF}_\backslash$,BF_d) are called the *twin balance laws for forces*, while the pair of balance laws ($\text{BM}_\backslash$, BM_d) are called the *twin balance laws for moments*. With the exception of the inertial forces, all of the terms appearing in the four balance laws are expressed directly in terms of the Piola–Kirchhoff stress, the external body force, and the structured motion under consideration. The detailed form of the inertial forces with and without disarrangements will be the subject of a subsequent analysis, based on second-order structured deformations [7].

It is customary in theories that treat motions at different length scales to introduce balance laws separately at each length scale, so that one postulates or derives, for example, both a law of balance of "macroforces" and a law of balance of "microforces". However, in the present theory, geometrical changes at smaller length scales affect *both* twins in each pair ($\cdot_\backslash$, $\cdot_d$) of twin balance laws, with smooth changes at smaller scales reflected in the first twin $\cdot_\backslash$ and non-smooth changes reflected in the second twin $\cdot_d$. Because the macroscopic motions treated here are assumed to be smooth, the disarrangements covered by the twin balance laws are "sub-macroscopic". The inclusion of macroscopic disarrangements can be accomplished by means of methods already well-established in the context of classical motions.

Included in Sect. 5 are derivations of the reduced formula for the power expended and of the classical balance of forces and moments. The reduced formula for the power gives the power as the sum of the *stress-power without disarrangements* and the *stress-power due to disarrangements*. The decomposition of the stress-power will play a central role in the analysis of dissipation that arises in structured motions.

2 Structured motions

Specification of a structured motion from a region $\mathcal{A}$ – in the Euclidean space $\mathcal{E}$ with translation space $\mathcal{V}$ – during a time interval $(0, T)$ includes the specification of three fields $\chi : \mathcal{A}\times(0, T) \to \mathcal{E}$, $G : \mathcal{A}\times(0, T) \to \text{Lin}\mathcal{V}$, and $\dot{\chi}_\backslash : \mathcal{A}\times(0, T) \to \mathcal{V}$ called the *macroscopic motion*, the *deformation without disarrangements* , and the *velocity without disarrangements*, respectively. The notation w instead of $\dot{\chi}_\backslash$ was used when structured motions were introduced in [3]; in addition, the "disarrangement site" $\kappa^{(4)}$ in [3] here is assumed to be the empty set. This assumption rules out in the present study discontinuities in the fields χ, G, and $\chi_\backslash$, although discontinuites are permitted in the piecewise-smooth approximating motions χ_n introduced below (1). To avoid technical issues that are not central to the present study, the precise smoothness assumptions on these fields and on the region $\mathcal{A}$ will not be specified here, but sufficient smoothness requirements on the fields can be inferred from the context. Other than smoothness requirements, the only conditions imposed on the fields χ, G, and $\dot{\chi}_\backslash$ are the injectivity of $x \mapsto \chi(x,t)$ for every $t \in (0,T)$, as well as the existence of a positive number m such that the inequalities

$$m < \det G(x,t) \leq \det \nabla\chi(x,t) \tag{1}$$

hold for all $(x,t) \in \mathcal{A}\times(0,T)$. The Approximation Theorem for structured motions [3, Sect. 6.5] assures us that there is a sequence $n \mapsto \chi_n$ of piecewise smooth motions defined almost everywhere on $\mathcal{A} \times (0,T)$ such that

$$\chi = \lim_{n\to\infty} \chi_n, \quad G = \lim_{n\to\infty} \nabla\chi_n, \quad \dot{\chi}_\backslash = \lim_{n\to\infty} \dot{\chi}_n, \tag{2}$$

with the limits taken in the sense of essentially uniform convergence (i.e., L^∞-convergence). The spatial derivatives $\nabla\chi_n$ and the time derivatives $\dot{\chi}_n$ are taken in the classical sense, and the limits G and $\dot{\chi}_\backslash$ of derivatives in $(2)_{2,3}$ need not equal the corresponding derivatives $\nabla\chi$ and $\dot{\chi}$ of the macroscopic motion (nor need G even be the gradient of some motion). Specific quantitative information about the differences $M := \nabla\chi - G$ and $\dot{\chi}_d := \dot{\chi} - \dot{\chi}_\backslash$ are provided in the first subsection below and justify the terminology *deformation due to disarrangements* for M and *velocity due to disarrangements* for $\dot{\chi}_d$.

2.1 Decompositions and identification relations

The algebraically trivial *additive decompositions*

$$\nabla\chi = G + M, \quad \dot{\chi} = \dot{\chi}_\backslash + \dot{\chi}_d \tag{3}$$

for the macroscopic deformation gradient and velocity are given deeper significance by means of the limit relations (2) for G and $\dot{\chi}_\backslash$ and by means of the following *identification relations* obtained in [3, Part Two, Sect. 6.6], for M and $\dot{\chi}_d$:

$$M(x,t) = \lim_{\delta\to 0}\lim_{n\to\infty} \mathrm{vol}\mathcal{B}(x,t;\delta)^{-1} \int_{\Gamma(\chi_n)\cap\mathcal{B}(x,t;\delta)} [\chi_n](y,\tau) \otimes \nu^{(3)}(y,\tau)\, dA_{(y,\tau)} \tag{4}$$

and

$$\dot{\chi}_d(x,t) = \lim_{\delta\to 0}\lim_{n\to\infty} \mathrm{vol}\mathcal{B}(x,t;\delta)^{-1}\int_{\Gamma(\chi_n)\cap\mathcal{B}(x,t;\delta)} \nu^{(1)}(y,\tau)[\chi_n](y,\tau)dA_{(y,\tau)}. \quad (5)$$

In these relations, $n \mapsto \chi_n$ is an arbitrary sequence of piecewise smooth motions that satisfies the limit relations (2). The symbol $\mathcal{B}(x,t;\delta) \subset \mathcal{E}\times\mathbb{R}$ denotes the space-time ball of radius $\delta > 0$ centered at a point (x,t) in $\mathcal{A}\times(0,T)$, and $\Gamma(\chi_n) \subset \mathcal{E}\times\mathbb{R}$, $[\chi_n](y,\tau) \in \mathcal{V}$, and $(\nu^{(3)}(y,\tau), \nu^{(1)}(y,\tau)) \in \mathcal{V}\times\mathbb{R}$ denote respectively the jump set of the piecewise smooth motion χ_n, the jump of χ_n at a point $(y,\tau) \in \Gamma(\chi_n)$, and the unit normal to the jump set $\Gamma(\chi_n)$ at the point (y,τ). The integrand in (4) is the tensor product of $[\chi_n](y,\tau)$ and $\nu^{(3)}(y,\tau)$, both vectors in $\mathcal{V}$, while the integrand in (5) is the scalar $\nu^{(1)}(y,\tau)$ multiplied by the vector $[\chi_n](y,\tau) \in \mathcal{V}$.

The precise interpretations now available for G, $M = \nabla\chi - G$, $\dot{\chi}_\backslash$ and $\dot{\chi}_d = \dot{\chi} - \dot{\chi}_\backslash$ permit us to understand and interpret various examples of structured motions as illustrated in the next two subsections.

2.2 Rigid motions without disarrangements

If $r : \mathcal{A}\times(0,T) \to \mathcal{E}$ is a smooth mapping that at each time preserves distances and orientation, then $\nabla r : \mathcal{A}\times(0,T) \to \mathrm{Lin}\mathcal{V}$ has proper orthogonal values and $\nabla\dot{r}\,\nabla r^T : \mathcal{A}\times(0,T) \to \mathrm{Lin}\mathcal{V}$ has skew values. The structured motion

$$\chi := r, \quad G := \nabla r,\ \dot{\chi}_\backslash := \dot{r} \quad (6)$$

satisfies $M = \nabla\chi - G = 0$ and $\dot{\chi}_d = \dot{\chi} - \dot{\chi}_\backslash = 0$. Here, there is no deformation due to disarrangements and no velocity due to disarrangements, and the identification relations above tell us that no jumps in approximating motions contribute to the macroscopic deformation and velocity. Accordingly, the triple $(r, \nabla r, \dot{r})$ will be called a *classical rigid motion* or a *rigid motion without disarrangements*.

2.3 Rigid motions caused by disarrangements

If r again is smooth, orientation- and distance-preserving, the structured motion

$$\chi := r,\ G := \nabla r,\ \dot{\chi}_\backslash := 0 \quad (7)$$

satisfies $M = \nabla\chi - G = 0$ and $\dot{\chi}_d = \dot{\chi} - \dot{\chi}_\backslash = \dot{r}$. Here, there is no deformation due to disarrangements, so that no space-like jumps in approximating motions contribute to the macroscopic deformation. However, the relation $\dot{r} = \dot{\chi}_d$ and the identification relations tell us that the entire macroscopic velocity field arises due to time-like jumps in approximating piecewise smooth motions. The triple $(r, \nabla r, 0)$ will be called a *rigid motion caused by (time-like) disarrangements*. An example in [3, Part Two, Sect. 6.3] (the "blinking motion") of approximating piecewise smooth motions provides instances of the time-like disarrangements required for a rigid motion caused by disarrangements.

2.4 Superposed rigid motions

The definition of composition of two structured deformations in [4] was shown in [3, Part Two, Sect. 6.5] to lead to the following notion of composition for structured motions:

$$(\eta, H, \dot{\eta}_\backslash) \circ (\chi, G, \dot{\chi}_\backslash) = \big(\eta \diamond \chi, (H \diamond \chi)G, (H \diamond \chi)\dot{\chi}_\backslash + \dot{\eta}_\backslash \diamond \chi\big). \tag{8}$$

Here, the symbol $\diamond$ denotes the "composition in the first variable"; for example, $(\eta \diamond \chi)(x, t) := \eta(\chi(x, t), t)$ for all $(x, t) \in \mathcal{A}\times(0, T)$. This formula provides the following formulas for a structured motion $(\chi, G, \dot{\chi}_\backslash)$ followed by rigid motions of the types just considered:

$$(r, \nabla r, \dot{r}) \circ (\chi, G, \dot{\chi}_\backslash) = \big(r \diamond \chi, (\nabla r \diamond \chi)G, (\nabla r \diamond \chi)\dot{\chi}_\backslash + \dot{r} \diamond \chi\big) \tag{9}$$

$$(r, \nabla r, 0) \circ (\chi, G, \dot{\chi}_\backslash) = \big(r \diamond \chi, (\nabla r \diamond \chi)G, (\nabla r \diamond \chi)\dot{\chi}_\backslash\big). \tag{10}$$

We interpret the compositions in (9) and (10) as *superpositions* of a rigid motion (without disarrangements or caused by disarrangements) on the given structured motion $(\chi, G, \dot{\chi}_\backslash)$. The superpositions result in new structured motions given by the right-hand sides of (9) and (10), and the velocity without disarrangements for $(r, \nabla r, \dot{r})\circ(\chi, G, \dot{\chi}_\backslash)$ is the third entry on the right-hand side of (9), $(\nabla r\diamond\chi)\dot{\chi}_\backslash+\dot{r}\diamond\chi$, while the velocity without disarrangements for $(r, \nabla r, 0)\circ(\chi, G, \dot{\chi}_\backslash)$ is the third entry of the right-hand side of (10), $(\nabla r \diamond \chi)\dot{\chi}_\backslash$. Moreover, we may compute the velocity due to disarrangements for each of the superpositions by subtracting the third entry from the time derivative of the first. For the superposition $(r, \nabla r, \dot{r}) \circ (\chi, G, \dot{\chi}_\backslash)$ the velocity due to disarrangements is

$$\begin{aligned}(r \diamond \chi)^{\cdot} - (\nabla r \diamond \chi)\dot{\chi}_\backslash - \dot{r} \diamond \chi &= \dot{r} \diamond \chi + (\nabla r \diamond \chi)(\dot{\chi} - \dot{\chi}_\backslash) - \dot{r} \diamond \chi \\ &= (\nabla r \diamond \chi)\dot{\chi}_d,\end{aligned} \tag{11}$$

while for $(r, \nabla r, 0) \circ (\chi, G, \dot{\chi}_\backslash)$ the velocity due to disarrangements is

$$\begin{aligned}(r \diamond \chi)^{\cdot} - (\nabla r \diamond \chi)\dot{\chi}_\backslash &= \dot{r} \diamond \chi + (\nabla r \diamond \chi)(\dot{\chi} - \dot{\chi}_\backslash) \\ &= \dot{r} \diamond \chi + (\nabla r \diamond \chi)\dot{\chi}_d.\end{aligned} \tag{12}$$

If r and $t_o \in (0, T)$ are such that $\nabla r(x, t_o) = I$ for all $x \in \mathcal{A}$, then the formulas just obtained may be simplified as follows when evaluated at t_o :

structured motion	*vel. without disarrangements*	*vel. due to disarrangements*
$(r, \nabla r, \dot{r}) \circ (\chi, G, \dot{\chi}_\backslash)$	$\dot{\chi}_\backslash(\cdot, t_o) + \dot{r}(\chi(\cdot, t_o), t_o)$	$\dot{\chi}_d(\cdot, t_o)$
$(r, \nabla r, 0) \circ (\chi, G, \dot{\chi}_\backslash)$	$\dot{\chi}_\backslash(\cdot, t_o)$	$\dot{\chi}_d(\cdot, t_o) + \dot{r}(\chi(\cdot, t_o), t_o)$.

In particular, if for every time t the mapping $r(\cdot, t)$ is a translation of $\mathcal{E}$, then $\nabla r = I$ holds at all times. The deformation without disarrangements for the superpositions in (9) and (10) are the second entries in each triple of the right-hand side, and the deformation due to disarrangements is obtained by subtracting the second entry on the right-hand sides from the gradient of the first. At a time $t_o \in (0, T)$ such that $\nabla r(x, t_o) = I$ for all $x \in A$, we find:

structured motion	*def. without disarrangments*	*def. due to disarrangments*
$(r, \nabla r, \dot{r}) \circ (\chi, G, \dot{\chi}_\backslash)$	$G(\cdot, t_o)$	$M(\cdot, t_o)$
$(r, \nabla r, 0) \circ (\chi, G, \dot{\chi}_\backslash)$	$G(\cdot, t_o)$	$M(\cdot, t_o)$.

3 Contact forces and body forces

3.1 Decompositions

Earlier studies of balance laws for bodies undergoing structured *deformations* ([5, Part Two, Sect. 1], [6]) – corresponding in the present context to the conditions $\dot{\chi} = \dot{\chi}_\backslash = 0$ at all times – showed that the classical law of balance of forces in the reference configuration is equivalent to the "refined balance law":

$$\operatorname{div}(SK^*) - S \operatorname{div} K^* + \nabla S[(\det K)I - K^*] + (\det K)\, b_{\text{ref}} = 0. \tag{13}$$

Here, $S : \mathcal{A}\times(0, T) \to \text{Lin}\mathcal{V}$ is the Piola–Kirchhoff stress field, $K := (\nabla\chi)^{-1}G$, b_{ref} is the body force per unit volume in the reference configuration, and $A^* := (\det A)A^{-T}$ for all invertible $A \in \text{Lin}\mathcal{V}$. Moreover, identification relations derived in the earlier studies permit us to call SK^* the stress without disarrangements, $\operatorname{div}(SK^*)$ the volume density of contact forces without disarrangements, and $-S \operatorname{div} K^* + \nabla S[(\det K)I - K^*]$ the volume density of contact forces due to disarrangements. The availability through structured deformations of a broader class of configurations to use as reference permits one to view (13) as balance of forces in a "virgin configuration" that differs from the given reference configuration by a "purely microscopic deformation" and that contains none of the disarrangements associated with a given structured deformation. Moreover, the scalar field $\det K$ may be thought of as the volume fraction associated with the given structured deformation.

As noted in [6], when $\det K = \text{const} = 1$ the volume density of contact forces due to disarrangements is the divergence of the tensor field $S(I - K^*)$, but when $\det K$ is not a constant, the volume density of contact forces due to disarrangements has no obvious representation as a divergence. We note that the contact forces due to disarrangements always obey the identity

$$-S \operatorname{div} K^* + \nabla S[(\det K)I - K^*] = \operatorname{div}(S[(\det K)I - K^*]) - S\nabla \det K \tag{14}$$

which permits us to write the refined balance of forces as:

$$\operatorname{div}(SK^*) + \operatorname{div}(S[(\det K)I - K^*]) - S\,\nabla \det K + (\det K)\, b_{\text{ref}} = 0. \tag{15}$$

We call $S[(\det K)I - K^*]$ the *stress due to disarrangments,* while we call $-S\,\nabla \det K$ the *body force due to disarrangements* and $(\det K)\, b_{\text{ref}}$ the *body force without disarrangements.* It is immediate that when $M = 0$, the case in which there is no deformation due to disarrangements, the stress and the body force due to disarrangements vanish, the body force without disarrangements reduces to b_{ref}, and the stress without disarrangements reduces to S. Actually, the body force due to disarrangements vanishes whenever the volume fraction $\det K$ does not vary with position.

If we define the body force per unit volume in the virgin cofiguration b_v to be the sum of the body force without disarrangements and the body force due to disarrangements, the considerations above lead us to the following decompositions of stress and body force:

$$(\det K)S = SK^* + S[(\det K)I - K^*] \tag{16}$$

$$b_v = (\det K)\, b_{\text{ref}} - S\,\nabla \det K. \tag{17}$$

We remark that the stress tensor $(\det K)S$ is an analogue of the "weighted Cauchy tensor" $(\det \nabla\chi)T \diamond \chi$ discussed in [8], and (16) shows that it is this measure of stress that readily decomposes into a part without disarrangements plus a part due to disarrangements.

When the special case $\dot{\chi} = \dot{\chi}_\backslash = 0$ is set aside in favor of the general case of structured motions, the measures of stress and body force and their decompositions remain relevant, but inertial forces must be taken into account. In the context of second-order structured deformations [7], it will be possible to decompose the acceleration additively into a part due to disarrangements and a part without disarrangements, so that an analogous decomposition of inertial forces is possible. Unfortunately, the derivation of such a decomposition is beyond the scope of the present article. Consequently, we proceed simply by postulating a decomposition of the inertial force (measured per unit volume in the virgin configuration)

$$b_v^{in} = (b_v^{in})_\backslash + (b_v^{in})_d \tag{18}$$

into an inertial force without disarrangements $(b_v^{in})_\backslash$ plus an inertial force due to disarrangements $(b_v^{in})_d$.

3.2 Transformation laws

Consider now a rigid motion r and a time t_o at which $\nabla r = I$. From the analysis in Sect. 2.4, both $\nabla\chi$ and G are unchanged at time t_o, and it follows that the tensor field $K = (\nabla\chi)^{-1}G$ is unchanged at time t_o under superposition of $(r, \nabla r, \dot{r})$ and $(r, \nabla r, 0)$. Moreover, we assume that the Piola–Kirchhoff stress S at time t_o also is unchanged under these superpositions, and we conclude that the stress without disarrangements SK^*, the stress due to disarrangements $S[(\det K)I - K^*]$, and the body force due to disarrangements $-S\,\nabla \det K$ in a given structured motion all remain unchanged at time t_o under superposition of $(r, \nabla r, \dot{r})$ and $(r, \nabla r, 0)$.

We specify the transformation law for the *total body force* $b^* := b_v + b_v^{in}$ first by noting that the decompositions (17) and (18) yield

$$\begin{aligned} b^* &= b_v + b_v^{in} \\ &= (\det K)\, b_{\text{ref}} - S\,\nabla \det K + (b_v^{in})_\backslash + (b_v^{in})_d \\ &= (\det K)\, b_{\text{ref}} + (b_v^{in})_\backslash - S\,\nabla \det K + (b_v^{in})_d, \end{aligned} \tag{19}$$

then by defining $b_\backslash^*$, the *total body force without disarrangements*, and b_d^*, the *total body force due to disarrangements,* through the formulas:

$$b_\backslash^* := (\det K)\, b_{\text{ref}} + (b_v^{in})_\backslash \tag{20}$$

$$b_d^* := -S\,\nabla \det K + (b_v^{in})_d, \tag{21}$$

and finally by requiring that *both $b_\backslash^*$ and b_d^* are unchanged at time t_o under superposition of $(r, \nabla r, \dot{r})$ and $(r, \nabla r, 0)$.*

4 Power expended

4.1 Postulate on the form of the power

Equipped now with decompositions of velocity, of contact forces, and of total body forces that reflect smooth versus non-smooth geometrical changes at smaller length scales, we postulate that the power expended at each time $t \in (0, T)$ on each subbody $\mathcal{S} \subset \mathcal{A}$ of the body in a structured motion $(\chi, G, \dot{\chi}_\backslash)$ be given by the formula

$$\begin{aligned} P(\mathcal{S}, t) = &\int_{\text{bdy}\mathcal{S}} \big(S_\backslash(x,t)\nu(x)\cdot\dot{\chi}_\backslash(x,t) + S_d(x,t)\nu(x)\cdot\dot{\chi}_d(x,t)\big)\, dA_x \\ &+ \int_{\mathcal{S}} \Big(b_\backslash^*(x,t)\cdot\dot{\chi}_\backslash(x,t) + b_d^*(x,t)\cdot\dot{\chi}_d(x,t)\Big)\, dV_x. \end{aligned} \tag{22}$$

In this formula, $\nu(x)$ denotes the unit outer normal to bdy$\mathcal{S}$ at x, and

$$S_\backslash := SK^*, \quad S_d := S[(\det K)I - K^*] \tag{23}$$

are the stress without disarrangements and the stress due to disarrangements introduced in Sect. 3. If the structured motion is a classical motion, i.e., if $G = \nabla\chi$ and $\dot{\chi}_\backslash = \dot{\chi}$, and, further, if $(b_v^{in})_\backslash = b_v^{in}$, then (18) - (23) imply the formula

$$P(\mathcal{S}, t) = \int_{\text{bdy}\mathcal{S}} S(x,t)\nu(x)\cdot\dot{\chi}(x,t)dA_x + \int_{\mathcal{S}} b^*(x,t)\cdot\dot{\chi}(x,t)dV_x,$$

which is the standard formula for the power expended.

4.2 Postulate of invariance of the power

We broaden the classical invariance assumptions affecting the power [1, 2] by requiring: *for every rigid motion* r *and time* t_o *satisfying* $\nabla r(\cdot, t_o) = I$, *for every part* $\mathcal{S}$, *and for every structured motion* $(\chi, G, \dot{\chi}_\backslash)$, *not only is the power expended at time* t_o *in the superposition* $(r, \nabla r, \dot{r}) \circ (\chi, G, \dot{\chi}_\backslash)$ *equal to the power expended in* $(\chi, G, \dot{\chi}_\backslash)$ *at time* t_o, *but also the power expended at time* t_o *in the superposition* $(r, \nabla r, 0) \circ (\chi, G, \dot{\chi}_\backslash)$ *equals the power expended at time* t_o *in* $(\chi, G, \dot{\chi}_\backslash)$.

5 Twin balance laws

Let $a \in \mathcal{V}$ be given, and put $r(x, t) := x + at$ for all $x \in \mathcal{E}$ and $t \in \mathbb{R}$. We then can use the results in the table below (12), with t_o an arbitrary time $t \in \mathbb{R}$, and apply the invariance of the power, of the stresses with and without disarrangements, and of the total body forces with and without disarrangements first for the superposition $(r, \nabla r, \dot{r}) \circ (\chi, G, \dot{\chi}_\backslash)$ to obtain the relation

$$\int_{\mathrm{bdy}\mathcal{S}} S_\backslash(x,t)\nu(x) \cdot a \, dA_x + \int_{\mathcal{S}} b_\backslash^*(x,t) \cdot a \, dV_x = 0 \tag{24}$$

and next for the superposition $(r, \nabla r, 0) \circ (\chi, G, \dot{\chi}_\backslash)$ to obtain the relation

$$\int_{\mathrm{bdy}\mathcal{S}} S_d(x,t)\nu(x) \cdot a \, dA_x + \int_{\mathcal{S}} b_d^*(x,t) \cdot a \, dV_x = 0. \tag{25}$$

Using the arbitrariness of a and of the part $\mathcal{S}$ as well as the Divergence Theorem, we obtain from (24) and (25) the *twin balance laws for forces* $(\text{BF}_\backslash, \text{BF}_d)$:

$$\text{BF}_\backslash : \ \mathrm{div}\, S_\backslash + b_\backslash^* = 0; \quad \text{BF}_d : \ \mathrm{div}\, S_d + b_d^* = 0. \tag{26}$$

In view of the twin balance laws for forces, the power (22) reduces to:

$$\begin{aligned} P(\mathcal{S}, t) = & \int_{\mathrm{bdy}\mathcal{S}} \big(S_\backslash(x,t)\nu(x) \cdot \dot{\chi}_\backslash(x,t) + S_d(x,t)\nu(x) \cdot \dot{\chi}_d(x,t)\big)\, dA_x \\ & - \int_{\mathcal{S}} \big(\mathrm{div}\, S_\backslash(x,t) \cdot \dot{\chi}_\backslash(x,t) + \mathrm{div}\, S_d(x,t) \cdot \dot{\chi}_d(x,t)\big)\, dV_x. \end{aligned}$$

Applying the Divergence Theorem to the surface integral and the product rule

$$\mathrm{div}(A^T w) = \mathrm{div}\, A \cdot w + A \cdot \nabla w$$

to the integrand in the resulting volume integral, we obtain the *reduced formula for the power expended*:

$$P(\mathcal{S}, t) = \int_{\mathcal{S}} \big(S_\backslash(x,t) \cdot \nabla \dot{\chi}_\backslash(x,t) + S_d(x,t) \cdot \nabla \dot{\chi}_d(x,t)\big)\, dV_x. \tag{27}$$

The integrand is the total stress-power decomposed as the sum of the *stress-power without disarrangements* $S_\backslash \cdot \nabla \dot{\chi}_\backslash$ and the *stress-power due to disarrangements* $S_d \cdot \nabla \dot{\chi}_d$.

For each $t \in \mathbb{R}$, $x_o \in \mathcal{E}$, and skew tensor $W \in \mathrm{Lin}\mathcal{V}$, we put

$$r(y, \tau) := x_o + e^{W(\tau - t)}(y - x_o) \text{ for all } y \in \mathcal{E}, \tau \in \mathbb{R}, \tag{28}$$

and note that $\dot{r}(\chi(x,t),t) = W(\chi(x,t) - x_o)$ and $\nabla(\dot{r} \diamond \chi)(x,t) = W\ \nabla\chi(x,t)$ for all $x \in \mathcal{A}$. We again can use the results below (12), with t_o the given time $t \in \mathbb{R}$, and apply the invariance of the power in the reduced form (27), as well as the invariance of the stresses with and without disarrangements, first for the superposition $(r, \nabla r, \dot{r}) \circ (\chi, G, \dot{\chi}_\backslash)$ to obtain the relation

$$\int_{\mathcal{S}} S_\backslash(x,t) \cdot W\ \nabla\chi(x,t)\, dV_x = 0 \tag{29}$$

and next for the superposition $(r, \nabla r, 0) \circ (\chi, G, \dot{\chi}_\backslash)$ to obtain the relation

$$\int_{\mathcal{S}} S_d(x,t) \cdot W\ \nabla\chi(x,t)\, dV_x = 0. \tag{30}$$

The arbitrariness of the skew tensor W and of the part $\mathcal{S}$ imply that $S_\backslash(x,t)\nabla\chi(x,t)^T$ and $S_d(x,t)\nabla\chi(x,t)^T$ are symmetric for each $(x,t) \in \mathcal{A} \times (0,T)$ and yield the *twin balance laws for moments* (BM$_\backslash$, BM$_d$):

$$\mathrm{BM}_\backslash : \ S_\backslash \nabla\chi^T = \nabla\chi\, S_\backslash^T; \ \mathrm{BM}_d : \ S_d \nabla\chi^T = \nabla\chi\, S_d^T. \tag{31}$$

Because (16) and the definitions $S_\backslash = SK^*$ and $S_d = S[(\det K)I - K^*]$ yield the relation

$$(\det K)S = S_\backslash + S_d, \tag{32}$$

the twin balance laws for moments (31) then imply the *classical balance law for moments*:

$$S\,\nabla\chi^T = \nabla\chi\, S^T.$$

Adding the twin balance laws for force $\mathrm{div}\, S_\backslash + b_\backslash^* = 0$ and $\mathrm{div}\, S_d + b_d^* = 0$, we obtain – in view of (32), (20), (21), and (18) – the relation

$$\det K\ (\mathrm{div}\, S + b_{\mathrm{ref}}) + b_v^{in} = 0. \tag{33}$$

This represents the classical balance law of forces written with respect to the virgin configuration. When the inertial forces take the form $b_v^{in} = -(\det K)\rho_{\mathrm{ref}}\,\ddot{\chi}$, we obtain the *classical balance of forces with respect to the given reference configuration*:

$$\mathrm{div}\, S + b_{\mathrm{ref}} - \rho_{\mathrm{ref}}\,\ddot{\chi} = 0. \tag{34}$$

References

[1] Noll, W. (1963): La mécanique classique, basée sur un axiome d'objectivité. In: La méthode axiomatique dans les mécaniques classiques et nouvelles. Gauthier-Villars, Paris, pp. 47–56

[2] Green, A.E., Rivlin, R.S. (1964): On Cauchy's equations of motion. Z. Angew. Math. Phys. **15**, 290–292

[3] Del Piero, G., Owen, D.R. (2000): Structured deformations. (GNFM-INDAM, no. 58). Gruppo Nazionale per la Fisica Matematica, Quaderni dell'Istituto Nazionale di Alta Matematica, Firenze

[4] Del Piero, G., Owen, D.R. (1993): Structured deformations of continua. Arch. Rational Mech. Anal. **124**, 99–155

[5] Del Piero, G., Owen, D.R. (1995): Integral-gradient formulae for structured deformations. Arch. Rational Mech. Anal. **131**, 121–138

[6] Owen, D.R. (1998): Structured deformations and the refinements of balance laws induced by microslip. Internat. J. Plasticity **14**, 289–299

[7] Owen, D.R., Paroni, R. (2000): Second-order structured deformations. Arch. Rational Mech. Anal. **155**, 215–235

[8] Haupt, P. (2000): Continuum mechanics and theory of materials. Springer Verlag, Berlin

On the Hysteresis in Martensitic Transformations

Miroslav Šilhavý

Abstract. The paper proposes an explanation of the hysteresis in shape memory alloys using energy minimization in nonlinear elasticity and the entropy criterion. The stored energy has two wells describing the two phases. At elongations from some interval, the minimum energy is realized on two-phase mixtures. If the loaded phases are incompatible, that minimum energy is a *concave* function, and so we have a phase equilibrium curve of negative slope. The quasistatic evolution during loading experiments is realized in the class of mechanically, but not thermodynamically, equilibrated mixtures. This family contains states of elongation and force covering the whole area of the hysteresis loop. The evolution must satisfy the entropy criterion for moving phase interfaces which implies that, in the region above the phase equilibrium line, only processes with nondecreasing amount of the second phase are possible while below the situation is the opposite. This picture provides all the elements necessary for the explanation of the hysteresis, including the internal hysteresis loops.

1 Introduction

The cyclic loading experiments with single crystals of shape memory alloys exhibit hysteresis. A solid/solid phase transition is involved, either from the austenite to martensite or between variants of the martensite.

The paper proposes an explanation of the hysteresis within the nonlinear elasticity based on stability considerations (energy minimization) and the entropy criterion for moving phase boundaries. The approach shares general ideas with Gibbsian thermostatics of phase transitions in fluids: Maxwell's relation, the common tangent construction to determine coexistent phases, and the convexity properties of the energy. The difference that eventually leads to the hysteresis is the necessity of taking into account the compatibility of coexistent solid phases. The hysteresis-free phase transitions occur only if, exceptionally, compatibility prevails in the picture. This is always the case in fluids, and phase transitions in fluids constitute a special "degenerate" case within the present scheme.

We briefly summarize the main ingredients of the approach. See [1–5] for references, origins and background; a self-contained derivation under the present assumptions is given below. For simplicity, the stored energy f is expressed as a function of the infinitesimal deformation tensors $\mathbf{E}$; f has two preferred (relative) minima surrounded by the wells W_0, W_1 describing the two phases involved in the experiment; see Fig. 1. The stress function is $\hat{\mathbf{S}} = \hat{\mathbf{S}}^T = Df$.

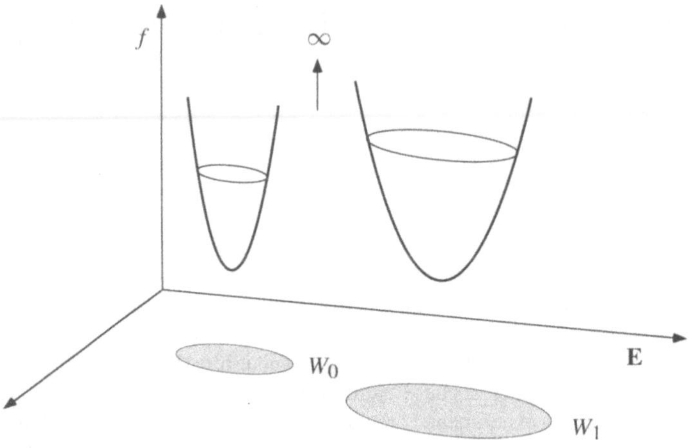

Fig. 1. The stored energy

The minimum energy is sought at constant elongation within the class consisting of homogeneous states and two phase states with homogeneous phases $\mathbf{E}_0, \mathbf{E}_1$ in the form of laminates; see Fig. 2. For the latter, the values $\mathbf{E}_0, \mathbf{E}_1$ must satisfy the compatibility condition

$$\mathbf{E}_1 - \mathbf{E}_0 = \mathbf{a} \odot \mathbf{n}, \quad \mathbf{a} \odot \mathbf{n} := \tfrac{1}{2}(\mathbf{a} \otimes \mathbf{n} + \mathbf{n} \otimes \mathbf{a}), \tag{1}$$

where $\mathbf{n}$ is the unit referential normal to the interface and $\mathbf{a}$ a vector called the amplitude. We say that $\mathbf{E}_0, \mathbf{E}_1$ are compatible if (1) holds for some $\mathbf{a}, \mathbf{n}$. Compatibility means that no dislocations are created at the interface, i.e., that the crystal lattices of the two phases match. The Maxwell relation is the equality

$$f(\mathbf{E}_1) - f(\mathbf{E}_0) = \hat{\mathbf{S}}(\mathbf{E}_0)\mathbf{n} \cdot \mathbf{a} \tag{2}$$

at each point of a solid/solid interface satisfying (1) and the balance of forces

$$\hat{\mathbf{S}}(\mathbf{E}_1)\mathbf{n} = \hat{\mathbf{S}}(\mathbf{E}_0)\mathbf{n}. \tag{3}$$

Equation (2) is obtained, e.g., if the laminate provides minimum energy amongst the class of laminates of the same elongation (see below). Equations (2) and (3) are the conditions of thermodynamical and mechanical equilibrium at the phase interface. The same ideas lead to the rank 1 convexity condition. The energy f is said to be rank 1 convex at $\mathbf{E}$ if

$$f(\mathbf{E}) \leq (1 - z) f(\mathbf{E}_0) + z f(\mathbf{E}_1) \tag{4}$$

whenever $0 \leq z \leq 1$ and $\mathbf{E}_0, \mathbf{E}_1$ are compatible deformation tensors such that

$$\mathbf{E} = (1 - z)\mathbf{E}_0 + z\mathbf{E}_1. \tag{5}$$

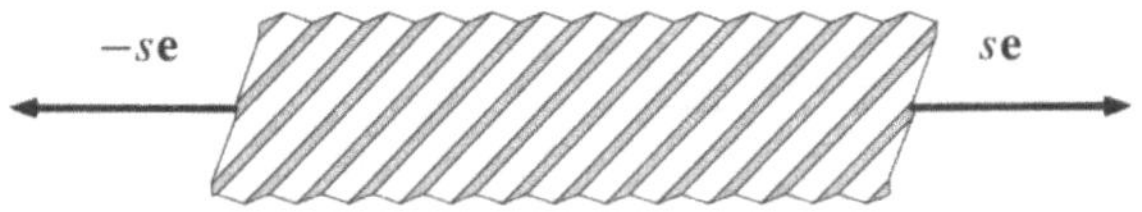

Fig. 2. A mixture of phases (laminate of order 1)

This is equivalent to

$$f(\mathbf{E}_1) \geq f(\mathbf{E}) + \hat{\mathbf{S}}(\mathbf{E}) \cdot (\mathbf{E}_1 - \mathbf{E})$$

for each $\mathbf{E}_1$ compatible with $\mathbf{E}$. If f is strictly rank 1 convex at $\mathbf{E}_0$, there is no other phase in mechanical and thermodynamical equilibrium with it; otherwise, if this is not the case, then Maxwell's relation determines pairs $\mathbf{E}_0, \mathbf{E}_1$ of phases capable of a stable coexistence in a laminate. In fact Maxwell's relation and the assumption of rank 1 convexity at $\mathbf{E}_0, \mathbf{E}_1$ is a generalization of the common tangent construction for the phase diagrams in fluids.

Introducing the sequence of functions [6] $\mathsf{R}_0 f \geq \mathsf{R}_1 f \geq \cdots \geq \mathsf{R}_k f \geq \ldots$ by $\mathsf{R}_0 f = f$,

$$\mathsf{R}_{k+1} f(\mathbf{E}) = \inf\{(1 - z)\mathsf{R}_k f(\mathbf{E}_0) + z\mathsf{R}_k f(\mathbf{E}_1)\},$$

where the infimum is taken over all z, $0 \leq z \leq 1$, and over all compatible $\mathbf{E}_0, \mathbf{E}_1$ such that (5) holds, we see that minimization in the class of homogeneous phases and (first order) laminates amounts to examining $\mathsf{R}_1 f$. Minimization in the class of laminates of arbitrary order corresponds to

$$\mathsf{R} f(\mathbf{E}) = \lim_{k \to \infty} \mathsf{R}_k f(\mathbf{E}),$$

which is the rank 1 convex hull of f, the largest rank 1 convex function not exceeding f. More generally, one can consider the quasiconvex hull $\mathsf{Q} f$. While the relaxations $\mathsf{R} f, \mathsf{Q} f$ are more satisfactory from the mathematical point of view, the relaxation $\mathsf{R}_1 f$ within the class of first order laminates seems to be more appropriate physically, since laminates of order > 1 are never observed in such experiments. This is due to the effects related to length scales, such as the width of the interface.

The reader is referred to [7] for an alternative view on the hysteresis within the framework of nonlinear elasticity.

2 The experiment and summary of the results

Consider a single crystal of a shape memory alloy in a hard loading device (Fig. 3). The device measures the force s at the ends of the specimen against the controlled elongation w in a fixed direction $\mathbf{e}$, $|\mathbf{e}| = 1$.

The schematic force-elongation diagram of Fig. 4 ([8–10]) distinguishes the two phases by the subscripts 0, 1 (e.g., austenite, martensite) and the two horizontal lines

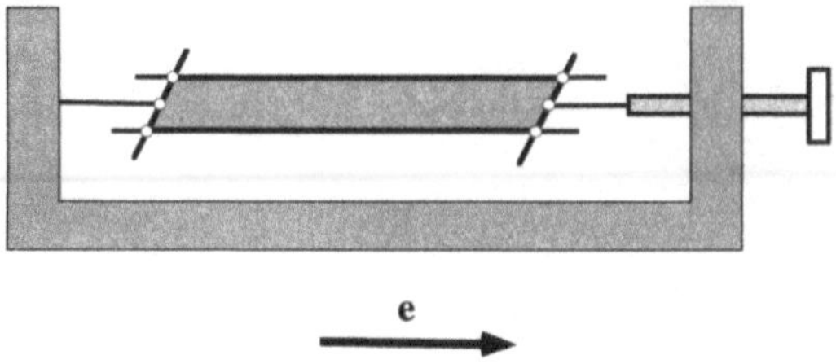

Fig. 3. The hard loading device

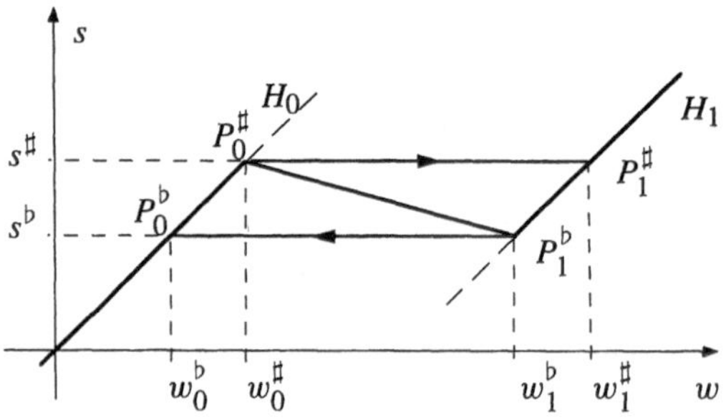

Fig. 4. Hysteresis in austenite martensite transformations

by the superscripts $\sharp, \flat$. The experiment starts at $w = 0, s = 0$ in the austenite phase. As the elongation increases, the specimen remains in a stable homogeneous austenite phase up to the point of loss of stability (the yield point) $P_0^\sharp = (w_0^\sharp, s^\sharp)$. At elongations greater than $w_0^\sharp$ the specimen exhibits an austenite/martensite mixture (Fig. 2). The mass fraction of the martensite z increases with elongation and the process follows the yield segment $[P_0^\sharp, P_1^\sharp]$ up to $P_1^\sharp$, where the austenite vanishes and a further elongation produces a stable homogeneous martensite phase. If the direction is reversed, the process passes through $P_1^\sharp$ undisturbed, and the austenite phase emerges at some lower point $P_1^\flat = (w_1^\flat, s^\flat)$. The process follows the recovery segment $[P_1^\flat, P_0^\flat]$ until $P_0^\flat$ where the specimen exhibits an austenite phase again. Call the parallelogram $P_0^\flat P_0^\sharp P_1^\sharp P_1^\flat$ the (maximal) hysteresis loop.

The phase equilibrium curve $[P_0^\sharp, P_1^\flat]$ is connected with internal hysteresis loops; see Fig. 5. If one reduces the elongation at some point on $[P_0^\sharp, P_1^\sharp]$, the process enters the interior of the maximal hysteresis loop, proceeds steeply down until it reaches $[P_0^\sharp, P_1^\flat]$, there it changes its direction into horizontal, and proceeds until it reaches the homogeneous phase; see Fig. 5(a). Alternatively (see Fig. 5(b)), if one starts loading at some point after crossing $[P_0^\sharp, P_1^\flat]$, the process goes up until it reaches $[P_0^\sharp, P_1^\flat]$, where it again changes direction to horizontal. The same occurs if one starts reloading at some point of $[P_1^\flat, P_0^\flat]$. Theoretically, $[P_0^\sharp, P_1^\flat]$ plots the force of the states of minimum energy (at a given elongation). For simplicity, rather strong hypotheses are made on the stored energy and on the orientation of the device axis **e**, namely, Assumptions 1–3. The stored energy f is assumed to be a class C^2 function on $U = W_0 \cup W_1$, where W_i, $i = 0, 1$, are bounded open convex subsets of the set of symmetric deformation tensors $\mathbf{E}$. For $i = 0, 1$ let

$$\alpha_i := \inf\{\mathbf{Ee} \cdot \mathbf{e} : \mathbf{E} \in W_i\}, \quad \beta_i := \sup\{\mathbf{Ee} \cdot \mathbf{e} : \mathbf{E} \in W_i\}. \tag{6}$$

Assumption 1. (a) *$f \geq 0$ on U, $D^2 f$ is positive definite on U, and $f(\mathbf{E}) \to \infty$ if $\mathbf{E} \to \partial U$.*
(b) $\beta_0 < \alpha_1$.

Formally we further assume that f is infinite outside U. Thus the wells are infinitely deep. However, the bottoms of the wells, i.e., the minima of f on W_0, W_1, can be different. The hypotheses lead to:

- the existence of the single phase equilibrium curves H_0, H_1 defined for all s (Proposition 4).
- If W_0, W_1 are incompatible (contain no pairs of compatible deformation tensors $\mathbf{E}_i \in W_i$, $i = 0, 1$), then the single phases H_0, H_1 are stable for all values of s (cf. the discussion before Theorem 1). If the wells are compatible, H_0 is stable below some critical value $s^\sharp$ and unstable above it while H_1 is stable above some critical value $s^\flat$ and unstable below it (Theorem 1). The values $s^\sharp$, $s^\flat$ are identified with the points so denoted in Fig. 4.
- Theorem 2 shows that $s^\sharp$ and $s^\flat$ are points of loss of rank 1 convexity, i.e., f is rank 1 convex at the corresponding deformation tensors, but not strictly – there is another compatible deformation tensor satisfying Maxwell's relation. In principle this gives the possibility of determining the values $s^\sharp$ and $s^\flat$ from the form of f.
- Theorem 3 proves that $s^\sharp \geq s^\flat$ and that the equality $s^\sharp = s^\flat$ holds only if the pure phases at $P_0^\sharp$ and $P_1^\flat$ are compatible. In the gap interval $(w_0^\sharp, w_1^\flat)$ the minimum energy is realized in the class of two-phase laminates. Theorem 3 shows that the minimum energy is a concave function of elongation on $[w_0^\sharp, w_1^\flat]$. The theory thus gives the nonpositive slope of the phase equilibrium line. Our proof of this general assertion has nothing to do with our stringent hypotheses on f. A related concavity property was established previously by Kohn [11] in the model case of a quadratic double-well energy with *equal elastic moduli* by evaluating the relaxed energy explicitly.

The slow quasistatic processes during the experiments are realized in the class of mechanically equilibrated metastable states. Metastability means that they generally violate Maxwell's relation (although they satisfy conditions of mechanical equilibrium). The evolution of the states is determined by the interface kinetics; this question is not addressed here. However, a priori, the direction of processes is restricted by the entropy criterion for moving interfaces (Sect. 6). This introduces the irreversibility.

- Proposition 5 shows that the area between H_0 and H_1 is covered by the force–elongation pairs of such states.
- Above the phase equilibrium line only processes with $\dot{z} \geq 0$ are possible while below only those with $\dot{z} \leq 0$. The processes are thus irreversible. In particular:

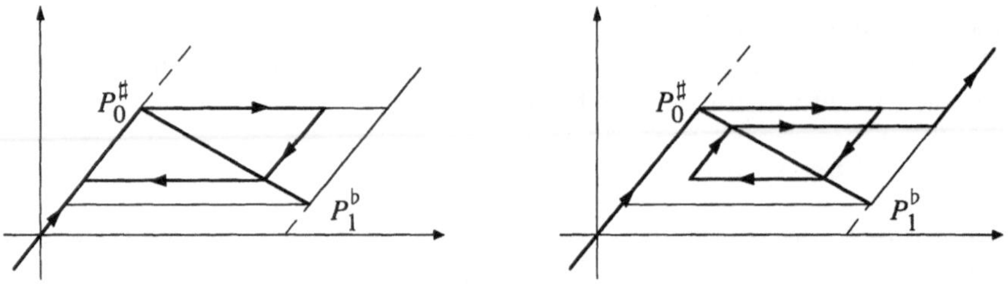

(a) (b)

Fig. 5. Internal hysteresis loops: (a) energy, (b) force

- $P_0^\sharp$ is the first point on H_0 where the process can change its direction from the steep upward direction into horizontal (but nothing says that this necessarily happens), and analogously for $P_1^\flat$. If the unloading occurs at some point of the segment $[P_0^\sharp, P_1^\sharp]$ the process cannot follow this segment in the backward direction; rather, it has to enter the interior of the maximal hysteresis loop in such a way as to satisfy $\dot{z} \geq 0$ (but nothing says that $\dot{z} = 0$, which appears to be the case). The point of crossing the phase equilibrium line is the first point where the process can change direction to the horizontal.

Sections 3–5 treat the states statically; the main tool is the minimization of energy. Processes are introduced only in Sect. 6 where the main tool is the entropy criterion for moving phase boundaries.

3 States, equilibrium and stability

This section reviews equilibrium equations for homogeneous phases or mixtures and introduces stability via energy minimization at fixed elongation.

We assume that the reference configuration Ω of the specimen is a cylinder with axis $\mathbf{e}$, height h and a basis B perpendicular to $\mathbf{e}$, i.e.,

$$\Omega = \{\mathbf{x} : 0 < \mathbf{x} \cdot \mathbf{e} < h, \mathbf{P}\mathbf{x} \in B\}$$

where $\mathbf{P} = \mathbf{1} - \mathbf{e} \otimes \mathbf{e}$ is the projection onto the orthogonal complement of $\mathbf{e}$.

Since the states occurring in the loading experiment are either homogeneous or two-phase mixtures, we define the state space Σ (the class of competitors in the minimum energy principle) by

$$\Sigma = U \cup \Sigma_m,$$

where the elements of U are identified with homogeneous states, and Σ_m is the set of mixtures, identified with triples $(\mathbf{E}_0, \mathbf{E}_1, z)$ where $\mathbf{E}_0, \mathbf{E}_1 \in U$ are compatible and $0 < z < 1$. We interpret z as the mass fraction of the phase $\mathbf{E}_1$ and identify $(\mathbf{E}_0, \mathbf{E}_1, z)$ with $(\mathbf{E}_1, \mathbf{E}_0, 1 - z)$. The energy Φ and the elongation $\bar{w}$ are functions on Σ defined by

$$\Phi(\sigma) = f(\mathbf{E}), \quad \bar{w}(\sigma) = \mathbf{Ee} \cdot \mathbf{e}$$

if $\sigma = \mathbf{E}$ is a homogeneous state and by

$$\Phi(\sigma) = (1 - z) f(\mathbf{E}_0) + z f(\mathbf{E}_1), \quad \bar{w}(\sigma) = [(1 - z)\mathbf{E}_0 + z\mathbf{E}_1]\mathbf{e} \cdot \mathbf{e}$$

if $\sigma = (\mathbf{E}_0, \mathbf{E}_1, z)$ is a mixture.

A homogeneous state $\mathbf{E}$ is called an equilibrium state if

$$\hat{\mathbf{S}}(\mathbf{E}) = s\mathbf{e} \otimes \mathbf{e} \tag{7}$$

for some $s \in \mathbb{R}$, called the force. The elongation w is given by

$$w = \mathbf{Ee} \cdot \mathbf{e}. \tag{8}$$

Equation (7) is motivated by the construction of the device: at the end faces B, $B+h\mathbf{e}$ (of the reference normal $\pm\mathbf{e}$), the device exerts a force $\pm s\mathbf{E}$ while the force vanishes on the mantel. A mixture $(\mathbf{E}_0, \mathbf{E}_1, z)$ is said to be an equilibrium mixture if

$$\hat{\mathbf{S}}(\mathbf{E}_0)\mathbf{n} = \hat{\mathbf{S}}(\mathbf{E}_1)\mathbf{n}, \tag{9}$$

$$(1 - z)\hat{\mathbf{S}}(\mathbf{E}_0) + z\hat{\mathbf{S}}(\mathbf{E}_1) = s\mathbf{e} \otimes \mathbf{e} \tag{10}$$

for some $s \in \mathbb{R}$ where $\mathbf{n}$ is the interface normal. Here (9) is the balance of force on the interface, and (10) is the balance of force between the specimen and the device. The elongation w is

$$w = (1 - z)\mathbf{E}_0\mathbf{e} \cdot \mathbf{e} + z\mathbf{E}_1\mathbf{e} \cdot \mathbf{e}. \tag{11}$$

For fine mixtures as in Fig. 2, Eq. (10) expresses the fact that at the end faces B, $B + h\mathbf{e}$ the device exerts the average force $\pm s\mathbf{e}$ while the average force vanishes on the mantel.

For each $w \in \mathbb{R}$ we consider the set of all states of elongation w,

$$\Sigma_w = \{\sigma \in \Sigma : \hat{w}(\sigma) = w\},$$

and say that a state σ is a minimizer (of energy), or equivalently that σ is stable, if

$$\Phi(\sigma) \leq \Phi(\tau) \tag{12}$$

for all $\tau \in \Sigma_{\hat{w}(\sigma)}$. We say that σ is unstable if it is not a minimizer. We say that σ is a strict minimizer (of energy) if (12) is strict for each $\tau \in \Sigma_{\hat{w}(\sigma)}$ that is not equivalent to σ, i.e., for each $\tau \neq \sigma$ if σ is homogeneous and for each $\tau \neq (\mathbf{E}_0, \mathbf{E}_1, z)$, $\tau \neq (\mathbf{E}_1, \mathbf{E}_0, 1 - z)$ if $\sigma = (\mathbf{E}_0, \mathbf{E}_1, z)$ is a mixture. For each $w \in \mathbb{R}$ we define the minimum energy

$$\hat{\phi}(w) = \inf\{\Phi(\sigma) : \sigma \in \Sigma_w\}. \tag{13}$$

Each minimizer is an equilibrium state of some force s. Indeed, (12) is a minimum subject to the constraint $\hat{w}(\tau) = w$, and s is the Lagrange multiplier. This is obvious for homogeneous minimizers. If $\sigma = (\mathbf{E}_0, \mathbf{E}_1, z)$ is a mixture minimizer, one seeks the stationary point of

$$(1-z)f(\mathbf{E}_0) + zf(\mathbf{E}_1) - s[(1-z)\mathbf{E}_0 + z\mathbf{E}_1]\mathbf{e}\cdot\mathbf{e},$$

where $\mathbf{E}_1 = \mathbf{E}_0 + \mathbf{a} \odot \mathbf{n}$, with respect to $\mathbf{E}_0$, $\mathbf{n}$, z, $\mathbf{a}$. The derivatives with respect to $\mathbf{E}_0$ and $\mathbf{a}$ lead to (10) and (9) while those with respect to z and $\mathbf{n}$ give additionally

$$f(\mathbf{E}_0) - f(\mathbf{E}_1) = s(\mathbf{E}_0 - \mathbf{E}_1)\mathbf{e}\cdot\mathbf{e}, \tag{14}$$

$$\hat{\mathbf{S}}(\mathbf{E}_0)\mathbf{a} = \hat{\mathbf{S}}(\mathbf{E}_1)\mathbf{a}. \tag{15}$$

Also, if a homogeneous state $\mathbf{E}$ is a minimizer then f is rank 1 convex at $\mathbf{E}$. Indeed, let $\mathbf{E}_0, \mathbf{E}_1, z, 0 < z < 1$, be as in the definition of rank 1 convexity (4). Then $\tau := (\mathbf{E}_0, \mathbf{E}_1, z) \in \Sigma_m$ and by (5) the elongation of τ is the same as that of $\sigma := \mathbf{E}$. Since τ is a minimizer we have (4).

4 Single phase curves and loss of stability

This section proves the existence of homogeneous phases and analyses equations of loss of stability which determine the yield/recovery points $P_0^\sharp/P_1^\flat$. Recall the notation (6).

Proposition 4. *There exist smooth functions* $\mathbf{E}_i^* : \mathbb{R} \to W_i$, $i = 0, 1$, *such that for each* $s \in \mathbb{R}$, $\mathbf{E}_i^*(s)$ *is an equilibrium state of force* s. *The elongations* $w_i^*(s) := \mathbf{E}_i^*(s)\mathbf{e}\cdot\mathbf{e}$ *satisfy*

$$\frac{dw_i^*(s)}{ds} > 0, \quad s \in \mathbb{R}, \tag{16}$$

and the range of $w_i^*(\cdot)$ *is the interval* (α_i, β_i). *If* s_i^* *denotes the inverse of* w_i^*, *then the function* f_i^* *defined by* $f_i^*(w) := f(\mathbf{E}_i^*(s_i^*(w)))$ *is strictly convex and*

$$\frac{df_i^*(w)}{dw} = s_i^*(w), \quad \frac{ds_i^*(w)}{dw} > 0, \quad w \in (\alpha_i, \beta_i).$$

The homogeneous curves H_0, H_1 in Fig. 4 are graphs of s_0^*, s_1^*. By Assumption 1(b) the intervals of elongations of the two phases (α_0, β_0) and (α_1, β_1) do not overlap.

Proof. To prove the existence of $\mathbf{E}_i^*(s)$, one looks for the minimum of the strictly convex function $g_s(\mathbf{E}) = f(\mathbf{E}) - s\mathbf{E}\mathbf{e}\cdot\mathbf{e}$ on W_i. Since $g_s(\mathbf{E}) \to \infty$ if $\mathbf{E} \to \partial W_i$, the point of minimum exists and satisfies (7). Inequality (16) follows from the fact that $D^2 f$ is positive definite. Next we prove that if $w \in (\alpha_i, \beta_i)$ then f has a minimum on the set $\{\mathbf{E} \in W_i : \mathbf{E}\mathbf{e}\cdot\mathbf{e} = w\}$ by the convexity and growth as before. The point $\mathbf{E}$ of minimum is a homogeneous equilibrium state of some force; see Sect. 3. This proves the assertion about the range of $w_i^*(\cdot)$ and the rest is again a consequence of the positive definiteness of $D^2 f$. □

The stability of the states $\mathbf{E}_i^*(s)$ is related to the compatibility of the wells W_0, W_1. Note first that, if the wells W_0, W_1 are not rank 1 connected in the sense that there is no pair $\mathbf{E}_0 \in W_0$, $\mathbf{E}_1 \in W_1$ of compatible tensors, then the homogeneous equilibrium states are stable for all $s \in \mathbb{R}$ and these are the only stable states. Indeed, in this case the class of mixtures Σ_m is void and thus the minimization occurs in the class of homogeneous states. The convexity of f on W_0 and on W_1 implies that $\mathbf{E}_i^*(s)$, $i = 0, 1$, are minimizers for all $s \in \mathbb{R}$ in the class $\{\mathbf{E} \in W_i : \mathbf{Ee} \cdot \mathbf{e} = w_i^*(s)\}$. By Assumption 1(b), for each $w \in \mathbb{R}$ either $\{\mathbf{E} \in W_0 : \mathbf{Ee} \cdot \mathbf{e} = w\} = \emptyset$ or $\{\mathbf{E} \in W_1 : \mathbf{Ee} \cdot \mathbf{e} = w\} = \emptyset$ and thus $\mathbf{E}_i^*(s)$, $i = 0, 1$, is a minimizer at constant elongation.

Assumption 2. *There exist compact sets $K_i \subset W_i, i = 0, 1$, such that, for each $s \in \mathbb{R}$, $\mathbf{E}_0^*(s)$ is compatible with some $\mathbf{E} \in K_1$ and $\mathbf{E}_1^*(s)$ is compatible with some $\mathbf{E} \in K_0$.*

Theorem 1. *There exist $s^\sharp, s^\flat \in \mathbb{R}$ such that*

(a) *$\mathbf{E}_0^*(s)$ is stable for all $s \leq s^\sharp$ and unstable for all $s > s^\sharp$;*
(b) *$\mathbf{E}_1^*(s)$ is stable for all $s \geq s^\flat$ and unstable for all $s < s^\flat$.*

We identify $s^\sharp, s^\flat$ with the points so denoted in Fig. 4 and write

$$\mathbf{E}_i^\sharp = \mathbf{E}_i^*(s^\sharp), \quad \mathbf{E}_i^\flat = \mathbf{E}_i^*(s^\flat), \quad w_i^\sharp = w_i^*(s^\sharp), \quad w_i^\flat = w_i^*(s^\flat), \tag{17}$$

for the deformation tensors and elongations of the homogeneous phases at $s^\sharp, s^\flat$.

Proof. We only prove (a); (b) is similar. The first part of the proof will show that $\mathbf{E}_0^*(s)$ is stable for all sufficiently small s and unstable for all sufficiently large s. Let

$$m_i = \min\{f(\mathbf{E}) : \mathbf{E} \in W_i\}, \quad i = 0, 1; \tag{18}$$

we show that, if

$$s \leq \min\{(m_1 - m_0)/(\alpha_1 - \beta_0), 0\}, \tag{19}$$

then $\mathbf{E}_0^*(s)$ is stable by proving that it is a point of convexity of f on $U = W_0 \cup W_1$ in the sense that

$$f(\mathbf{E}) \geq f(\mathbf{E}_0^*(s)) + s[\mathbf{E} - \mathbf{E}_0^*(s)]\mathbf{e} \cdot \mathbf{e} \tag{20}$$

for all $\mathbf{E} \in U$. Inequality (20) is clear if $\mathbf{E} \in W_0$ since $D^2 f$ is positive definite on the convex set W_0. Hence let $\mathbf{E} \in W_1$. Let $\mathbf{M_0} \in W_0$ be such that $m_0 = f(\mathbf{M_0})$. By the convexity of f on W_0,

$$m_0 = f(\mathbf{M_0}) \geq \mathbf{f(E_0^*(s))} + \mathbf{s[M_0} - \mathbf{E_0^*(s)]e} \cdot \mathbf{e}, \tag{21}$$

and, by Assumption 1(b), $\mathbf{M_0 e} \cdot \mathbf{e} < \beta_0$, $\alpha_1 < \mathbf{Ee} \cdot \mathbf{e}$. Using this and Eqs. (21) and (19) one finds that

$$f(\mathbf{E}_0^*(s)) + s[\mathbf{E} - \mathbf{E}_0^*(s)]\mathbf{e} \cdot \mathbf{e} \leq m_0 + s(\alpha_1 - \beta_0) \leq f(\mathbf{E})$$

and thus (20) holds generally. A standard argument, given below, implies that $\mathbf{E}_0^*(s)$ is stable, i.e.,

$$\Phi(\tau) \geq \Phi(\mathbf{E}_0^*(s)) = f(\mathbf{E}_0^*(s)) \tag{22}$$

for any state τ of elongation $w = w_0^*(s)$. If $\tau = (\mathbf{E}_0, \mathbf{E}_1, z)$ is a mixture, (22) is obtained by inserting $\mathbf{E} = \mathbf{E}_0$ and $\mathbf{E} = \mathbf{E}_1$ into (20) and taking a convex combination of the resulting inequalities with weights $1 - z, z$. The case of a homogeneous state is similar. Next we show that $\mathbf{E}_0^*(s)$ is unstable for all sufficiently large s. This part of the proof uses Assumption 2. Let

$$N_i = \max\{f(\mathbf{E}) : \mathbf{E} \in K_i\} + 1, \quad i = 0, 1;$$

we prove that, if

$$s \geq \max\{(N_1 - m_0)/(\alpha_1 - \beta_0), 0\}, \tag{23}$$

then f is not rank 1 convex at $\mathbf{E}_0^*(s)$ and hence $\mathbf{E}_0^*(s)$ is unstable. Indeed, if s satisfies (23) and $\mathbf{E} \in K_1 \subset U$ is compatible with $\mathbf{E}_0^*(s)$ (Assumption 2), then

$$f(\mathbf{E}) < N_1 \leq m_0 + s(\alpha_1 - \beta_0) \leq f(\mathbf{E}_0^*(s)) + s[\mathbf{E} - \mathbf{E}_0^*(s)]\mathbf{e} \cdot \mathbf{e}.$$

The failure of rank 1 convexity implies instability by the assertion at the end of Sect. 3. Let S_0 be the set of all $\bar{s} \in \mathbb{R}$ such that $\mathbf{E}_0^*(s)$ is stable for all $s \leq \bar{s}$. By the above, S_0 is nonempty and bounded from above and let $s^\sharp = \sup S_0$. A continuity argument shows that $\mathbf{E}_0^*(s^\sharp)$ is still stable. We complete the proof by proving that $\mathbf{E}_0^*(s)$ is unstable for all $s > s^\sharp$. There exist sequences s^n and σ_n such that $s^n \to s^\sharp$, $s^n > s^\sharp$, $\sigma_n \in \Sigma_{w^n}$, $w^n := w_0^*(s^n)$, and $\Phi(\mathbf{E}_0^*(s^n)) > \Phi(\sigma_n)$. Using the strict convexity of f on W_0 we can show that σ_n is a mixture with phases belonging to different wells. Thus $\sigma_n = (\mathbf{E}_0^n, \mathbf{E}_1^n, z^n) \in W_0 \times W_1 \times (0, 1)$ for all n. Since the wells are bounded, we have, for a subsequence, $(\mathbf{E}_0^n, \mathbf{E}_1^n, z^n) \to (\bar{\mathbf{E}}_0, \bar{\mathbf{E}}_1, u) \in \bar{W}_0 \times \bar{W}_1 \times [0, 1]$. Necessarily $u < 1$ since, with $u = 1$, the limit in $w^n = [(1 - z^n)\mathbf{E}_0^n + z^n\mathbf{E}_1^n]\mathbf{e} \cdot \mathbf{e}$ gives $\mathbf{E}_0^*(s^\sharp)\mathbf{e} \cdot \mathbf{e} = \bar{\mathbf{E}}_1\mathbf{e} \cdot \mathbf{e}$ and $\mathbf{E}_0^*(s^\sharp) \in W_0, \bar{\mathbf{E}}_1 \in \bar{W}_1$ which contradicts Assumption 1(b). Combining

$$f(\mathbf{E}_0^*(s^n)) > (1 - z^n)f(\mathbf{E}_0^n) + z^n f(\mathbf{E}_1^n), \tag{24}$$

with $f \geq 0$ and $u < 1$, we see that the sequence $f(\mathbf{E}_0^n)$ is bounded and hence $\bar{\mathbf{E}}_0 \in W_0$ by the growth of f near ∂U. By the convexity of f on W_0,

$$f(\mathbf{E}_0^n) \geq f(\mathbf{E}_0^*(s^n)) + s^n(\mathbf{E}_0^n\mathbf{e} \cdot \mathbf{e} - w^n)$$

and thus, from (24) and $w^n - \mathbf{E}_0^n\mathbf{e} \cdot \mathbf{e} = z^n(\mathbf{E}_1^n - \mathbf{E}_0^n)\mathbf{e} \cdot \mathbf{e}$ we obtain

$$s^n(\mathbf{E}_1^n - \mathbf{E}_0^n)\mathbf{e} \cdot \mathbf{e} \geq f(\mathbf{E}_1^n) - f(\mathbf{E}_0^n). \tag{25}$$

Since $f(\mathbf{E}_0^n)$, s^n, $\mathbf{E}_1^n$ and $\mathbf{E}_0^n$ are bounded we see from (25) that $f(\mathbf{E}_1^n)$ is also bounded and hence $\bar{\mathbf{E}}_1 \in W_1$ by the growth of f near ∂U. Thus $\tau := (\bar{\mathbf{E}}_0, \bar{\mathbf{E}}_1, u)$ is a state of elongation $w_0^\sharp$. Combining the fact that it is a minimizer with the limit in (24) gives

$$\Phi(\tau) = \Phi(\mathbf{E}_0^*(s^\sharp)). \tag{26}$$

Further, the limit in (25) gives

$$f(\bar{\mathbf{E}}_1) \le f(\bar{\mathbf{E}}_0) + s^\sharp(\bar{\mathbf{E}}_1 - \bar{\mathbf{E}}_0)\mathbf{e}\cdot\mathbf{e}. \tag{27}$$

For any $z, 0 < z < 1$, let $\bar{\sigma} = (\bar{\mathbf{E}}_0, \bar{\mathbf{E}}_1, z)$. For $z = u$ this is a state of elongation $w_0^\sharp$ and energy $f(\mathbf{E}_0^\sharp)$; if z ranges the interval $u < z < 1$ the elongation w of $\bar{\sigma}$ ranges the interval $w_0^\sharp < w < \bar{\mathbf{E}}_1\mathbf{e}\cdot\mathbf{e}$. Combining Assumption 1(b) with the assertion of Proposition 4 about the range of w_0^*, one finds that the interval $w_0^\sharp < w < \bar{\mathbf{E}}_1\mathbf{e}\cdot\mathbf{e}$ covers the range of $w_0^*(s)$, $s^\sharp < s < \infty$. Further,

$$\Phi(\bar{\sigma}) = \Phi(\tau) + (z-u)(f(\bar{\mathbf{E}}_1) - f(\bar{\mathbf{E}}_0)). \tag{28}$$

Then for $w > w_0^\sharp$ we obtain from (26)–(28) that

$$\Phi(\bar{\sigma}) \le f(\mathbf{E}_0^\sharp) + s^\sharp(w - w_0^\sharp) < f_0^*(w)$$

since f_0^* is strictly convex. □

The following assumption is made mainly for convenience; an alternative development without it is possible at the expense of more complicated forms of the results.

Assumption 3. $\mathbf{E}_0^\sharp$ *and* $\mathbf{E}_1^\flat$ *are strict minimizers.*

The following proposition identifies $s^\sharp, s^\flat$ with the points of loss of rank 1 convexity.

Theorem 2. (a) *There exists* $\bar{\mathbf{E}}_1^\sharp \in W_1$ *compatible with* $\mathbf{E}_0^\sharp$ *such that*

$$f(\bar{\mathbf{E}}_1^\sharp) - f(\mathbf{E}_0^\sharp) = s^\sharp(\bar{\mathbf{E}}_1^\sharp - \mathbf{E}_0^\sharp)\mathbf{e}\cdot\mathbf{e}. \tag{29}$$

(b) *There exists* $\bar{\mathbf{E}}_0^\flat \in W_0$ *compatible with* $\mathbf{E}_1^\flat$ *such that*

$$f(\mathbf{E}_1^\flat) - f(\bar{\mathbf{E}}_0^\flat) = s^\flat(\mathbf{E}_1^\flat - \bar{\mathbf{E}}_0^\flat)\mathbf{e}\cdot\mathbf{e}. \tag{30}$$

Proof. We only prove (a); (b) is similar. Let τ be the state derived in the proof of Theorem 1. As the elongation of the homogeneous state $\mathbf{E}_0^\circ$ and of τ is $w_0^\sharp$, Eq. (26) and an appeal to the strict nature of the minimizer $\mathbf{E}_0^\sharp$ (Assumption 3) shows that $u = 0$ and $\bar{\mathbf{E}}_0 = \mathbf{E}_0^\sharp$. Denoting $\bar{\mathbf{E}}_1^\sharp := \bar{\mathbf{E}}_1$ and combining (27) with the rank 1 convexity of f at $\mathbf{E}_0^\sharp$ establishes (29). □

Theorem 2(a) asserts the existence of $\bar{\mathbf{E}}_0^\sharp \in W_0$ compatible with $\mathbf{E}_0^\sharp$, written

$$\bar{\mathbf{E}}_1^\sharp - \mathbf{E}_0^\sharp = \mathbf{a}^\sharp \odot \mathbf{n}^\sharp, \quad \mathbf{a}^\sharp \ne \mathbf{0}, \quad |\mathbf{n}^\sharp| = 1, \tag{31}$$

such that (29) holds; moreover, since f is rank 1 convex at $\mathbf{E}_0^\sharp$, we also have

$$\hat{\mathbf{S}}(\bar{\mathbf{E}}_1^\sharp)\mathbf{n}^\sharp = \hat{\mathbf{S}}(\mathbf{E}_0^\sharp)\mathbf{n}^\sharp, \quad \hat{\mathbf{S}}(\bar{\mathbf{E}}_1^\sharp)\mathbf{a}^\sharp = \hat{\mathbf{S}}(\mathbf{E}_0^\sharp)\mathbf{a}^\sharp \tag{32}$$

by the considerations at the end of Sect. 3. Mechanically, $\bar{\mathbf{E}}_1^\sharp$ is a second phase that can coexist with $\mathbf{E}_0^\sharp$ in a thermodynamic equilibrium, a germ of the second phase emerging at the onset of the phase transformation at the yield point $P_0^\sharp$. For $s < s^\sharp$, no such phase exists, because the argument at the end of the proof of Theorem 1 would show that $\mathbf{E}_0^*(t)$ is unstable for all $t > s$. If $\mathbf{E}_0^*(\cdot)$ is known, the system (31), (29), (32) can be used to determine the value $s^\sharp$ and the second phase $\bar{\mathbf{E}}_1$. For the quadratic double well material with equal elastic moduli, that system and a similar system for $\mathbf{E}_1^\flat$ can be solved explicitly [12]. The main difference between solid phase transitions and those in fluids is that there is no reason to expect that

$$\mathbf{E}_1^\sharp = \bar{\mathbf{E}}_1^\sharp, \quad \mathbf{E}_0^\flat = \bar{\mathbf{E}}_0^\flat. \tag{33}$$

Indeed, expressed in one way, $(33)_1$ means that the second phase $\bar{\mathbf{E}}_1^\sharp$ is not only in equilibrium with the phase $\mathbf{E}_0^\sharp$, but also with the device at the force $s^\sharp$. Expressed in another way, $(33)_1$ says that the homogeneous phase $\mathbf{E}_1^*(w_1^\sharp)$ is geometrically compatible with $\mathbf{E}_0^*(s^\sharp)$. Both can occur only exceptionally. In the exceptional case when $(33)_1$ holds, the proof of Theorem 3 will show that

$$\mathbf{E}_0^\sharp = \mathbf{E}_0^\flat = \bar{\mathbf{E}}_0^\flat, \quad \mathbf{E}_1^\flat = \mathbf{E}_1^\sharp = \bar{\mathbf{E}}_1^\sharp.$$

In this case, $s^\sharp = s^\flat$, no hysteresis occurs and the mixtures along the phase equilibrium line are mixtures of the phases at the endpoints. Otherwise, one has to solve the system of Eqs. (9), (10) to determine the exact values of the phases $\mathbf{E}_0, \mathbf{E}_1$ at the composition z and force $s = s^\sharp$. See Proposition 5.

5 Minimum energy and the phase equilibrium curve

The following statement is one of the main results of the paper. Recall the notation (6) and the minimum energy $\hat{\phi}$ (13).

Theorem 3. *The minimum energy $\hat{\phi}$ is convex on $(\alpha_0, w_0^\sharp]$, concave on $[w_0^\sharp, w_1^\flat]$ and convex on $[w_1^\flat, \beta_1)$. For each $w \in (w_0^\sharp, w_1^\flat)$ there exists a mixture minimizer of elongation w and force s that satisfies* (14) *and* (15)*; s is a supergradient of $\hat{\phi}$ on $[w_0^\sharp, w_1^\flat]$ at w and*

$$s^\sharp \geq s \geq s^\flat. \tag{34}$$

Thus $s^\sharp \geq s^\flat$; equality holds if and only if $\mathbf{E}_0^\sharp$ and $\mathbf{E}_1^\flat$ are compatible.

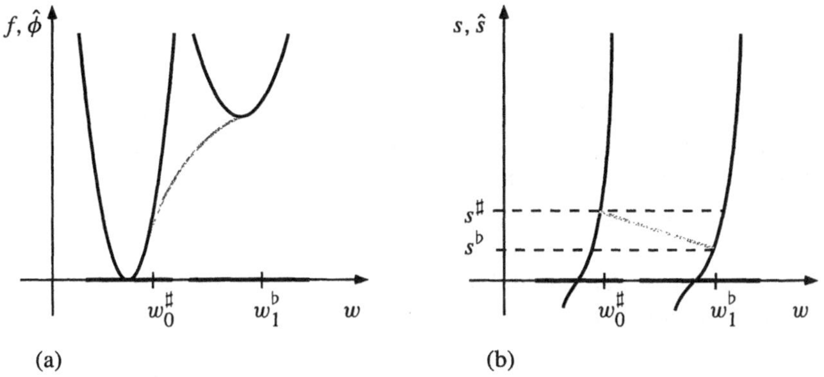

Fig. 6. Single-phase and effective quantities: (a) energy, (b) force

If $\hat{\phi}$ is twice continuously differentiable on $[w_0^\sharp, w_1^\flat]$, concavity says that the effective force $\hat{s}(w) := d\hat{\phi}(w)/dw$ is a nonincreasing function:

$$\frac{d\hat{s}(w)}{dw} \leq 0.$$

The graph of $\hat{s}(w)$ is the phase equilibrium curve. See Fig. 6, where the bold lines denote the energy of the single phase states and the shadow lines the effective quantities. (The bold line intervals on the w-axis denote the nonoverlapping intervals (α_0, β_0) and (α_1, β_1).)

Proof. The convexity of $\hat{\phi}$ on $(\alpha_0, w_0^\sharp]$ and $[w_1^\flat, \beta_1)$ follows from Proposition 4. We show that, for each $w \in (w_0^\sharp, w_1^\flat)$, there exists a minimizer of elongation w. Note first that Assumption 2 implies that, for each $w \in [w_0^\sharp, w_1^\flat]$, $\Sigma_w \neq \emptyset$. Let $w \in (w_0^\sharp, w_1^\flat)$ now be fixed; we show that there exists a minimizer of Φ in Σ_w. Let $\sigma^n \in \Sigma_w$ be a minimizing sequence for Φ on Σ_w. Since the homogeneous equilibrium states are not minimizers in this interval of elongations by Theorem 1, we can assume that all members of σ^n are mixtures $\sigma^n = (\mathbf{E}_0^n, \mathbf{E}_1^n, z^n)$, $0 < z^n < 1$. The strict convexity of f on W_0 and W_1 shows that we can further modify the sequence so that it consists only of mixtures with $\mathbf{E}_0^n \in W_0$, $\mathbf{E}_1^n \in W_1$. Since the wells are bounded, we have, for a subsequence,

$$(\mathbf{E}_0^n, \mathbf{E}_1^n, z^n) \to (\mathbf{E}_0, \mathbf{E}_1, z),$$

where $\mathbf{E}_0 \in \bar{W}_0$, $\mathbf{E}_1 \in \bar{W}_1$, $0 \leq z \leq 1$. We show that $0 < z < 1$. Indeed, assume $z = 0$. Since

$$\Phi(\sigma^n) = (1 - z^n) f(\mathbf{E}_0^n) + z^n f(\mathbf{E}_1^n) \tag{35}$$

is bounded, $f(\mathbf{E}_0^n)$ is also bounded and consequently $\mathbf{E}_0 \in W_0$. As $f \geq 0$,

$$\Phi(\sigma^n) = (1 - z^n) f(\mathbf{E}_0^n) + z^n f(\mathbf{E}_1^n) \geq (1 - z^n) f(\mathbf{E}_0^n)$$

and thus $\lim \Phi(\sigma^n) \geq \lim(1-z^n) f(\mathbf{E}_0^n) = f(\mathbf{E}_0)$. This shows that the homogeneous state $\mathbf{E}_0$ of elongation w is a minimizer of Φ on Σ_w, a contradiction to Theorem 1. Thus $z > 0$ and the same argument shows that $z < 1$. Then the boundedness of (35) yields the boundedness of $f(\mathbf{E}_0^n), f(\mathbf{E}_0^n)$; consequently, $\mathbf{E}_0 \in W_0$, $\mathbf{E}_1 \in W_1$. Thus $\sigma := (\mathbf{E}_0, \mathbf{E}_1, z)$ is a mixture and by continuity, $\lim \Phi(\sigma^n) = \Phi(\sigma)$; this shows that σ is a minimizer satisfying (14) and (15) by Sect. 3.

We next prove the concavity of $\hat{\phi}$ on $[w_0^\sharp, w_1^\flat]$ and the assertion about the supergradient. Let $w \in (w_0^\sharp, w_1^\flat)$ be fixed and let $(\mathbf{E}_0, \mathbf{E}_1, z) \in \Sigma_w$ be any minimizer of Φ, which by the above is an equilibrium mixture of some force s satisfying (14). For any $z' \in (0, 1)$, $\tau := (\mathbf{E}_0, \mathbf{E}_1, z')$ is a state of elongation

$$w' = w + (z' - z)(\mathbf{E}_1 - \mathbf{E}_0)\mathbf{e} \cdot \mathbf{e} \tag{36}$$

and energy

$$\Phi(\tau) = \hat{\phi}(w) + s(w' - w),$$

where (14) has been used. We note that $(\mathbf{E}_1 - \mathbf{E}_0)\mathbf{e} \cdot \mathbf{e} > 0$ by Assumption 1(b) and as z' ranges the interval $[0, 1]$, w' ranges the nondegenerate interval

$$J := [w - z(\mathbf{E}_1 - \mathbf{E}_0)\mathbf{e} \cdot \mathbf{e},\, w + (1 - z)(\mathbf{E}_1 - \mathbf{E}_0)\mathbf{e} \cdot \mathbf{e}]$$

which contains w as an interior point. Thus if $w' \in J$, the minimum energy $\hat{\phi}(w')$ at elongation w' must satisfy

$$\hat{\phi}(w') \leq \Phi(\tau) = \hat{\phi}(w) + s(w' - w). \tag{37}$$

Hence s is a local supergradient of $\hat{\phi}$ of w in the sense that (37) holds for all w' sufficiently close to w. As one can easily prove that $\hat{\phi}$ is continuous, [13, Proposition 6.1] can be invoked to assert that $\hat{\phi}$ is concave on $(w_0^\sharp, w_1^\flat)$ and s is a supergradient at w. The proof of Theorem 1 (namely, the construction of the state $\bar{\sigma}$ and an analogous construction for $s^\flat$) gives

$$\hat{\phi}(w) \leq f_0^*(w_0^\sharp) + s^\sharp(w - w_0^\sharp), \quad \hat{\phi}(w) \leq f_1^*(w_1^\flat) + s^\flat(w - w_1^\flat),$$

for all $w \in [w_0^\sharp, w_1^\flat]$ sufficiently close to $w_0^\sharp$ and $w_1^\flat$, respectively. Thus $s^\sharp, s^\flat$ are supergradients at $w_0^\sharp, w_1^\flat$, respectively. The decreasing character of the supergradient yields (34).

Assume that $s^\sharp = s^\flat$; we prove that $\mathbf{E}_0^\sharp$ and $\mathbf{E}_1^\flat$ are compatible. By (34), $s := s^\sharp = s^\flat$ is a supergradient of $\hat{\phi}$ at each point of $[w_0^\sharp, w_1^\flat]$. Thus $\hat{\phi}$ is affine:

$$\hat{\phi}(w) = f(\mathbf{E}_0^\sharp) + s^\sharp(w - w_0^\sharp), \quad w \in [w_0^\sharp, w_1^\flat]. \tag{38}$$

Let $w \in (w_0^\sharp, w_1^\flat)$ be arbitrary. By the above there exists an equilibrium mixture minimizer $\sigma = (\mathbf{E}_0, \mathbf{E}_1, z)$, $\mathbf{E}_0 \in W_0$, $\mathbf{E}_1 \in W_1$, of force s; as, moreover, (14) holds, the energy is

$$\Phi(\sigma) = (1 - z) f(\mathbf{E}_0) + z f(\mathbf{E}_1) = f(\mathbf{E}_0) + s^\sharp(w - \mathbf{E}_0\mathbf{e} \cdot \mathbf{e}).$$

Combining this with (38) provides

$$f(\mathbf{E}_0) = f(\mathbf{E}_0^\sharp) + s^\sharp(\mathbf{E}_0 - \mathbf{E}_0^\sharp)\mathbf{e}\cdot\mathbf{e}. \tag{39}$$

This implies that $\mathbf{E}_0 = \mathbf{E}_0^\sharp$ since $\mathbf{E}_0 \neq \mathbf{E}_0^\sharp$ and strict convexity of f on W_0 would imply strict inequality in (39). Similarly, $\mathbf{E}_1 = \mathbf{E}_1^\flat$ and since $\mathbf{E}_0 = \mathbf{E}_0^\sharp, \mathbf{E}_1 = \mathbf{E}_1^\flat$ belong to the same mixture, they are compatible. Conversely, let $\mathbf{E}_0^\sharp$ and $\mathbf{E}_1^\flat$ be compatible. By Theorem 2, f is rank 1 convex at $\mathbf{E}_0^\sharp$ and $\mathbf{E}_1^\flat$ and thus the compatibility gives

$$f(\mathbf{E}_1^\flat) \geq f(\mathbf{E}_0^\sharp) + s^\sharp(\mathbf{E}_1^\flat - \mathbf{E}_0^\sharp)\mathbf{e}\cdot\mathbf{e}, \quad f(\mathbf{E}_0^\sharp) \geq f(\mathbf{E}_1^\flat) + s^\flat(\mathbf{E}_0^\sharp - \mathbf{E}_1^\flat)\mathbf{e}\cdot\mathbf{e},$$

from which $(s^\flat - s^\sharp)(\mathbf{E}_1^\flat - \mathbf{E}_0^\sharp)\mathbf{e}\cdot\mathbf{e} \geq 0$ and hence $s^\flat \geq s^\sharp$ by Assumption 1(b). Thus $s^\sharp = s^\flat$. □

6 Equilibrium mixtures, irreversibility, hysteresis

As we saw at the end of Sect. 4, the phase mixtures along the yield segment $[P_0^\sharp, P_1^\sharp]$ are determined by the equilibrium equations (9), (10). This section starts with a proof of solvability of (9), (10). To this end, we slightly extend the terminology and consider mixtures $(\mathbf{E}_0, \mathbf{E}_1, z)$ with $\mathbf{E}_i \in U$ compatible and $0 \leq z \leq 1$, instead of the previous $0 < z < 1$. A mixture $\sigma = (\mathbf{E}_0, \mathbf{E}_1, z)$ is called an equilibrium mixture of force s, mass fraction z and interface normal $\mathbf{n}$ if (9), (10) hold. Let N be the set of all unit vectors $\mathbf{n}$ such that there exist $\mathbf{E}_0 \in W_0$, $\mathbf{E}_1 \in W_1$ and a vector $\mathbf{a}$ satisfying (1). By Assumption 2, this is a nonempty open subset of the unit sphere.

Proposition 5. *For each $\alpha = (s, z, \mathbf{n}) \in D := \mathbb{R} \times [0, 1] \times N$ there exists an equilibrium mixture $\sigma(\alpha) = (\tilde{\mathbf{E}}_0(\alpha), \tilde{\mathbf{E}}_1(\alpha), z)$ of force s, mass fraction z and interface normal* $\mathbf{n}$; *moreover,*

$$\tilde{\mathbf{E}}_i(s, i, \mathbf{n}) = \mathbf{E}_i^*(s), \tag{40}$$

and the values $\tilde{\mathbf{E}}_i(\alpha)$ are uniquely determined within W_i.

We introduce the free energy $\tilde{\phi}$ and the excess function $\tilde{E}$ as functions on D by

$$\tilde{\phi}(\alpha) = (1 - z)f(\tilde{\mathbf{E}}_0) + zf(\tilde{\mathbf{E}}_1), \tag{41}$$

$$\tilde{E}(\alpha) = f(\tilde{\mathbf{E}}_0) - f(\tilde{\mathbf{E}}_1) - s[\tilde{\mathbf{E}}_0(\alpha) - \tilde{\mathbf{E}}_1(\alpha)]\mathbf{e}\cdot\mathbf{e}, \tag{42}$$

where $\tilde{\mathbf{E}}_i = \tilde{\mathbf{E}}_i(\alpha)$, $i = 0, 1$.

Proof. For each $s \in \mathbb{R}$, let $G_s : \Sigma_m \to \mathbb{R}$ be defined by

$$G_s(\mathbf{E}_0, \mathbf{E}_1, z) = (1 - z)g_s(\mathbf{E}_0) + zg_s(\mathbf{E}_1), \quad (\mathbf{E}_0, \mathbf{E}_1, z) \in \Sigma_m,$$

where $g_s(\mathbf{E}) = f(\mathbf{E}) - s\mathbf{E}\mathbf{e}\cdot\mathbf{e}$. Let $s \in \mathbb{R}$, $0 < z < 1, \mathbf{n} \in N$ be fixed and let $H : C \to \mathbb{R}$ be defined by

$$C := \{(\mathbf{E}, \mathbf{a}) : (\mathbf{E}, \mathbf{E} + \mathbf{a} \odot \mathbf{n}, z) \in \Sigma\}$$
$$H(\mathbf{E}, \mathbf{a}) = G_s(\mathbf{E}, \mathbf{E} + \mathbf{a} \odot \mathbf{n}, z), \quad (\mathbf{E}, \mathbf{a}) \in C.$$

Since the restrictions of g_s to W_0 and to W_1 are strictly convex, H is a strictly convex function on an open convex set C. We have

$$\partial C = \{(\mathbf{E}, \mathbf{a}) : \mathbf{E} \in \partial W_0, \mathbf{E} + \mathbf{a} \odot \mathbf{n} \in \bar{W}_1\} \cup \{(\mathbf{E}, \mathbf{a}) : \mathbf{E} \in \bar{W}_0, \mathbf{E} + \mathbf{a} \odot \mathbf{n} \in \partial W_1\}$$

and thus $H(\mathbf{E}, \mathbf{a}) \to \infty$ if $(\mathbf{E}, \mathbf{a}) \to \partial C$ by Assumption 1(a). Hence H has a unique point of minimum on C at which $D_{\mathbf{E}} H(\mathbf{E}, \mathbf{a}) = \mathbf{0}$, $D_{\mathbf{a}} H(\mathbf{E}, \mathbf{a}) = \mathbf{0}$; these give the equations for the equilibrium mixtures. This completes the proof in the case $0 < z < 1$. Now let $z = 0$. Let $s \in \mathbb{R}$, $\mathbf{n} \in N$ be fixed and let $H : C \to \mathbb{R}$ be defined by

$$C := \{\mathbf{a} : \mathbf{E}_0^*(s) + \mathbf{a} \odot \mathbf{n} \in W_1\},$$
$$H(\mathbf{a}) = g_s(\mathbf{E}_0^*(s) + \mathbf{a} \odot \mathbf{n}), \quad \mathbf{a} \in C.$$

The function H is strictly convex on an open convex set. We again have $H(\mathbf{a}) \to \infty$ as $\mathbf{a} \to \partial C$ and thus H has a unique point of minimum where $DH(\mathbf{a}) = \mathbf{0}$; this equation shows that $(\mathbf{E}_0^*(s), \mathbf{E}_0^*(s) + \mathbf{a} \odot \mathbf{n}, 0)$ is an equilibrium mixture. Similar considerations apply to the case $z = 1$. The assertion about uniqueness follows from the strict convexity and the assertion about continuous differentiability from an application of the implicit function theorem. The details are omitted. □

We next return to Fig. 4 and assume that the phases $\mathbf{E}_0^\sharp$ and $\mathbf{E}_1^\flat$ are incompatible. The preceding sections show that $s^\sharp > s^\flat$, $\mathbf{E}_0^*(s)$ is a point of rank 1 convexity for all $s \le s^\sharp$, $\mathbf{E}_1^*(s)$ is a point of rank 1 convexity for all $s \ge s^\flat$ and there exist $\bar{\mathbf{E}}_1^\sharp$, $\bar{\mathbf{E}}_0^\flat$ compatible with $\mathbf{E}_0^\sharp$, $\mathbf{E}_1^\flat$, respectively, such that (29) and (30) hold. We have the following picture of the excess function on the boundary of D :

$$\left.\begin{array}{lll} s < s^\sharp \Rightarrow \tilde{E}(s, 0, \mathbf{n}) < 0 & \text{for all } \mathbf{n} \in N, & \\ s = s^\sharp \Rightarrow \tilde{E}(s, 0, \mathbf{n}) \le 0 & \text{for all } \mathbf{n} \in N, & \text{and } \tilde{E}(s, 0, \mathbf{n}^\sharp) = 0, \\ s > s^\flat \Rightarrow \tilde{E}(s, 1, \mathbf{n}) > 0 & \text{for all } \mathbf{n} \in N, & \\ s = s^\flat \Rightarrow \tilde{E}(s, 1, \mathbf{n}) \ge 0 & \text{for all } \mathbf{n} \in N, & \text{and } \tilde{E}(s, 1, \mathbf{n}^\flat) = 0. \end{array}\right\} \quad (43)$$

Moreover, $\tilde{E} = 0$ along the phase equilibrium line and it can be further shown that $\tilde{E}$ is decreasing in z at constant s, $\mathbf{n}$.

Consider quasistatic processes $(s(t), z(t), \mathbf{n}(t))$ in D interpreted as sequences of equilibrium mixtures. Changing z amounts to changing positions of phase interfaces which is necessarily accompanied by the production of entropy (e.g., [5, Chap. 22]).

For a fixed specimen at fixed temperature, the total production of entropy per unit time in the specimen is proportional to $\dot{z}\tilde{E}(\alpha)$ and hence real processes must satisfy

$$\dot{z}\tilde{E}(\alpha) \geq 0. \tag{44}$$

Thus if we denote

$$A_{\pm} = \{\alpha \in D : \pm\tilde{E}(\alpha) > 0\}, \quad A_{\circ} = \{\alpha \in D : \tilde{E}(\alpha) = 0\},$$

then real processes must satisfy $\dot{z} \geq 0$ on A_+, $\dot{z} \leq 0$ on A_- while on $A_\circ$ the sign of $\dot{z}$ is unrestricted. Hence, referring to (43), we see that the processes starting from the single phase austenite segment $[P_0^\flat, P_0^\sharp]$ and entering the hysteresis loop $P_0^\flat P_0^\sharp P_1^\sharp P_1^\flat$ are inadmissible. At $P_0^\sharp$ we have $\tilde{E} = 0$ and thus $P_0^\sharp$ is the first point where the process can change direction to horizontal. Nature uses this possibility since the homogeneous phase beyond the point $w_0^\sharp$ is unstable; thus for elongations $w > w_0^\sharp$ the material exhibits an equilibrium mixture. If we accept that the process proceeds horizontally along $[P_0^\sharp, P_1^\sharp]$, then the horizontal part of the process necessarily terminates at $P_1^\sharp$. If the change of elongation is reversed, the resulting process is not the reversal of the direct process since that process violates (44). The function $\tilde{E}$ remains positive on $[P_1^\sharp, P_1^\flat)$ until the point $P_1^\flat$ and thus the process cannot enter the hysteresis loop which indicates the necessity of the hysteresis.

The entropy criterion leaves the processes otherwise unrestricted. However, it is known that in many problems connected with shock waves and propagating phase boundaries, the entropy criterion alone is insufficient to single out admissible processes. Thus, for example, a process starting from $P_0^\sharp$ and following the phase equilibrium line $[P_0^\sharp, P_1^\flat]$ is admissible (and reversible) according to (44). Another candidate for an admissibility criterion (for combination with the entropy criterion) is

$$\dot{s}\dot{w} \geq 0. \tag{45}$$

This excludes the processes along the phase equilibrium line. Note that the observed processes, including the internal hysteresis loops, satisfy both (44) and (45). Also consistent with (44) and (45) is the requirement that

$$D_{\mathbf{n}}\tilde{\phi} = 0$$

which may be interpreted as a condition determining the unknown interface normal, and as a consequence of the energy minimization with respect to $\mathbf{n}$ at fixed w, z. Furthermore, one may appeal to an additional minimization. (See [10] for an application of the principle of least work in a model of hysteresis.)

A similar situation arises if we reduce the elongation at some point of the yield segment. The reverse motion along the yield segment is impossible because $\tilde{E} > 0$ there, and thus the process must enter the interior of the hysteresis loop. Consider the processes within the hysteresis loop. Horizontal processes towards the left above the phase equilibrium curve are impossible, and the phase equilibrium curve is the first possibility where the process can change direction to horizontal. The possibility of changing direction, however, does not explain why the change of direction on the phase equilibrium curve really occurs.

Acknowledgements. This work was supported by the Weierstraß-Institut für Angewandte Analysis und Stochastik, Grant A2019603 of the AV ČR and Grant 201/00/1516 of the GA ČR. This support is gratefully acknowledged. The author also thanks Professor K. Wilmański for discussions on the topic of martensitic transformations and for comments on an early draft of the manuscript.

References

[1] Pedregal, P. (1997): Parametrized measures and variational principles. Birkhäuser, Basel

[2] Pedregal, P. (2000): Variational methods in nonlinear elasticity. SIAM, Philadelphia

[3] Müller, S. (1999): Variational models for microstructure and phase transitions. In: Bethuel, F. et al. (eds.) Calculus of variations and geometric evolution problems. (Lecture Notes in Mathematics, vol. 1713). Springer, Berlin, pp. 85–210

[4] Roubíček, T. (1997): Relaxation in optimization theory and variational calculus. W. de Gruyter, Berlin

[5] Šilhavý, M. (1997): The mechanics and thermodynamics of continuous media. Springer, Berlin

[6] Kohn, R.V., Strang, G. (1986): Optimal design and relaxation of variational problems., I, II, III. Comm. Pure Appl. Math. **39**, 113–137, 139–182, 353–377

[7] Ball, J.M., Chu, C., James, R.D. (1995): Hysteresis during stress-induced variant rearrangement. J. Physique IV **C8**, 245–251

[8] Müller, I. (1989): On the size of the hysteresis in pseudoelasticity. Contin. Mech. Thermodyn. **1**, 125–142

[9] Müller, I., & Xu, H. (1991): On the pseudoelastic hysteresis. Acta Metall. Mater. **39**, 263–271

[10] Wilmański, K. (1993): Symmetric model of stress-strain hysteresis loops in shape memory alloys. Internat. J. Engrg. Sci. **31**, 1121–1138

[11] Kohn, R.V. (1991): The relaxation of a double-well energy. Contin. Mech. Thermodyn. **3**, 193–236

[12] Šilhavý, M. (2001): On the pseudoelastic hysteresis in a double well material. In preparation

[13] Šilhavý, M. (2001): Rotationally invariant rank 1 convex functions. Appl. Math. Optim. **44**, 1–15

Dissipative Fluids with Microstructure

André M. Sonnet, Epifanio G. Virga

Abstract. A variational principle is proposed which allows us to derive the equations of motion for a dissipative fluid with general microstructure. The only constitutive ingredients are the densities of the free energy and the dissipation, both subject to appropriate invariance requirements. The strict interplay between the microstructures considered here and those studied by Capriz is also examined in some detail.

1 Introduction

Many *soft* materials can be modeled as fluids with order parameters. The clearest examples of these are perhaps liquid crystals, which in their different phases are rich in microstructures. Nematic liquid crystals, for example, which are usually uniaxial, can also exhibit biaxial states around defects or under shear. Moreover, intermediate states that are optically uniaxial, and yet reveal different degrees of molecular order, are also observed. When the nematic is uniaxial, it can be described by a *director*, and its dynamics has long been understood in terms of balances of linear and angular momenta, also including those of the microstructure [1, 2]. The biaxial states are instead described by a second-rank order tensor (see, e.g., [3, Chap. 2] and [4]), for which a fairly complete dynamical model is analyzed in [5]. Moreover, the intermediate uniaxial states are described by supplementing the director with a variable scalar order parameter, whose dynamics was treated in depth by Ericksen [6]. However, a rational systematic theory to arrive at the evolution equations for more general dissipative microstructures is still missing. Presumably, this is so because simply balancing linear and angular momenta does not suffice to predict the evolution of the body when the microstructure fails to be vectorial.

Here, inspired by a paper of Leslie [7] where he rederived the classical theory of nematodynamics from a dissipation principle, we propose a general theory for dissipative fluids with microstructure. Rather than positing the evolution equations by analogy, we arrive at them from a variational principle, which, when applied to mass-point dynamics, gives the classical Lagrange- Rayleigh equations. For continua, this theory leads to the evolution equations along with the appropriate boundary conditions. A special application of this theory to dissipative fluids with arbitrary tensorial order is illustrated in [8].

The variational principle adopted here is formulated in Sect. 2. In Sect. 3, we apply it to a class of continua with dissipative microstructure: this class is chosen so as to encompass nematic liquid crystals, among many other materials (see [5, 8]). We arrive at a set of general dynamical equations and we show in Sect. 4 how these can subsume the balance equations for linear and angular momenta. We close the paper by illuminating the points of contact between this theory and Capriz's general theory for continua with microstructure (see [9]). One advantage of our method is that it introduces few constitutive laws: both elastic and viscous stresses, for example, are obtained from the elastic and dissipated energies.

2 Variational principle

We aim at extending to a class of dissipative ordered media the method originally put forward by Rayleigh to describe dissipative discrete systems [10]. The essence of Rayleigh's approach is to balance all generalized forces in Lagrange equations, including inertia, against frictional forces which derive from a dissipation function and which are linear in the generalized velocities. We first illustrate this basic balance by taking a holonomic dynamical system as a paradigm and we then confine attention to systems subject to conservative active forces. In general, a variational principle allows one to arrive directly at the equations of motion including the appropriate dissipative terms. This principle will be applied in Sect. 3 to a variety of dissipative ordered fluids.

2.1 Basic balance

Essentially, our development rests on two assumptions: first, that the total mechanical power, excluding dissipation, can be written as the product of generalized forces by generalized velocities, and, second, that these forces are balanced by frictional forces that possess a quadratic velocity potential.

Consider a holonomic dynamical system described by m generalized co-ordinates $q_1, \dots, q_m$. We denote by q and $\dot{q}$ the vectors in $\mathbb{R}^m$ of the generalized co-ordinates and the generalized velocities. We assume that the total mechanical power $\mathcal{W}$ can be written as

$$\mathcal{W} = X \cdot \dot{q} = \sum_{i=1}^{m} X_i \dot{q}_i, \tag{1}$$

where X_i are the generalized forces, including inertia. To help identify the generalized forces, we may suppose that X remains unchanged under time reversal, while $\dot{q}$ changes its sign.

When, in addition to the mechanical forces X, the system is subject to dissipation, frictional generalized forces Y are also at work which satisfy the balance equation

$$X + Y = 0. \tag{2}$$

Here we make the constitutive assumption that the forces Y are linear in the velocities $\dot{q}$ and that they can be derived from a positive-definite quadratic form $\mathcal{R}$ as

$$Y = \frac{\partial \mathcal{R}}{\partial \dot{q}}. \tag{3}$$

The velocity potential $\mathcal{R}$ is called the Rayleigh *dissipation function*. The equations of motion are then obtained by inserting (3) into Eq. (2):

$$X + \frac{\partial \mathcal{R}}{\partial \dot{q}} = 0. \tag{4}$$

Taking the inner product of both sides of Eq. (4) with $\dot{q}$ yields the balance of energy in the form

$$\mathcal{W} + 2\mathcal{R} = 0, \tag{5}$$

since $\mathcal{R}$ is a homogeneous function of degree two, for which

$$\frac{\partial \mathcal{R}}{\partial \dot{q}} \cdot \dot{q} = 2\mathcal{R}\,. \tag{6}$$

2.2 Conservative forces

When all active forces are conservative the total mechanical power $\mathcal{W}$ can be written as the rate of change of the total mechanical energy $\mathcal{F}$. Let $V = V(q)$ be the potential energy of the system and $T = T(q, \dot{q})$ its kinetic energy, which is assumed to be a positive-definite quadratic form in the velocities $\dot{q}$. A standard computation yields

$$\dot{\mathcal{F}} = \dot{T} + \dot{V} = \sum_{i=1}^{m} \left(\frac{\mathrm{d}}{\mathrm{d}t} \frac{\partial L}{\partial \dot{q}_i} - \frac{\partial L}{\partial q_i} \right) \dot{q}_i, \tag{7}$$

with the Lagrange function $L := T - V$, and so the generalized forces conjugate to the velocities $\dot{q}$ are here found to be

$$X = \frac{\mathrm{d}}{\mathrm{d}t} \frac{\partial L}{\partial \dot{q}} - \frac{\partial L}{\partial q}. \tag{8}$$

The equations of motion (4) then read simply as

$$\frac{\mathrm{d}}{\mathrm{d}t} \frac{\partial L}{\partial \dot{q}} - \frac{\partial L}{\partial q} + \frac{\partial R}{\partial \dot{q}} = 0, \tag{9}$$

which is the standard text-book form (cf., e.g., [11, p. 231]).

Clearly, the energy balance (5) still holds with $\mathcal{W} = \dot{F}$.

It should be noted that, since T is a positive-definite quadratic form in $\dot{q}$ and V is independent of $\dot{q}$, by Eq. (8) X is linear in $\ddot{q}$ and

$$\det \frac{\partial X}{\partial \ddot{q}} \neq 0;$$

thus $\ddot{q}$ can also be expressed in terms of X, q and $\dot{q}$ as

$$\ddot{q} = B(q, \dot{q})X + a(q, \dot{q}) \tag{10}$$

(cf. [11, p. 40]), where B is an invertible matrix in $\mathbb{R}^{m \times m}$ and a a vector in $\mathbb{R}^m$.

2.3 Variational formulation

We next show how the equations of motion (4) can be derived from an appropriate variational principle for $\mathcal{R}$, where the configuration remains unchanged, while both the velocities $\dot{q}$ and the accelerations $\ddot{q}$ are subject to judiciously constrained variations.

For a given configuration we imagine a system of variations $\delta\dot{q}$ of the actual velocity vector $\dot{q}$ that leaves both the generalized forces X and their power $\mathcal{W}$ unchanged. This implies that variations $\delta\ddot{q}$ of $\ddot{q}$ are chosen accordingly: they eventually result in linear combinations of the variations $\delta\dot{q}$ (cf. Eq. (10)). The constraint on the power input $\mathcal{W}$ can then be treated in the standard way through a Lagrange multiplier, so that $\dot{q}$ may be arbitrarily perturbed.

In the actual evolution of the system through a given configuration, $\dot{q}$ is such that $\mathcal{R}$ attains its minimum relative to all virtual values it can achieve, once both the forces X and their power $\mathcal{W}$ are *frozen* in their actual state. This is a principle of *minimum restrained dissipation*, which in a similar form has also been used in irreversible thermodynamics [12, 13]. We first require $\mathcal{R}$ to be stationary with respect to this special class of variations. This leads to

$$\delta\mathcal{R} + \lambda\delta\mathcal{W} = \left(\frac{\partial\mathcal{R}}{\partial\dot{q}} + \lambda X\right)\cdot\delta\dot{q} = 0 \tag{11}$$

for all variations $\delta\dot{q}$, where λ is a Lagrange multiplier. Since $\delta\dot{q}$ is arbitrary, Eq. (11) amounts to

$$\lambda X + \frac{\partial\mathcal{R}}{\partial\dot{q}} = 0. \tag{12}$$

The value of λ can be determined by taking the inner product of both sides of Eq. (12) with $\dot{q}$ and requiring the energy balance (5) to hold. This shows that $\lambda = 1$, and the stationarity conditions on $\mathcal{R}$ in Eq. (12) just become the equations of motion (4). These are indeed minimality conditions on $\mathcal{R}$ constrained to the linear subspace where both X and $\mathcal{W}$ are prescribed, because $\mathcal{R}$ is a positive-definite quadratic form of $\dot{q}$.

3 Dissipative microstructures

In this section we apply the variational principle stated in the preceding section to ordered fluids. Our aim is to arrive at the evolution equations for a class of dissipative microstructures by positing in each case the appropriate energy and dissipation functionals, which in our development are the only relevant constitutive quantities for these materials.

We consider a material bearing an internal microstructure that occupies the region $\mathcal{B}$ in the three-dimensional space $\mathcal{E}$, with smooth boundary $\partial\mathcal{B}$. Following Capriz [9] (cf., in particular, §§ 1–5), we describe a microstructure through the elements $\boldsymbol{\nu} \in \mathcal{M}$ of a differentiable manifold M of dimension m. Often we represent a single element

$\boldsymbol{\nu} \in \mathsf{M}$ by means of its coordinates ν^α $(\alpha = 1, 2, \ldots, m)$ in a local chart of an atlas of M; they are the *order parameters* of the microstructure. When a body endowed with such a structure is subject to a rotation $\mathbf{Q}$, a given microstate represented by $\boldsymbol{\nu} \in \mathsf{M}$ appears in the same frame as being represented by a different element $\boldsymbol{\nu}^*$ of M. For $\mathbf{Q}$ close to the identity,

$$\boldsymbol{\nu}^* = \boldsymbol{\nu} + \mathbf{A}(\boldsymbol{\nu})\boldsymbol{q} + o(\boldsymbol{q}), \tag{13}$$

where, for every $\boldsymbol{\nu} \in \mathsf{M}$, $\mathbf{A}(\boldsymbol{\nu})$ is a linear operator from the translation space $\mathcal{V}$ of $\mathcal{E}$ into the tangent space $\mathcal{T}_{\boldsymbol{\nu}}\mathsf{M}$ to M at $\boldsymbol{\nu}$, and $\boldsymbol{q}$ is the vector associated with the infinitesimal generator of $\mathbf{Q}$. The operator $\mathbf{A}$ represents the local action on M of the group of rotations.

In general, the microstate changes both in space and time, and so $\boldsymbol{\nu}$ is thought of as a field $\boldsymbol{\nu} : \mathcal{B} \times \mathbb{R}^+ \to \mathsf{M}$, which associates an element of M to each position $p \in \mathcal{B}$ and time $t \in \mathbb{R}^+$. We denote by $\dot{\boldsymbol{\nu}}$ the material time derivative of $\boldsymbol{\nu}$ along a motion of the body:

$$\dot{\boldsymbol{\nu}} = \frac{\partial}{\partial t}\boldsymbol{\nu} + (\nabla \boldsymbol{\nu})\boldsymbol{v}, \tag{14}$$

where $\boldsymbol{v}$ is the macrovelocity field in the actual placement of the body. Here $\boldsymbol{v}$ and $\dot{\boldsymbol{\nu}}$ are the appropriate generalized velocities pertaining to the macro- and micro-motion, respectively.

The operator $\mathbf{A}$ defined by (13) plays a special rôle in Capriz's theory, as it enters all invariance requirements for the microstructure. For example, a function φ of $\boldsymbol{\nu}$ and $\dot{\boldsymbol{\nu}}$ is invariant under the whole Galilean group whenever the following condition is satisfied (cf. [9, §5]):

$$\frac{\partial \varphi}{\partial \nu^\alpha}\mathrm{A}_i^\alpha + \frac{\partial \varphi}{\partial \dot{\nu}^\alpha}\frac{\partial \mathrm{A}_i^\alpha}{\partial \nu^\beta}\dot{\nu}^\beta = 0, \tag{15}$$

which is written in components to avoid any ambiguity. Here and in what follows we adopt the summation convention for repeated indices.

The dissipation function must be invariant under all changes of frame, that is, under the mappings acting on the position vector $\boldsymbol{x}$ as

$$\boldsymbol{x}^* = \mathbf{Q}(t)\boldsymbol{x} + \boldsymbol{b}(t), \tag{16}$$

where $\mathbf{Q}$ is an orthogonal tensor and $\boldsymbol{b}$ a vector, both arbitrarily depending on time. Equivalently, the elementary measures of dissipation must not change when a rigid motion is superimposed on the actual motion. The invariants we shall consider here are the *shearing* tensor

$$\mathbf{D} := \frac{1}{2}(\nabla \boldsymbol{v} + (\nabla \boldsymbol{v})^{\mathrm{T}})$$

and the co-rotational time derivative $\overset{\circ}{\boldsymbol{\nu}}$ of $\boldsymbol{\nu}$ defined as (cf. [9, Eq. (6.7)])

$$\overset{\circ}{\boldsymbol{\nu}} := \dot{\boldsymbol{\nu}} - \mathbf{A}\boldsymbol{w}, \tag{17}$$

where $\boldsymbol{w} = \frac{1}{2}\operatorname{curl}\boldsymbol{v}$ is the axial vector associated with the *vorticity* tensor

$$\mathbf{W} := \frac{1}{2}(\nabla\boldsymbol{v} - (\nabla\boldsymbol{v})^{\mathrm{T}}).$$

3.1 Power input

We write the total energy stored in $\mathcal{B}$ as

$$\mathcal{F} = \int_{\mathcal{B}} F \mathrm{d}V$$

with

$$F = \varrho\left(\frac{1}{2}v^2 + \phi + \sigma(\varrho) + \kappa(\boldsymbol{\nu}, \dot{\boldsymbol{\nu}}) + \chi(\boldsymbol{\nu})\right) + W(\boldsymbol{\nu}, \nabla\boldsymbol{\nu}), \tag{18}$$

where ϱ is the mass density, ϕ is the potential energy per unit mass of a body force $\boldsymbol{f} = -\nabla\phi$, σ is the potential energy associated with the compressibility of the material, κ is the kinetic energy of the microstructure, taken to be a quadratic form of $\dot{\boldsymbol{\nu}}$, χ is the potential energy for the external actions exerted on $\boldsymbol{\nu}$ and W is the elastic energy related to the microstructure. In general W will be required to be frame indifferent (see Sect. 4 below). While the elastic energy might in principle also depend on higher gradients of $\boldsymbol{\nu}$, to keep the following discussion simple, we take into account only first gradients. The inclusion of higher gradients is feasible within our setting, but it would also require the introduction of higher order forces and moments in the bulk [14]. The main features of our method are already evident with an energy density of the form (18).

We write the balance of mass in the usual way, that is,

$$\dot{\varrho} + \varrho \operatorname{div}\boldsymbol{v} = 0. \tag{19}$$

Then, by the transport theorem, the time rate of $\mathcal{F}$ is

$$\begin{aligned}\dot{\mathcal{F}} = \int_{\mathcal{B}} \Big\{ &\varrho(\dot{\boldsymbol{v}} - \boldsymbol{f})\cdot\boldsymbol{v} + \varrho\,\boldsymbol{\alpha}\cdot\dot{\boldsymbol{\nu}} + \left(\varrho\frac{\partial\chi}{\partial\boldsymbol{\nu}} + \frac{\partial W}{\partial\boldsymbol{\nu}}\right)\cdot\dot{\boldsymbol{\nu}} \\ &+ \frac{\partial W}{\partial\nabla\boldsymbol{\nu}}\cdot(\nabla\boldsymbol{\nu})^{\cdot} + (W - \varrho^2\sigma')\operatorname{div}\boldsymbol{v}\Big\}\,\mathrm{d}V,\end{aligned} \tag{20}$$

where $\sigma' := \mathrm{d}\sigma/\mathrm{d}\varrho$, and where use has been made of the fact that κ is a quadratic function of $\dot{\boldsymbol{\nu}}$ in arranging the microinertia per unit mass in the form

$$\boldsymbol{\alpha} := \left(\frac{\partial\kappa}{\partial\dot{\boldsymbol{\nu}}}\right)^{\cdot} - \frac{\partial\kappa}{\partial\boldsymbol{\nu}} \tag{21}$$

(cf. (7) and also [9, p. 19]). At variance with the Lagrangian paradigm, here $\dot{\mathcal{F}}$ fails to be the total power input for the system; it must be supplemented with the *surface* power $\mathcal{W}^s$, which for a movable boundary $\partial\mathcal{B}$ takes the general form

$$\mathcal{W}^s = \int_{\partial\mathcal{B}} \{\boldsymbol{X}^s\cdot\boldsymbol{v} + \boldsymbol{\Xi}^s\cdot\dot{\boldsymbol{\nu}}\}\,\mathrm{d}A, \tag{22}$$

where $\boldsymbol{X}^s$, Ξ^s are generalized external forces associated with the velocities $\boldsymbol{v}$ and $\dot{\boldsymbol{\nu}}$, respectively. While $\boldsymbol{X}^s$ is a vector in the translation space $\mathcal{V}$, Ξ^s belongs to the cotangent space $\mathcal{T}_{\boldsymbol{\nu}}^*\mathsf{M}$ of M to $\boldsymbol{\nu}$. Often the surface power derives from a surface potential, i.e., it can be represented as

$$\mathcal{W}^s = \frac{\mathrm{d}}{\mathrm{d}t}\int_{\partial\mathcal{B}} W^s(\boldsymbol{x}, \boldsymbol{\nu})\mathrm{d}A;$$

clearly, in such a case,

$$\boldsymbol{X}^s = \frac{\partial W^s}{\partial \boldsymbol{x}} \quad \text{and} \quad \Xi^s = \frac{\partial W^s}{\partial \boldsymbol{\nu}}.$$

After some integration by parts in Eq. (20), by use of

$$(\nabla\boldsymbol{\nu})^{\cdot} = \nabla\dot{\boldsymbol{\nu}} - (\nabla\boldsymbol{\nu})\nabla\boldsymbol{v}, \tag{23}$$

the total power input can be cast in the form

$$\begin{aligned}\dot{\mathcal{F}} + \mathcal{W}^s = &\int_{\mathcal{B}} \{\boldsymbol{X}\cdot\boldsymbol{v} + \Xi\cdot\dot{\boldsymbol{\nu}}\}\,\mathrm{d}V \\ &+ \int_{\partial\mathcal{B}} \left\{(\boldsymbol{X}^b + \boldsymbol{X}^s)\cdot\boldsymbol{v} + (\Xi^b + \Xi^s)\cdot\dot{\boldsymbol{\nu}}\right\}\mathrm{d}A,\end{aligned} \tag{24}$$

where $\boldsymbol{X}$, Ξ and $\boldsymbol{X}^b$, Ξ^b are the generalized internal forces in the body and on its boundary, respectively. The forces Ξ and Ξ^b belong to $\mathcal{T}_{\boldsymbol{\nu}}^*\mathsf{M}$, while $\boldsymbol{X}$ and $\boldsymbol{X}^b$ are vectors in $\mathcal{V}$. In particular,

$$\boldsymbol{X} = \varrho(\dot{\boldsymbol{v}} - \boldsymbol{f}) - \nabla(W - \varrho^2\sigma') + \operatorname{div}\left((\nabla\boldsymbol{\nu})^{\mathrm{T}}\frac{\partial W}{\partial\nabla\boldsymbol{\nu}}\right) \tag{25}$$

$$\Xi = \varrho\left(\boldsymbol{\alpha} + \frac{\partial\chi}{\partial\boldsymbol{\nu}}\right) + \frac{\partial W}{\partial\boldsymbol{\nu}} - \operatorname{div}\frac{\partial W}{\partial\nabla\boldsymbol{\nu}} \tag{26}$$

and

$$\boldsymbol{X}^b = (W - \varrho^2\sigma')\boldsymbol{n} - \left((\nabla\boldsymbol{\nu})^{\mathrm{T}}\frac{\partial W}{\partial\nabla\boldsymbol{\nu}}\right)\boldsymbol{n} \tag{27}$$

$$\Xi^b = \frac{\partial W}{\partial\nabla\boldsymbol{\nu}}\boldsymbol{n}, \tag{28}$$

where $\boldsymbol{n}$ is the outer unit normal to $\partial\mathcal{B}$ and

$$(\nabla\boldsymbol{\nu})^{\mathrm{T}}\frac{\partial W}{\partial\nabla\boldsymbol{\nu}} \tag{29}$$

is a tensor bearing a microstructural contribution to the generalized force $\boldsymbol{X}$ conjugated to $\boldsymbol{v}$. It will become apparent in Sect. 4 how this tensor can be interpreted as an elastic stress. Henceforth a comma will denote differentiation with respect to space variables so that, for example,

$$\left((\nabla \boldsymbol{\nu})^{\mathrm{T}} \frac{\partial W}{\partial \nabla \boldsymbol{\nu}}\right)_{ij} = \nu^{\alpha}_{,i} \frac{\partial W}{\partial \nu^{\alpha}_{,j}}.$$

It should be noted that, when the velocity field $\boldsymbol{v}$ is subject to a possibly differential constraint, both $\boldsymbol{X}$ and $\boldsymbol{X}^b$ may not be uniquely determined by Eq. (24) and additional terms may arise in Eqs. (25) and (28) (see, e.g., [8]).

The special constrained variation of the total power defined in the preceding section here becomes

$$\begin{aligned} \delta(\dot{\mathcal{F}} + \mathcal{W}^s) = &\int_{\mathcal{B}} \{\boldsymbol{X} \cdot \delta \boldsymbol{v} + \Xi \cdot \delta \dot{\boldsymbol{v}}\} \, \mathrm{d}V \\ &+ \int_{\partial \mathcal{B}} \left\{(\boldsymbol{X}^b + \boldsymbol{X}^s) \cdot \delta \boldsymbol{v} + (\Xi^b + \Xi^s) \cdot \delta \dot{\boldsymbol{v}}\right\} \mathrm{d}A. \end{aligned} \tag{30}$$

3.2 Dissipation

In the present setting $\mathcal{R}$ is indeed a functional invariant under all changes of frame (16):

$$\mathcal{R} = \int_{\mathcal{B}} R \mathrm{d}V.$$

For any analytic dissipation function R, this requirement amounts to the requirement that R be the sum of invariant homogeneous terms. Thus we can assume that R is bilinear in $\mathring{\boldsymbol{\nu}}$ and $\mathbf{D}$:

$$R = R(\boldsymbol{\nu}, \mathring{\boldsymbol{\nu}}, \mathbf{D}).$$

The variation of the dissipation function here takes the form

$$\delta \mathcal{R} = \int_{\mathcal{B}} \left\{ \frac{\partial R}{\partial \dot{\boldsymbol{v}}} \cdot \delta \dot{\boldsymbol{v}} + \frac{\partial R}{\partial \nabla \boldsymbol{v}} \cdot \nabla(\delta \boldsymbol{v}) \right\} \mathrm{d}V, \tag{31}$$

since $\delta(\nabla \boldsymbol{v}) = \nabla(\delta \boldsymbol{v})$. An integration by parts then changes Eq. (31) into

$$\begin{aligned} \delta \mathcal{R} = &\int_{\mathcal{B}} \left\{ \frac{\partial R}{\partial \dot{\boldsymbol{v}}} \cdot \delta \dot{\boldsymbol{v}} - \operatorname{div}\left(\frac{\partial R}{\partial \nabla \boldsymbol{v}}\right) \cdot \delta \boldsymbol{v} \right\} \mathrm{d}V \\ &+ \int_{\partial \mathcal{B}} \left(\frac{\partial R}{\partial \nabla \boldsymbol{v}} \boldsymbol{n}\right) \cdot \delta \boldsymbol{v} \mathrm{d}A. \end{aligned} \tag{32}$$

By Eq. (17) and the chain rule,

$$\frac{\partial R}{\partial \dot{\boldsymbol{\nu}}} = \frac{\partial R}{\partial \mathring{\boldsymbol{\nu}}} \tag{33}$$

and

$$\frac{\partial R}{\partial \nabla \boldsymbol{v}} = \frac{\partial R}{\partial \mathbf{D}} + \frac{\partial R}{\partial \overset{\circ}{\boldsymbol{\nu}}} \frac{\partial \overset{\circ}{\boldsymbol{\nu}}}{\partial \nabla \boldsymbol{v}}. \tag{34}$$

It is readily seen that the tensors on the right-hand side of (34) are symmetric and skew-symmetric, respectively, and moreover

$$\left(\frac{\partial R}{\partial \overset{\circ}{\boldsymbol{\nu}}} \frac{\partial \overset{\circ}{\boldsymbol{\nu}}}{\partial \nabla \boldsymbol{v}} \right) \boldsymbol{u} = -\frac{1}{2} \left(\mathbf{A}^{\mathrm{T}} \frac{\partial R}{\partial \overset{\circ}{\boldsymbol{\nu}}} \right) \times \boldsymbol{u} \quad \forall \boldsymbol{u} \in \mathcal{V}. \tag{35}$$

3.3 Evolution equations and boundary conditions

Combining Eqs. (33) and (34) with Eqs. (30) and (32), and with the aid of (25)–(28), we arrive at the generalized form of Eq. (4) valid for a dissipative ordered fluid. Hence we obtain the evolution equations for both the macro- and micro-motions along with the associated boundary conditions:

$$\begin{cases} \varrho(\dot{\boldsymbol{v}} - \boldsymbol{f}) - \operatorname{div}\left((W - \varrho^2 \sigma')\mathbf{I} - (\nabla \boldsymbol{\nu})^{\mathrm{T}} \dfrac{\partial W}{\partial \nabla \boldsymbol{\nu}} + \dfrac{\partial R}{\partial \mathbf{D}} + \dfrac{\partial R}{\partial \overset{\circ}{\boldsymbol{\nu}}} \dfrac{\partial \overset{\circ}{\boldsymbol{\nu}}}{\partial \nabla \boldsymbol{v}} \right) = \mathbf{0} \\ \varrho\left(\boldsymbol{\alpha} + \dfrac{\partial \chi}{\partial \boldsymbol{\nu}} \right) + \dfrac{\partial W}{\partial \boldsymbol{\nu}} - \operatorname{div} \dfrac{\partial W}{\partial \nabla \boldsymbol{\nu}} + \dfrac{\partial R}{\partial \overset{\circ}{\boldsymbol{\nu}}} = 0 \end{cases} \tag{36}$$

in $\mathcal{B}$, and

$$\begin{cases} \left((W - \varrho^2 \sigma')\mathbf{I} - (\nabla \boldsymbol{\nu})^{\mathrm{T}} \dfrac{\partial W}{\partial \nabla \boldsymbol{\nu}} + \dfrac{\partial R}{\partial \mathbf{D}} + \dfrac{\partial R}{\partial \overset{\circ}{\boldsymbol{\nu}}} \dfrac{\partial \overset{\circ}{\boldsymbol{\nu}}}{\partial \nabla \boldsymbol{v}} \right) \boldsymbol{n} + \boldsymbol{X}^s = \mathbf{0} \\ \dfrac{\partial W}{\partial \nabla \boldsymbol{\nu}} \boldsymbol{n} + \Xi^s = 0 \end{cases} \tag{37}$$

on $\partial\mathcal{B}$, where $\mathbf{I}$ is the identity tensor. In particular, the boundary Eqs. (37) deserve a comment. When the region $\mathcal{B}$ is not free to change in time, all admissible $\delta \boldsymbol{v}$ vanish on $\partial\mathcal{B}$, and so Eq. $(37)_1$ would not properly follow from our reasoning; however, it can still be regarded as valid, provided that $\boldsymbol{X}^s$ is interpreted as a *reactive* generalized force exerted by the boundary. A similar conclusion applies to Eq. $(37)_2$ and the generalized force Ξ^s, when $\boldsymbol{\nu}$ is prescribed on $\partial\mathcal{B}$.

Finally, it should be noted that, to arrive at Eqs. (36) and (37), we have set equal to unity the Lagrange multiplier which here plays the rôle of λ in Sect. 2. That this is indeed justified follows from Euler's theorem on homogeneous functions, by which

$$\frac{\partial R}{\partial \overset{\circ}{\boldsymbol{\nu}}} \cdot \overset{\circ}{\boldsymbol{\nu}} + \frac{\partial R}{\partial \mathbf{D}} \cdot \mathbf{D} = 2R, \tag{38}$$

and the linearity of the relation between the pairs $(\overset{\circ}{\boldsymbol{\nu}}, \mathbf{D})$ and $(\dot{\boldsymbol{\nu}}, \nabla \boldsymbol{v})$, which together with Eq. (38) ensures that

$$\frac{\partial R}{\partial \dot{\boldsymbol{\nu}}} \cdot \dot{\boldsymbol{\nu}} + \frac{\partial R}{\partial \nabla \boldsymbol{v}} \cdot \nabla \boldsymbol{v} = 2R.$$

4 Balance equations

Equations (36) and (37) are the basic equations of this theory. Classically this rôle is played by the balance equations for linear and angular momenta. Though in the present setting it would be illusory to derive from these balances the evolution of microstructures described by a manifold M with dimension $m > 3$, it remains crucial to ascertain that the evolution predicted by this theory does not violate the classical balance equations. Here we show how these equations can indeed be recognized as valid.

The balance of linear momentum requires that

$$\frac{\mathrm{d}}{\mathrm{d}t}\int_C \varrho \boldsymbol{v}\,\mathrm{d}V = \int_C \varrho \boldsymbol{f}\,\mathrm{d}V + \int_{\partial C} \boldsymbol{t}\,\mathrm{d}A \tag{39}$$

for any sub-body C, where $\boldsymbol{f}$ is the body force per unit mass and $\boldsymbol{t}$ is the contact force per unit area. For unstructured continua, only the moments of the same forces entering Eq. (39) contribute to the balance of angular momentum. When, however, the body possesses a microstructure somehow related to an internal rotation of the material element, additional couples in the form of body and surface torques must be accounted for. The general form of the balance of angular momentum is then

$$\frac{\mathrm{d}}{\mathrm{d}t}\int_C \varrho(\boldsymbol{m} + \boldsymbol{x} \times \boldsymbol{v})\,\mathrm{d}V = \int_C \varrho(\boldsymbol{x} \times \boldsymbol{f} + \boldsymbol{k})\,\mathrm{d}V + \int_{\partial C} (\boldsymbol{x} \times \boldsymbol{t} + \boldsymbol{l})\,\mathrm{d}A,$$

where $\varrho\boldsymbol{m}$ is the intrinsic angular momentum, $\boldsymbol{k}$ the body moment per unit mass, and $\boldsymbol{l}$ the surface contact moment per unit area.

When $\boldsymbol{t}$ and $\boldsymbol{l}$ at a given point are assumed to depend only on the local surface normal $\boldsymbol{n}$, by Cauchy's tetrahedron argument one can show that they can be expressed in terms of a stress tensor $\mathbf{T}$ and a couple stress tensor $\mathbf{L}$ as

$$\boldsymbol{t} = \mathbf{T}\boldsymbol{n}, \quad \boldsymbol{l} = \mathbf{L}\boldsymbol{n}. \tag{40}$$

Also, by use of the conservation of mass (19), it is possible to write the classical balance equations in point form as

$$\varrho\dot{\boldsymbol{v}} = \varrho\boldsymbol{f} + \operatorname{div}\mathbf{T} \tag{41}$$

and

$$\varrho\dot{\boldsymbol{m}} = \varrho\boldsymbol{k} + 2\boldsymbol{s} + \operatorname{div}\mathbf{L}, \tag{42}$$

where $\boldsymbol{s}$ is the axial vector associated with the skew symmetric part of $\mathbf{T}$. In the absence of internal rotational degrees of freedom $\boldsymbol{m}$, $\boldsymbol{k}$ and $\mathbf{L}$ vanish and Eq. (42) reduces to the usual requirement that the stress tensor be symmetric.

One readily sees that the equation of motion $(36)_1$ takes the form (41) when the stress tensor is set equal to

$$\mathbf{T} = (W - \varrho^2\sigma')\mathbf{I} - (\nabla\boldsymbol{\nu})^{\mathsf{T}}\frac{\partial W}{\partial\nabla\boldsymbol{\nu}} + \frac{\partial R}{\partial\mathbf{D}} + \frac{\partial R}{\partial\,\mathring{\boldsymbol{\nu}}}\frac{\partial\,\mathring{\boldsymbol{\nu}}}{\partial\nabla\boldsymbol{v}}. \tag{43}$$

It further follows from Eq. $(37)_1$ that the traction on $\partial\mathcal{B}$ is balanced by the surface force $\boldsymbol{X}^s$. In (43) we can clearly identify both elastic and viscous components of $\mathbf{T}$ and write $\mathbf{T} = \mathbf{T}^{(e)} + \mathbf{T}^{(v)}$ accordingly, where

$$\mathbf{T}^{(v)} := \frac{\partial R}{\partial \mathbf{D}} + \frac{\partial R}{\partial \overset{\circ}{\boldsymbol{\nu}}} \frac{\partial \overset{\circ}{\boldsymbol{\nu}}}{\partial \nabla \boldsymbol{v}}. \tag{44}$$

Correspondingly, the vector $\boldsymbol{s}$ can be written as $\boldsymbol{s} = \boldsymbol{s}^{(e)} + \boldsymbol{s}^{(v)}$, where by (35)

$$\boldsymbol{s}^{(v)} = -\frac{1}{2}\mathbf{A}^{\mathrm{T}} \frac{\partial R}{\partial \overset{\circ}{\boldsymbol{\nu}}}. \tag{45}$$

To identify within this theory the appropriate expression for the couple stress $\mathbf{L}$, we find it convenient to digress slightly and first compare our theory with that of Capriz.

4.1 Capriz's theory

In his tract [9] Capriz proposed a general theory for continua with microstructure which encompasses various special models. Here we show that our theory for dissipative fluids is compatible with that of Capriz. Moreover, our way of writing the balance of angular momentum in (42) illuminates an equivalent way of reading this balance within Capriz's theory.

According to this theory, the equations that govern the dynamical evolution of a body with microstructure are the balance of linear macromomentum, written in the local form (41), and the balance of *micromomentum*, which by analogy with Lagrangian dynamics is posited in § 8 of [9] in the form

$$\varrho\boldsymbol{\alpha} = \varrho\boldsymbol{\beta} - \boldsymbol{\zeta} + \operatorname{div}\mathbf{S}, \tag{46}$$

where $\boldsymbol{\alpha}$ is as in (21) when the microkinetic energy is quadratic in $\dot{\boldsymbol{\nu}}$, $\boldsymbol{\beta}$ is the density per unit mass of the external microstructural actions, $\boldsymbol{\zeta}$ is the density per unit volume of the internal microstructural actions and $\mathbf{S}$ is the *microstress*. For a given microstate $\boldsymbol{\nu} \in \mathsf{M}$, $\boldsymbol{\beta}$ and $\boldsymbol{\zeta}$ are elements of the cotangent space $\mathcal{T}^*_{\boldsymbol{\nu}}\mathsf{M}$, while $\mathbf{S}$ is a linear operator of $\mathcal{V}$ into $\mathcal{T}^*_{\boldsymbol{\nu}}\mathsf{M}$. In accordance with (40)

$$\boldsymbol{\sigma} = \mathbf{S}\boldsymbol{n}$$

represents the internal contact actions on the microstructure.

Capriz expresses the balance of angular momentum through an invariance requirement: he derives the balance of energy from the balances of macro- and micromomenta and requires the total power of the internal actions to vanish identically on all possible rigid motions. Specifically, this amounts to the following restriction on the constitutive laws for $\mathbf{T}$, $\boldsymbol{\zeta}$, and $\mathbf{S}$:

$$2\boldsymbol{s} + \mathbf{A}^{\mathrm{T}}\boldsymbol{\zeta} + (\nabla\mathbf{A}^{\mathrm{T}})\mathbf{S} = \mathbf{0}, \tag{47}$$

which, in components, reads as

$$2s_i + A_i^\alpha \zeta_\alpha + \frac{\partial A_i^\alpha}{\partial \nu^\beta} \nu^\beta_{,j} S_{\alpha j} = 0.$$

Also in view of $(37)_2$, Eq. $(36)_2$ can be given the form (46), provided we set

$$\boldsymbol{\beta} = -\frac{\partial \chi}{\partial \boldsymbol{\nu}}, \quad \boldsymbol{\zeta} = \frac{\partial W}{\partial \boldsymbol{\nu}} + \frac{\partial R}{\partial \overset{\circ}{\boldsymbol{\nu}}} \quad \text{and} \quad \mathbf{S} = \frac{\partial W}{\partial \nabla \boldsymbol{\nu}}. \tag{48}$$

As for the torque $\boldsymbol{s}$ and the stress $\mathbf{T}$, we can easily identify both elastic and viscous components for the forces and stress in (48): they are indeed all purely elastic except for $\boldsymbol{\zeta}$, for which

$$\boldsymbol{\zeta}^{(\mathrm{v})} = \frac{\partial R}{\partial \overset{\circ}{\boldsymbol{\nu}}}. \tag{49}$$

It was already shown in [9, §13] that, for perfect fluids, the balance of angular momentum in the form (47) is identically satisfied by the elastic components of $\boldsymbol{s}$, $\boldsymbol{\zeta}$ and $\mathbf{S}$. That (47) is also satisfied by the viscous components of these fields readily follows from (45) and (49).

Within Capriz's theory the balance of angular momentum can equivalently be given the form (42). By requiring κ to obey (15) and using (46) in (47) to eliminate $\boldsymbol{\zeta}$, one arrives at

$$\varrho \left(\mathbf{A}^{\mathrm{T}} \frac{\partial \kappa}{\partial \dot{\boldsymbol{\nu}}} \right)^{\cdot} = \varrho \mathbf{A}^{\mathrm{T}} \boldsymbol{\beta} + 2\boldsymbol{s} + \operatorname{div}(\mathbf{A}^{\mathrm{T}} \mathbf{S})$$

(cf. [9, (9.11)]), which reduces to (42) upon setting

$$\boldsymbol{m} = \mathbf{A}^{\mathrm{T}} \frac{\partial \kappa}{\partial \overset{\circ}{\boldsymbol{\nu}}}, \quad \boldsymbol{k} = \mathbf{A}^{\mathrm{T}} \boldsymbol{\beta}, \quad \text{and} \quad \mathbf{L} = \mathbf{A}^{\mathrm{T}} \mathbf{S}. \tag{50}$$

In particular, it follows from $(48)_3$ and $(50)_3$ that the couple stress $\mathbf{L}$ is purely elastic. The absence of any viscous contribution to the couple stress is a consequence of $\nabla \dot{\boldsymbol{\nu}}$ being excluded from the dissipation (cf. [7] for an analogous result in a restricted setting).

We have already shown in [8] that the classical dynamical theory for uniaxial liquid crystals put forward by Ericksen and Leslie in [1] and [2] is indeed a special case of our theory for dissipative fluids with general tensorial order. The classical expression for the couple stress tensor is

$$L_{ij} = \varepsilon_{ihk} d_h \frac{\partial W}{\partial d_{k,j}},$$

where ε_{ihk} are the components of Ricci's alternator and d_h those of the nematic director. This is indeed an immediate consequence of $(50)_3$ and $(48)_3$, once the operator $\mathbf{A}$ for this microstructure is recognized as the skew-symmetric tensor with components

$$A_{ij} = \varepsilon_{ijk} d_k$$

(cf. [9, (19.3)]).

References

[1] Ericksen, J.L. (1961): Conservation laws for liquid crystals. Trans. Soc. Rheology **5**, 23–34
[2] Leslie, F.M. (1968): Some constitutive equations for liquid crystals. Arch. Rational Mech. Anal. **28**, 265–283
[3] de Gennes, P.G., Prost, J. (1993): The physics of liquid crystals. Second ed. Clarendon Press, Oxford
[4] Hess, S. (1975): Irreversible thermodynamics of non-equilibrium alignment phenomena in molecular liquids and in liquid crystals. I; II. Z. Naturforsch. **30a**, 728–738; 1224–1232
[5] Sonnet, A.M., Maffettone, P.L., Virga, E.G. (2002): Continuum theory for nematic liquid crystals with tensorial order. J. Non-Newton. Fluid Mech. To appear
[6] Ericksen, J.L. (1990): Liquid crystals with variable degree of orientation. Arch. Rational Mech. Anal. **113**, 97–120
[7] Leslie, F.M. (1992): Continuum theory for nematic liquid crystals. Contin. Mech. Thermodyn. **4**, 167–175
[8] Sonnet, A.M., Virga, E.G. (2001): Dynamics of dissipative ordered fluids. Phys. Rev. E **64**, 031705
[9] Capriz, G. (1989): Continua with microstructure. Springer, New York
[10] Strutt, J.W.S. (Lord Rayleigh) (1873): Some general theorems relating to vibrations. Proc. London Math. Soc. **4**, 357–368
[11] Whittaker, E.T. (1937): A treatise on the analytical dynamics of particles and rigid bodies. Fourth ed. Cambridge University Press, Cambridge
[12] Biot, M.A. (1955): Variational principles in irreversible thermodynamics with application to viscoelasticity. Phys. Rev. (2) **97**, 1463–1469
[13] Lavenda, B.H. (1978): Thermodynamics of irreversible processes. Macmillan, London
[14] Green, A.E., Rivlin, R.S. (1964): Multipolar continuum mechanics. Arch. Rational Mech. Anal. **17**, 113–147

Some Questions on Material Objectivity Arising in Models of Porous Materials

Krzysztof Wilmanski

Abstract. The paper is devoted to the analysis of nonobjective contributions to continuous models of porous materials. Explicit contributions appearing in momentum balance equations in noninertial reference systems influence Darcy's law considered as a limit case of a multicomponent model, and render it nonobjective. On the other hand an implicit constitutive contribution of relative accelerations to partial momentum balance equations of multicomponent models violates the principle of material objectivity. These two types of contribution bear physically different significances. The former seems to be a natural consequence of body forces on relative motions of components. The latter is incompatible with principles of a macroscopical continuous model. Even if we ignore the principle of material objectivity, contributions of relative accelerations are either indistinguishable from other contributions or yield unacceptable modes of propagation of sound waves or both. Consequently, in contrast to the nonobjectivity of Darcy's law, we conclude that such contributions should be ignored completely in the construction of macroscopical models of porous materials.

1 Introduction

In his classical work on two-component macroscopical models of porous materials M.A. Biot [1] considered interactions of components described by a relative acceleration (see formula (3.21)). Such contributions are still being used in various applications of Biot's model, e.g., in order to account for a tortuosity.

We argue in this note that such models violate the principle of material objectivity under rather natural assumptions concerning transformation properties of a continuous model. In addition they do not seem to reflect any essential microscopical properties which should be transferred to the macroscopical level of description. In contrast to theories of rarified gases (e.g. [2, p. 147]), nonobjective macroscopical contributions in theories of porous materials seem to be some orders of magnitude smaller than many other effects not appearing at all on the macroscopical level of description or appearing solely in a very crude manner. A typical example is an influence of microscopical vorticities in flows of a fluid component which yields an essential contribution to a form of macroscopical boundary condition on permeable boundaries (seepage conditions), and should be incorporated by an additional macroscopical field of – say – tortuosity.

It is known that interactions between a flow of fluid and an obstacle are influenced much more strongly by vorticities than by time dependent relative velocities (compare the classical problem of Euler–d'Alembert paradox). The latter lead in classical fluid dynamics to an explicit solution, and to the notion of added mass, and consequently, to speculations on a macroscopical dependence on relative accelerations in theories of porous materials. Such an argument does not hold water, and not only due to the fact that vorticities, and similar effects, are ignored. In addition, in the process of averaging microscopical properties to obtain a macroscopical model of porous materials, microscopical interactions through contact surfaces are replaced by an extension of the set of constitutive variables, and this does not require explicit solutions of any microscopical boundary value problems which are never available anyway.

In Sect. 2 of this note we present some classical transformation properties of continuum mechanics related to the form of momentum balance equations in noninertial frames of reference. In Sect. 3 we discuss the form of momentum sources under the assumption of material objectivity. We show that in such a model a contribution of relative accelerations cannot appear. Section 4 contains an example of a model in which the dependence on relative accelerations is incorporated in a way similar to that of Biot. By means of a solution of a simple steady state flow we show that the influence of this contribution is so small that it cannot be observed in any experiment on a porous material which can be described by a classical continuum model. In Sect. 5 we present propagation conditions for sound waves within such a model. It is shown that corrections due to relative accelerations cannot be distinguished from those of constitutive relations for partial stresses, and, in addition, the transversal wave contains a contribution of a fluid component which is assumed to be ideal. This does not seem to be plausible, and it is, certainly, not observable.

The general conclusion of these considerations is that contributions of relative accelerations in continuum models of porous materials should not appear at all.

2 Transformation properties of momentum balance

We check invariance properties of partial momentum balance equations under Euclidean transformations. Such invariance follows in classical mechanics from the assumption that the space of configuration is isometric, and the time space is homogeneous (see, e.g., [3]). The latter is of no interest for our purposes. Such transformations are of the form

$$\mathbf{x}^* = \mathbf{O}(t)\,\mathbf{x} + \mathbf{d}(t)\,, \quad \mathbf{O}^T = \mathbf{O}^{-1}, \quad \mathbf{x}^*, \mathbf{x} \in \Re^3, \quad \mathbf{O} \in \mathrm{Orth}, \quad \mathbf{d} \in V^3, \tag{2.1}$$

where $\mathbf{x}^*$, $\mathbf{x}$ denote points of the space of configuration, t is time, Orth is the group of orthogonal transformations, and V^3 denotes the space of 3D vectors.

A scalar φ, a vector ϖ, a second order tensor $\mathbf{\Upsilon}$ are called *objective* if they satisfy the following transformation rules

$$\forall \mathbf{O} \in \mathrm{Orth}: \quad \varphi^* = \varphi, \quad \varpi^* = \mathbf{O}\varpi\,, \quad \mathbf{\Upsilon}^* = \mathbf{O}\mathbf{\Upsilon}\mathbf{O}^T. \tag{2.2}$$

It is assumed that the mass densities of continuum mechanics are objective scalars, and that contact forces are objective vectors. The latter together with the transformation rule for unit vectors perpendicular to material surfaces yields the objectivity of Cauchy stress tensors.

On the other hand kinematic quantities of continuum mechanics are not objective. In the case of a single component body B we have in the so-called Lagrangian description:

$$\mathbf{x} = \mathbf{f}(\mathbf{X}, t) \Longrightarrow \mathbf{f}^*(\mathbf{X}, t) = \mathbf{O}\mathbf{f}(\mathbf{X}, t) + \mathbf{d},$$

$$\frac{\partial \mathbf{f}^*}{\partial t}(X, t) = \mathbf{O}\frac{\partial \mathbf{f}}{\partial t}(\mathbf{X}, t) + \dot{\mathbf{O}} f(\mathbf{X}, t) + \dot{\mathbf{d}}, \quad \dot{\mathbf{O}} := \frac{d\mathbf{O}}{dt}, \quad \dot{\mathbf{d}} := \frac{d\mathbf{d}}{dt},$$

$$\frac{\partial^2 \mathbf{f}^*}{\partial t^2}(\mathbf{X}, t) = \mathbf{O}\frac{\partial^2 \mathbf{f}}{\partial t^2}(\mathbf{X}, t) + 2\dot{\mathbf{O}}\frac{\partial \mathbf{f}}{\partial t}(\mathbf{X}, t) + \ddot{\mathbf{O}} f(\mathbf{X}, t) + \ddot{\mathbf{d}},$$

$$\ddot{\mathbf{O}} := \frac{d^2\mathbf{O}}{dt^2}, \quad \ddot{\mathbf{d}} := \frac{d^2\mathbf{d}}{dt^2},$$

$$\operatorname{Grad} \mathbf{f}^*(\mathbf{X}, t) = \mathbf{O} \operatorname{Grad} \mathbf{f}(\mathbf{X}, t),$$

where $\mathbf{X}$ denotes a material point of the continuous body B and $\mathbf{f}$ is a function of motion (local configuration) in the Lagrangian description of motion of the body B. The latter satisfies the usual smoothness assumptions. After transformation to the Eulerian description we obtain from the above relations the following transformation rules for the velocity field $\mathbf{v}$, the acceleration field $\mathbf{a}$, and for the right Cauchy–Green deformation tensor $\mathbf{C} := (\operatorname{Grad} \mathbf{f})^T (\operatorname{Grad} \mathbf{f})$:

$$\begin{aligned} \mathbf{v}^*\left(\mathbf{x}^*, t\right) &= \mathbf{O}v(\mathbf{x}, t) + \dot{\mathbf{O}}x + \dot{\mathbf{d}}, \quad x^* = \mathbf{O}\mathbf{x} + \mathbf{d}, \\ \mathbf{a}^*\left(\mathbf{x}^*, t\right) &= \mathbf{O}a(\mathbf{x}, t) + 2\dot{\mathbf{O}}v(\mathbf{x}, t) + \ddot{\mathbf{O}}x + \ddot{\mathbf{d}}, \quad \mathbf{C}^* = \mathbf{C}. \end{aligned} \tag{2.3}$$

Hence the deformation tensor $\mathbf{C}$ is not objective, but its six independent components behave as if they were six objective scalars in relation to transformations in the space of configurations. The velocity $\mathbf{v}$ and the acceleration $\mathbf{a}$ are not objective as well but their transformation rules are more complicated because their components are directly related to reference systems in the space of configuration.

We proceed to investigate partial balance equations of a two-component porous body. We rely on the Eulerian description, i.e., fields are functions of the space point $\mathbf{x}$ and time t.

We assume that there is no mass exchange between components:

$$\frac{\partial \rho^F}{\partial t} + \operatorname{div}\left(\rho^F \mathbf{v}^F\right) = 0, \quad \frac{\partial \rho^S}{\partial t} + \operatorname{div}\left(\rho^S \mathbf{v}^S\right) = 0, \tag{2.4}$$

where ρ^F, ρ^S denote the partial mass densities and $\mathbf{v}^F$, $\mathbf{v}^S$ are velocities of components. This assumption solely simplifies arguments, and can be ignored, if needed, without much additional effort.

For the fluid component, and for the skeleton, respectively, the momentum balance equations have the following form in an inertial reference system:

$$\rho^F \mathbf{a}^F = \operatorname{div} \mathbf{T}^F + \hat{\mathbf{p}}^F + \rho^F \mathbf{b}^F, \tag{2.5}$$

$$\rho^S \mathbf{a}^S = \operatorname{div} \mathbf{T}^S + \hat{\mathbf{p}}^S + \rho^S \mathbf{b}^S, \qquad \hat{\mathbf{p}}^F + \hat{\mathbf{p}}^S = 0, \tag{2.6}$$

where $\mathbf{T}^F, \mathbf{T}^S$ are the partial Cauchy stress tensors, $\hat{\mathbf{p}}^F, \hat{\mathbf{p}}^S$ are the momentum sources, and $\mathbf{b}^F, \mathbf{b}^S$ denote partial body forces. After the transformation to a noninertial $*$-system the above balance equations have the form:

$$\begin{aligned} &\rho^F \left(\mathbf{O}a^F + 2\dot{\mathbf{O}}v^F + \ddot{\mathbf{O}}x + \ddot{\mathbf{d}} \right) = \mathbf{O} \left(\operatorname{div} \mathbf{T}^F \right) + \hat{\mathbf{p}}^{F*} + \rho^F \mathbf{b}^{F*}, \\ &\rho^S \left(\mathbf{O}\mathbf{a}^S + 2\dot{\mathbf{O}}\mathbf{v}^S + \ddot{\mathbf{O}}\mathbf{x} + \ddot{\mathbf{d}} \right) = \mathbf{O} \left(\operatorname{div} \mathbf{T}^S \right) + \hat{\mathbf{p}}^{S*} + \rho^S \mathbf{b}^{S*}, \\ &\hat{\mathbf{p}}^{F*} + \hat{\mathbf{p}}^{S*} = 0, \end{aligned} \tag{2.7}$$

where the following relations have been used:

$$\operatorname{div} {}^*\mathbf{T}^{F*} = \mathbf{O} \left(\operatorname{div} \mathbf{T}^F \right), \quad \operatorname{grad}{}^* (\cdots) = \mathbf{O}^T \operatorname{grad} (\cdots), \tag{2.8}$$

and similarily for the skeleton.

Bearing these relations in mind we obtain the following identities:

$$\begin{aligned} &\rho^F \left[\mathbf{b}^{F*} - \mathbf{i}^{F*} - \mathbf{O}\mathbf{b}^F \right] = - \left(\hat{\mathbf{p}}^{F*} - \mathbf{O}\hat{\mathbf{p}}^F \right), \\ &\rho^S \left[\mathbf{b}^{S*} - \mathbf{i}^{S*} - \mathbf{O}\mathbf{b}^S \right] = \hat{\mathbf{p}}^{F*} - \mathbf{O}\hat{\mathbf{p}}^F, \\ &\mathbf{i}^{F*} := 2\mathbf{W} \left(\mathbf{v}^{F*} - \dot{\mathbf{d}} \right) - \mathbf{W}^2 \left(\mathbf{x}^* - \mathbf{d} \right) + \dot{\mathbf{W}} \left(\mathbf{x}^* - \mathbf{d} \right) + \ddot{\mathbf{d}}, \\ &\mathbf{i}^{S*} := 2\mathbf{W} \left(\mathbf{v}^{S*} - \dot{\mathbf{d}} \right) - \mathbf{W}^2 \left(\mathbf{x}^* - \mathbf{d} \right) + \dot{\mathbf{W}} \left(\mathbf{x}^* - \mathbf{d} \right) + \ddot{\mathbf{d}}, \quad \mathbf{W} := \dot{O}\mathbf{O}^T, \end{aligned} \tag{2.9}$$

where the contributions on the right-hand side of $(2.9)_{3,4}$ correspond to the Coriolis acceleration, the centrifugal acceleration, the Euler acceleration and the relative translational acceleration in both components, respectively. Here $\mathbf{W}$ denotes the skew-symmetric tensor of angular velocities of two reference systems.

In the mechanics of single component systems the right-hand side of the relations $(2.9)_{1,2}$ is identically zero (conservation of momentum!). If we make a similar assumption for the system of two components we obtain the rules of transformation for body forces:

$$\mathbf{b}^{F*} - \mathbf{i}^{F*} = \mathbf{O}\mathbf{b}^F, \quad \mathbf{b}^{S*} - \mathbf{i}^{S*} = \mathbf{O}\mathbf{b}^S. \tag{2.10}$$

i.e., the combinations on the left-hand sides (the so-called apparent body forces) are objective. However, in the mechanics of multicomponent systems there is no argument based on a conservation law which would eliminate contributions from sources. The main aim of this work is to show that such contributions are not plausible. This means that we claim that the momentum sources are objective vectors, i.e.,

$$\hat{\mathbf{p}}^{F*} = \mathbf{O}\hat{\mathbf{p}}^F. \tag{2.11}$$

In the next section we show simple consequences of this assumption. In Sect. 4 we present an example of a model in which the relation (2.11) is not assumed to hold. Such is the case in the Biot model [1] in which the momentum sources depend on the relative acceleration of components.

3 Objective sources of momentum

We consider a simple case of the poroelastic material undergoing isothermal processes. A larger class of materials can be considered in a similar manner but calculations are more tedious. In a chosen inertial frame constitutive laws for momentum sources are assumed to have the following form

$$\hat{\mathbf{p}}^F \equiv -\hat{\mathbf{p}}^S = \mathfrak{p}\left(\rho^F, \mathbf{C}^S, \mathbf{w}, \mathbf{a}\right), \tag{3.1}$$

where

$$\mathbf{w} := \mathbf{v}^F - \mathbf{v}^S, \quad \mathbf{a} := \mathbf{a}^F - \mathbf{a}^S, \tag{3.2}$$

and $\mathbf{C}^S$ is the right Cauchy–Green deformation tensor of the skeleton.

We assume *material objectivity*, i.e., in any other frame obtained by an orthogonal transformation the constitutive relation must have the form

$$\hat{\mathbf{p}}^{F*} = \mathfrak{p}\left(\rho^{F*}, \mathbf{C}^{S*}, \mathbf{w}^*, \mathbf{a}^*\right), \tag{3.3}$$

with the transformation rules

$$\mathbf{C}^{S*} = \mathbf{C}^S, \quad \mathbf{w}^* = \mathbf{O}\mathbf{w}, \quad \mathbf{a}^* = \mathbf{O}\mathbf{a} + 2\dot{\mathbf{O}}\mathbf{w}. \tag{3.4}$$

The last two follow easily from (2.3) and definitions (3.2). Further, substitution in (2.11) yields

$$\forall \mathbf{O} \in \mathrm{Orth}: \quad \mathbf{O}\mathfrak{p}\left(\rho^F, \mathbf{C}^S, \mathbf{w}, \mathbf{a}\right) = \mathfrak{p}\left(\rho^F, \mathbf{C}^S, \mathbf{O}\mathbf{w}, \mathbf{O}\mathbf{a} + 2\dot{\mathbf{O}}\mathbf{w}\right). \tag{3.5}$$

We choose a particular instant of time in which $\mathbf{O} = 1$, and $\dot{\mathbf{O}}$ is arbitrary. Then

$$\mathfrak{p}\left(\rho^F, \mathbf{C}^S, \mathbf{w}, \mathbf{a}\right) = \mathfrak{p}\left(\rho^F, \mathbf{C}^S, \mathbf{w}, \mathbf{a} + 2\dot{\mathbf{O}}\mathbf{w}\right). \tag{3.6}$$

Certainly, this relation can hold only if the source is independent of $\mathbf{a}$. Then the relation (3.5) reduces to the following condition

$$\forall \mathbf{O} \in \mathrm{Orth}: \quad \mathbf{O}\mathfrak{p}\left(\rho^F, \mathbf{C}^S, \mathbf{w}\right) = \mathfrak{p}\left(\rho^F, \mathbf{C}^S, \mathbf{O}\mathbf{w}\right), \tag{3.7}$$

which means that the function $\mathfrak{p}$ should be isotropic with respect to the relative velocity $\mathbf{w}$. Consequently,

$$\hat{\mathbf{p}}^F = -\pi\left(\rho^F, \mathbf{C}^S, |\mathbf{w}|\right)\mathbf{w}, \tag{3.8}$$

where the minus sign has been introduced for historical reasons, and π is a scalar function not limited any further by objectivity arguments.

4 Linear dependence on the relative acceleration

We now ignore the assumption (2.11) and consider a constitutive law for the momentum source which, in the inertial frame of reference, is linear and isotropic with respect to the relative velocity $\mathbf{w}$ and the relative acceleration $\mathbf{a}$, i.e.,

$$\hat{\mathbf{p}}^F = -\pi \mathbf{w} + \mathfrak{b}\mathbf{a}, \tag{4.1}$$

where the coefficients π and $\mathfrak{b}$ for poroelastic materials may be functions of ρ^F and $\mathbf{C}^S$. In an arbitrary *-frame we have

$$\hat{\mathbf{p}}^{F*} = -\pi \mathbf{w}^* + \mathfrak{b}\mathbf{a}^* = \mathbf{O}\hat{\mathbf{p}}^F + 2\mathfrak{b}\dot{\mathbf{O}}\mathbf{w}, \tag{4.2}$$

because the mass density ρ^F and the deformation tensor $\mathbf{C}^S$ do not change under this transformation and, consequently, the coefficients π and $\mathfrak{b}$ remain unchanged as well. According to the relations (2.9) the transformation of body forces yields in this case the following relations:

$$\rho^F \left[\mathbf{b}^{F*} - \mathbf{i}^{F*} - \mathbf{O}b^F\right] = -2\mathfrak{b}\dot{\mathbf{O}}\mathbf{w},$$
$$\rho^S \left[\mathbf{b}^{S*} - \mathbf{i}^{S*} - \mathbf{O}\mathbf{b}^S\right] = 2\mathfrak{b}\dot{\mathbf{O}}\mathbf{w}.$$

In the explicit form, we have:

$$\mathbf{b}^{F*} - 2\mathbf{W}\left(\mathbf{v}^{F*} - \dot{\mathbf{d}} - \frac{\mathfrak{b}}{\rho^F}\mathbf{w}^*\right) - \mathbf{W}^2\left(\mathbf{x}^* - \mathbf{d}\right) + \dot{\mathbf{W}}\left(\mathbf{x}^* - \mathbf{d}\right) + \ddot{\mathbf{d}} = \mathbf{O}\mathbf{b}^F,$$
$$\mathbf{b}^{S*} - 2\mathbf{W}\left(\mathbf{v}^{S*} - \dot{\mathbf{d}} + \frac{\mathfrak{b}}{\rho^F}\mathbf{w}^*\right) - \mathbf{W}^2\left(\mathbf{x}^* - \mathbf{d}\right) + \dot{\mathbf{W}}\left(\mathbf{x}^* - \mathbf{d}\right) + \ddot{\mathbf{d}} = \mathbf{O}\mathbf{b}^S, \tag{4.3}$$

where the second relation follows easily from condition (2.6).

Hence, in contrast to the classical nonobjective contributions to body forces, the constitutive relation (4.1) leads to body forces in noninertial systems which depend on *material properties* of the system. This does not seem to be very plausible. However such a contribution can be relatively easily verified experimentally if it is at all essential.

In order to discuss this point in some detail we consider a simple example of stationary flow through a cylinder. The geometry of the system is shown in Fig. 1. It is assumed that a fluid flows into the porous material at the surface $r = r_i$, and it flows out at the surface $r = r_e$. Deformations of the skeleton are assumed to be small which means, approximately, that the radii of the cylinder do not change. The cylinder rotates with a constant angular velocity ω as shown in Fig. 1.

Under the assumption of a constant porosity, and a constant mass density of the fluid ρ^F (this corresponds to the incompressibility of a real fluid), the problem is described by the following fields:

$$\left\{p^F, v_r^F, v_\phi^F, u_r^S, u_\phi^S\right\}, \tag{4.4}$$

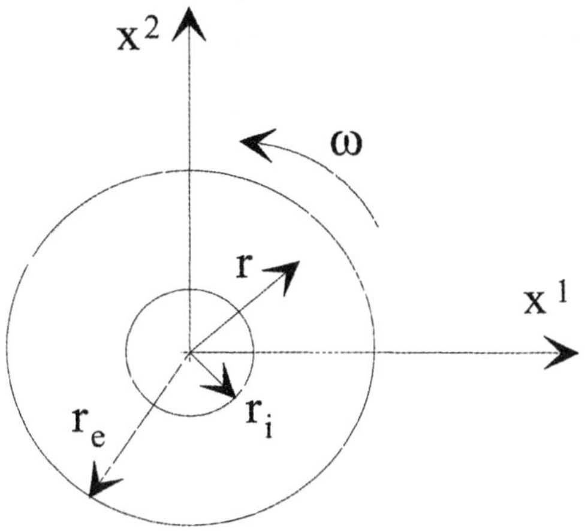

Fig. 1. Geometry of the flow through the cylinder

where p^F denotes the partial pressure of the fluid, v_r^F, v_ϕ^F are the radial and circumferential physical components of the velocity of the fluid, respectively, and u_r^S, u_ϕ^S describe the physical components of displacement of the skeleton. We use the noninertial cylindrical frame of reference rotating together with the cylinder with angular velocity ω. To simplify the notation we leave out the asterisk indicating the noninertial character of the reference system. All the above listed fields are functions of the radius r alone. Consequently the velocity of the skeleton $\mathbf{v}^S$ and the acceleration $\mathbf{a}^S$ are identically zero, and solely convective contributions to the acceleration $\mathbf{a}^F$ remain.

The field equations in the noninertial frame have the form:

– mass conservation of the fluid component

$$\frac{d\left(rv_r^F\right)}{dr} = 0, \tag{4.5}$$

– momentum balance for the fluid component

$$\begin{aligned} &\left(\rho^F - \mathfrak{b}\right)\left(\frac{1}{2}\frac{dv_r^{F2}}{dr} - \frac{v_\phi^{F2}}{r}\right) = -\frac{dp^F}{dr} - \pi v_r^F + 2\left(\rho^F - \mathfrak{b}\right)\omega v_\phi^F - \omega^2 r, \\ &\left(\rho^F - \mathfrak{b}\right) v_r^F \frac{1}{r}\frac{d}{dr}\left(rv_\phi^F\right) + \pi v_\phi^F = -2\omega v_r^F, \end{aligned} \tag{4.6}$$

and the momentum balance equation for the skeleton which is immaterial for our present argument.

We can easily construct solutions for components of the partial velocity of the fluid. From Eq. (4.5) we obtain

$$v_r^F = \frac{C}{r}, \quad C = \text{const.} \tag{4.7}$$

Equation $(4.6)_2$ then yields

$$\frac{C}{r^2}\frac{d}{dr}\left(rv_\phi^F\right) + \frac{\pi}{\rho^F - \mathfrak{b}}v_\phi^F = -2\omega\frac{C}{r}. \tag{4.8}$$

Hence we obtain

$$v_\phi^F = \frac{1}{r}\left(-\frac{2\omega}{\pi}C\left(\rho^F - \mathfrak{b}\right) + A\exp\left(-\frac{\pi}{2C\left(\rho^F - \mathfrak{b}\right)}r^2\right)\right), \quad A = \text{const.} \tag{4.9}$$

We need boundary conditions in order to find the constants A and C. One of them is obvious - the circumferencial velocity v_ϕ^F should be zero for $r = r_i$ because the fluid is assumed to enter the cylinder in the normal direction. A condition for the velocity v_r^F at this surface should follow from a condition of the third kind (seepage condition; see, e.g., [4]) which we shall not quote here. It describes the amount of fluid mass which flows into the cylinder per unit time and unit surface. We only need an order of magnitude of this quantity. The corresponding value of the velocity is assumed to be $v_{r_i}^F$. After easy manipulations we obtain

$$\frac{v_\phi^F}{v_r^F} = -\frac{\omega\left(\rho^F - \mathfrak{b}\right)}{\pi}\left[1 - \exp\left(-\frac{\pi r_i}{2v_{r_i}^F\left(\rho^F - \mathfrak{b}\right)}\left(\frac{r^2}{r_i^{F2}} - 1\right)\right)\right]. \tag{4.10}$$

For typical values of the mass density $\rho^F \backsim 10^3\,\frac{\text{kg}}{\text{m}^3}$ and permeability coefficient $\pi \backsim 10^8\,\frac{\text{kg}}{\text{m}^3\text{s}}$, we have to rotate the system with the angular speed $\omega \backsim 10^5\,\frac{1}{\text{s}}$ in order to be able to observe an influence of the relative acceleration on the flow in the cylinder. Certainly this is not reasonable in the case of classical porous and granular materials. Consequently, even if we accepted the lack of material objectivity, its influence could not be observed in steady state experiments. In the next section we consider the dynamical effects of such contributions.

5 Influence of nonobjectivity on the propagation of sound waves

Additional contributions of accelerations to momentum balance equations influence not only steady-state flows but primarily dynamical processes such as the propagation of sound waves. We check the propagation condition for such waves in the case of poroelastic materials. Propagation conditions determine speeds of propagation of wave fronts, and amplitudes of waves on these fronts.

Sound waves are also called weak discontinuity waves because their fronts are characterized by the discontinuity of derivatives of fields but not of fields themselves. This means in our case that

$$\left[\left[\rho^F\right]\right]=0, \quad \left[\left[\rho^S\right]\right]=0, \quad \left[\left[\mathbf{v}^F\right]\right]=0, \quad \left[\left[\mathbf{v}^S\right]\right]=0, \quad \left[\left[\mathbf{e}^S\right]\right]=0,$$
$$[[\cdots]] := (\cdots)^+ - (\cdots)^-, \tag{5.1}$$

where

$$\mathbf{e}^S := \operatorname{sym}\operatorname{grad}\mathbf{u}^S, \tag{5.2}$$

$\mathbf{u}^S$ denotes the displacement of the skeleton, and the values $(\cdots)^+$, $(\cdots)^-$ are estimated on the positive and negative side of the front, respectively. Time derivatives of the fields do not have to be continuous, and we introduce the following notation:

$$\begin{aligned} R^F &:= \left[\left[\frac{\partial \rho^F}{\partial t}\right]\right], \quad R^S := \left[\left[\frac{\partial \rho^S}{\partial t}\right]\right], \\ \mathbf{A}^F &:= \left[\left[\frac{\partial \mathbf{v}^F}{\partial t}\right]\right], \quad \mathbf{A}^S := \left[\left[\frac{\partial \mathbf{v}^S}{\partial t}\right]\right]. \end{aligned} \tag{5.3}$$

If we denote by c the speed of propagation of the front then we have the following Hadamard kinematical compatibility conditions (e.g., [5]):

$$\begin{aligned} &c\left[\left[\operatorname{grad}\rho^F\right]\right] = -R^F\mathbf{n}, \qquad c\left[\left[\operatorname{grad}\rho^S\right]\right] = -R^S\mathbf{n}, \\ &c\left[\left[\operatorname{grad}\mathbf{v}^F\right]\right] = -\mathbf{A}^F\otimes\mathbf{n}, \qquad c\left[\left[\operatorname{grad}\mathbf{v}^S\right]\right] = -\mathbf{A}^S\otimes\mathbf{n}, \\ &c\left[\left[\operatorname{grad}\mathbf{e}^S\right]\right] = -\left[\left[\frac{\partial \mathbf{e}^S}{\partial t}\right]\right]\otimes\mathbf{n}. \end{aligned} \tag{5.4}$$

In these relations $\mathbf{n}$ denotes the unit vector normal to the surface of the front, and its orientation defines the positive side of the surface.

We proceed to investigate conditions following from the field equations. In order to simplify arguments we consider the linear model in which stress tensors appearing in the balance equations (2.5) are given by the following constitutive relations:

$$\mathbf{T}^F = -\left[p_0^F + \kappa\left(\rho^F - \rho_0^F\right)\right]\mathbf{1}, \quad \mathbf{T}^S = \mathbf{T}_0^S + \lambda^S \operatorname{tr}\mathbf{e}^S\mathbf{1} + 2\mu^S\mathbf{e}^S, \tag{5.5}$$

where the material parameters κ, λ^S, μ^S are assumed to be constant, and $\mathbf{T}_0^S$, p_0^F, ρ_0^F denote constant reference values of the partial stress tensor in the skeleton, the partial pressure and the partial mass density, respectively. Hence the fluid component is ideal (no contributions of the velocity gradient), and the skeleton is elastic. Such porous materials are called linear poroelastic.

After substitution of the above constitutive relations as well as of the relation (4.1) in the momentum balance equations, we obtain field equations provided that we account for the following integrability condition:

$$\frac{\partial \mathbf{e}^S}{\partial t} = \text{sym grad } \mathbf{v}^S. \tag{5.6}$$

This relation leads to the following condition on the wave front:

$$c^2 \left[\left[\text{grad } \mathbf{e}^S\right]\right] = \frac{1}{2}\left(\mathbf{A}^S \otimes \mathbf{n} \otimes \mathbf{n} + \mathbf{n} \otimes \mathbf{A}^S \otimes \mathbf{n}\right). \tag{5.7}$$

We construct the limits of mass, and momentum balance equations on the wave front. This immediately yields the following set of algebraic relations for amplitudes:

$$\begin{aligned}
&R^F\left(c - \mathbf{v}^F \cdot \mathbf{n}\right) = \rho^F \mathbf{A}^F \cdot \mathbf{n}, \\
&R^S\left(c - \mathbf{v}^S \cdot \mathbf{n}\right) = \rho^S \mathbf{A}^S \cdot \mathbf{n}, \\
&\left(\rho^F - \mathfrak{b}\right)\left(c - \mathbf{v}^F \cdot \mathbf{n}\right)\mathbf{A}^F + \mathfrak{b}\left(c - \mathbf{v}^S \cdot \mathbf{n}\right)\mathbf{A}^S = \kappa R^F \mathbf{n},
\end{aligned} \tag{5.8}$$

$$\begin{aligned}
\mathfrak{b}c\left(c - \mathbf{v}^F \cdot \mathbf{n}\right)\mathbf{A}^F + \left(\rho^S - \mathfrak{b}\right)c\left(c - \mathbf{v}^S \cdot \mathbf{n}\right)\mathbf{A}^S \\
= \left(\lambda^S + \mu^S\right)\mathbf{A}^S \cdot \mathbf{n}\mathbf{n} + \mu^S \mathbf{A}^S,
\end{aligned} \tag{5.9}$$

This is a homogeneous set of equations for the amplitudes R^F, R^S, $\mathbf{A}^F$, $\mathbf{A}^S$. Consequently its determinant must vanish, and this yields the so-called propagation condition determining speeds of propagation, and some relations between components of amplitudes. We consider solutions of this condition under the assumption that the speed of propagation c is much larger than the normal velocities $\mathbf{v}^F \cdot \mathbf{n}$ and $\mathbf{v}^S \cdot \mathbf{n}$. Then we easily obtain:

$$\begin{aligned}
&\left[\left(\rho^F - \mathfrak{b}\right)c^2\mathbf{1} - \rho^F \kappa \mathbf{n} \otimes \mathbf{n}\right]\mathbf{A}^F + \mathfrak{b}c^2\mathbf{A}^S = 0, \\
&\mathfrak{b}c^2\mathbf{A}^F + \left[\left(\rho^S - \mathfrak{b}\right)c^2\mathbf{1} - \left(\lambda^S + \mu^S\right)\mathbf{n} \otimes \mathbf{n} - \mu^S\mathbf{1}\right]\mathbf{A}^S = 0.
\end{aligned} \tag{5.10}$$

To see the structure of the solutions it is convenient to split the amplitudes $\mathbf{A}^F$, $\mathbf{A}^S$ into normal and tangential components:

$$\begin{aligned}
&\mathbf{A}^F = \mathbf{A}^F \cdot \mathbf{n}\mathbf{n} + \mathbf{A}^F_\perp, \quad &&\mathbf{A}^S = \mathbf{A}^S \cdot \mathbf{n}\mathbf{n} + \mathbf{A}^S_\perp, \\
&\mathbf{A}^F_\perp \cdot \mathbf{n} \equiv 0, &&\mathbf{A}^S_\perp \cdot \mathbf{n} \equiv 0.
\end{aligned} \tag{5.11}$$

The set of two scalar equations for the normal components $\mathbf{A}^F \cdot \mathbf{n}$, $\mathbf{A}^S \cdot \mathbf{n}$ has nontrivial solutions if the following condition is satisfied:

$$\left[\left(\rho^F - \mathfrak{b}\right)c^2 - \rho^F\kappa\right]\left[\left(\rho^S - \mathfrak{b}\right)c^2 - \left(\lambda^S + 2\mu^S\right)\right] - \mathfrak{b}^2c^4 = 0. \tag{5.12}$$

Consequently we obtain two modes of propagation – the so-called P1- and P2-waves which propagate with speeds $\pm c_{P1}$ and $\pm c_{P2}$, the solutions of the above biquadratic equation for c. For these speeds to be real a condition on material coefficients must be satisfied which we do not need to present here.

This is the usual result for two-component porous materials [5]. However in the present case the speeds of propagation depend on the value of the coefficient $\mathfrak{b}$, and they differ from the following classical results:

$$\mathfrak{b} \equiv 0 \quad \Longrightarrow \quad c_{P1}^2 = \frac{\lambda^S + 2\mu^S}{\rho^S}, \quad c_{P2}^2 = \kappa. \tag{5.13}$$

Corrections predicted by the relation (5.12) would be still acceptable due to the fact that the speeds depend on multiplicative combinations of elastic properties of components with mass densities, and the coupling coefficient $\mathfrak{b}$. This means that measurements of speeds do not yield any information on the coupling coefficient alone, and we can obtain a good fit by correcting the material coefficients in (5.13), particularily by a low accuracy in the measurements of these speeds for porous materials. This fitting by means of the coupling coefficient is extensively discussed in [6,7]. We return to these papers in the next section.

In contrast to the above modes of propagation the result for the transversal mode seems to contradict all available experimental observations. Namely we obtain the following solution of the propagation condition:

$$\mathbf{A}_{\perp}^F = -\frac{\mathfrak{b}}{\rho^F - \mathfrak{b}}\mathbf{A}_{\perp}^S, \qquad c^2\left[\left(\rho^S - \mathfrak{b}\right) - \frac{\mathfrak{b}^2}{\rho^F - \mathfrak{b}}\right] = \mu^S. \tag{5.14}$$

Consequently, for $\mathfrak{b} \neq 0$, the speed of propagation of transversal waves would be dependent on the mass density of the fluid component in contrast to the classical result $c^2 = \frac{\mu^S}{\rho^S}$, and, in addition, there would exist a transversal component of the amplitude $\mathbf{A}_{\perp}^F$ of the wave carried by an ideal fluid. Such effects are very unlikely and they have never been reported by experimentalists.

6 Some heuristic remarks

Biot's correction of interactions in the momentum balance equations has recently been by results for flows through obstacles, and dynamical effects in suspensions. In these fields of research the notion of an *added mass* which should support Biot's corrections is well established and justified by reasonable physical arguments (see [8, p. 134ff]; in this book an extensive literature on the subject is quoted).

However, in contrast to multicomponent theories of porous materials, solutions of those flow problems are constructed, from the viewpoint of multicomponent theories, on a microscopical level of observation. Interactions through the pressure on surfaces of obstacles are eliminated by means of explicit solutions of mass and momentum conservation equations for the fluid. In this way one can construct an explicit momentum balance equation for obstacles where, as the result of elimination of surface interactions, corrections to the mass appear. Certainly, this is not the case in macroscopical theories of porous materials where microscopical interactions are smeared out by averaging. They contribute on the macroscopical level through the correction of the set of constitutive variables. Such corrections are reflected both by a simultaneous appearance of constitutive variables of all components in all constitutive relations, and by additional microstructural variables such as porosity or tortuosity.

It seems to be justified to assume that tortuosity should be the microstructural variable which replaces an added mass in macroscopical models. However the way in which it is introduced by Gajo and others [6,7] contradicts the classical definition of this notion (see [9, Sect. 4.8]). According to such a definition the tortuosity describes a geometrical property *additional* to the volume fraction of voids described by the porosity. Consequently it cannot be related to the porosity by any algebraic relation as is claimed in these, and some other, papers.

Tortuosity as an additional field seems to be a natural candidate to describe, on the macroscopical level, effects of dynamical interactions of flows of the fluid component through complex channels of the skeleton without contradicting the principle of material objectivity.

Bearing in mind the above analysis we have to accept the consequence that the relative accelerations cannot contribute to the momentum balance equations.

7 On the objectivity of Darcy's law

We present another aspect of the material objectivity which is related to the transition between two levels of description. It is easy to see that, for processes in which the acceleration $\mathbf{a}^F$ approximately vanishes in inertial reference systems, the momentum balance equation for the fluid implies the prototype of Darcy's law. It becomes identical with this law for incompressible fluids in which $\mathbf{T}^F = -p^F\mathbf{1}$, $p^F = -np$, where p is the pore pressure and n denotes the porosity. Then for the momentum source $\hat{\mathbf{p}}^F = -\pi\left(\mathbf{v}^F - \mathbf{v}^S\right) + p \operatorname{grad} n$ (see [10] for a discussion of the thermodynamical admissibility of this relation) the momentum balance equation yields

$$\operatorname{grad} p = -\frac{\pi}{n}\left(\mathbf{v}^F - \mathbf{v}^S\right). \tag{7.1}$$

However, in a noninertial reference frame, the momentum balance equation yields the following form of the relation for momentum sources:

$$\hat{\mathbf{p}}^{F*} = \operatorname{grad}{}^*\mathbf{p}^{F*} + \rho^{F*}\left[\dot{\mathbf{W}}\left(\mathbf{x}^* - d\right) - 2\mathbf{W}\left(\mathbf{v}^* - \dot{\mathbf{d}}\right) + \mathbf{W}^2\left(\mathbf{x}^* - \mathbf{d}\right)\right]. \tag{7.2}$$

Hence Darcy's law, considered as a constitutive relation between the momentum source (i.e., the relative velocity of components – see (7.1)) and the pressure gradient, is not objective. There is an explicit contribution of kinematics of the noninertial system in the form of the term in square brackets in relation (7.2). However, in contrast to corrections arising from the presence of the difference of accelerations, this contribution is independent of material parameters, and, consequently, universal.

Two questions arise whether such a form of nonobjectivity is admissible in macroscopical models, and, secondly, whether it has any practical bearing.

The second question is easy to answer. If we again substitute typical values of parameters appearing in geophysics (cf. Sect. 4), the influence of the nonobjective contributions becomes significant for angle velocities of noninertial systems of the order $\omega \sim 10^5 \frac{1}{s}$. Hence these contributions can be safely neglected.

The first question is more difficult as it concerns general principles of continuum thermodynamics. Without making any attempt to answer it in general we can only mention that such nonobjective contributions always appear if we simplify the hierarchy of equations of extended thermodynamics [11]. For instance the Cattaneo-type balance equation for the heat flux leads to a nonobjective Fourier's law relating the heat flux to the temperature gradient. Such effects may have a practical bearing in gas dynamics (e.g., [2]) but they seem to be immaterial for solids and fluids.

References

[1] Biot, M. A. (1956): Theory of propagation of elastic waves in a fluid-saturated porous solid. I. Low-frequency range. J. Acoust. Soc. Amer. **28**, 168–178

[2] Müller, I. (1985): Thermodynamics. Pitman, Boston

[3] Wilmański, K. (1998): Thermomechanics of continua. Springer, Berlin

[4] Albers, B., Wilmański, K. (1999): An axisymmetric steady-state flow through a poroelastic medium under large deformations. Arch. Appl. Mech. **69**, 121–132

[5] Wilmański, K. (1999): Waves in porous and granular materials. In: Hutter, K., Wilmański, K. (eds.) Kinetic and continuum theories of granular and porous media, Springer, Vienna, pp. 131–185

[6] Gajo, A. (1996): The effects of inertial coupling in the interpretation of dynamic soil tests. Geotech. **46**, 245–257

[7] Gajo, A., Fedel, A., Mongiovi, J. (1997): Experimental analysis of the effects of fluid-solid coupling on the velocity of elastic waves in saturated porous media. Geotech. **47**, 993–1008

[8] Brennen, C.E. (1995): Cavitation and bubble dynamics. Oxford Univ. Press, New York

[9] Bear, J. (1988): Dynamics of fluids in porous media. Dover, New York

[10] Wilmanski, K. (2001): Note on the notion of incompressibility in thermodynamic theories of porous and granular materials. Z. Angew. Math. Mech. **81**, 37–42

[11] Müller, I., Ruggeri, T. (1998): Rational extended thermodynamics. Second ed. Springer, New York

Simplifield Dynamics of a Continuous Peptide Chain

Henryk Zorski

Abstract. The chain is regarded as the continuum limit of rigid units-quadruples (their diameter tends to zero), connected at points called nods. The relative motion of the neighbouring units is described by two angles called dihedrals [1].The rigidity and planarity of the units is well established experimentally for most peptide chains. It is shown that the difficult kinematic relation (1.4) at every point of the chain is identically satisfied for the case of the travelling waves. This fact and the planarity of the units lead to a simplification of the inertia term in the equations of motion resulting in a system of nonlinear wave equations or even one equation, the nonlinearity being due now to the static term, i.e. the constitutive law for the continuous chain. In a particular periodic case the problem is governed by the pendulum equation. To examine the latter, in Sect. 2 we present a procedure applied to the discrete system of rigid units, introducing an average argument in the interaction energy. Thus, the continuum equations of motion are further simplified.

Introduction

The rigidity and planarity of the peptide unit are well established experimentally, with some exceptions. We call our continuum limit a locally rigid chain.

A general theory of motion of such peptide chains as presented in [1] has been considerably simplified, without further model assumptions, by introducing directly the basic physical variables, namely, the dihedral angles. The resulting system of equations is still both kinematically and physically nonlinear (possibly nonlinear constitutive laws) but in the case of travelling waves the form of the equations is familiar and therefore conventional methods of the theory of dynamical systems can now be employed. For instance, when the external moment is periodic and only the sine term is retained, the whole system of five partial differential equations is reduced to the pendulum equation.

In Sect. 2 the assumption of inextensibility is removed and a three-dimensional displacement, additional to the rigid motion, is introduced. By taking averages in the interaction between the units we find the argument of the internal energy. It turns out to be a linear combination of the angular variables (dihedral angles) and the displacement gradient. The resulting form of the final equations is thus simplified.

The kinematics of the discrete model is described in the Appendix.

1 Pure rotation of the chain

In the continuum case considered the only degrees of freedom are the dihedral angles $\psi(s,t)$ and $\phi(s,t)$. The two vectors defining the time and space rotations (Fig. 1)

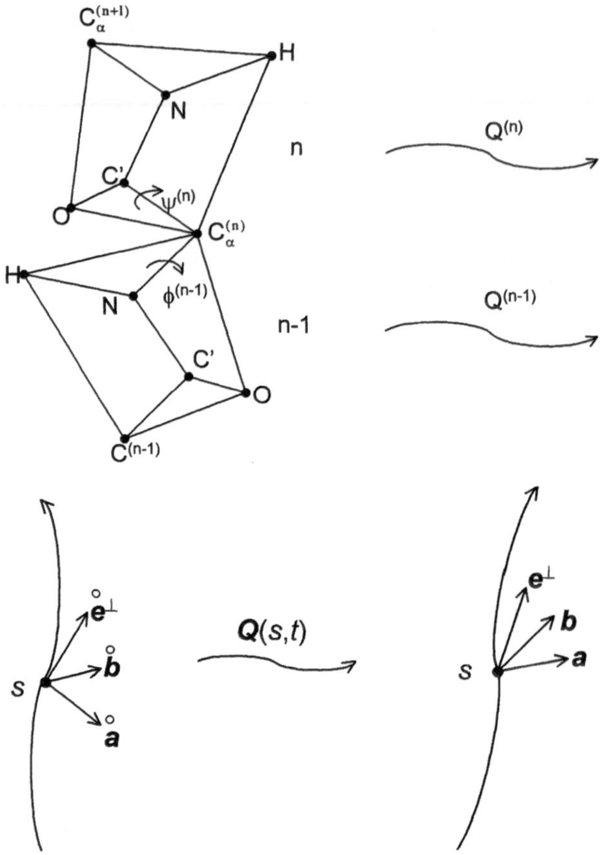

Fig. 1. (**Top**) The $(n-1)$st and nth peptide units. The dihedral angles are indicated at the joint (see Appendix). (**Bottom**) The continuum model (see Appendix). We shall also use the vectors transformed to the initial configuration

are (see [1] and the Appendix)

$$
\begin{aligned}
&\boldsymbol{\omega} = \text{vect}\,\boldsymbol{\Omega}, \quad \boldsymbol{\Omega} = \dot{\boldsymbol{Q}}\boldsymbol{Q}^{\top}; \qquad \omega = \frac{1}{2}\varepsilon_{ipq}\Omega_{pq}\\
&\boldsymbol{\gamma} = \text{vect}\,\boldsymbol{\Gamma}, \quad \boldsymbol{\Gamma} = \boldsymbol{Q}'\boldsymbol{Q}^{\top}; \quad \gamma_i = \frac{1}{2}\varepsilon_{ipq}\Gamma_{pq},
\end{aligned}
\tag{1.1}
$$

where $\boldsymbol{Q} = \frac{\partial}{\partial t}\boldsymbol{Q}(s,t)$ and $\boldsymbol{Q}' = \frac{\partial}{\partial s}\boldsymbol{Q}(s,t)$. In the discrete chain we are faced with the rotation matrices $\boldsymbol{Q}^{(n)}(t)$, $n = 1, 2, \ldots$, while in the continuum model we have $\boldsymbol{Q}(s,t)$. Introduce

$$
\tilde{\omega} = \boldsymbol{Q}^{\top}\cdot\omega, \quad \tilde{\gamma} = \boldsymbol{Q}^{\top}\cdot\gamma \tag{1.2}
$$

Cross-differentiating the definitions of $\boldsymbol{\Omega}$ and $\boldsymbol{\Gamma}$ in (1.1) leads to the compatibility conditions

$$\dot{\boldsymbol{\Gamma}} = \boldsymbol{\Omega}' + [\boldsymbol{\Omega}, \boldsymbol{\Gamma}], \tag{1.3}$$

where $[\boldsymbol{\Omega}, \boldsymbol{\Gamma}] = \boldsymbol{\Omega}\boldsymbol{\Gamma} - \boldsymbol{\Gamma}\boldsymbol{\Omega}$, or, in terms of the vectors $\boldsymbol{\omega}$ and $\boldsymbol{\gamma}$,

$$\dot{\boldsymbol{\gamma}} + \frac{1}{2}\boldsymbol{\omega} \times \boldsymbol{\gamma} = \boldsymbol{\omega}' + \frac{1}{2}\boldsymbol{\gamma} \times \boldsymbol{\omega}$$

or

$$\dot{\boldsymbol{\gamma}} + \frac{1}{2}\tilde{\boldsymbol{\gamma}} \times \tilde{\boldsymbol{\omega}} = \tilde{\boldsymbol{\omega}}' + \frac{1}{2}\tilde{\boldsymbol{\omega}} \times \tilde{\boldsymbol{\gamma}}. \tag{1.4}$$

It was shown in [1] that retaining only first derivatives in the Taylor expansion of the discrete rotation matrix $\boldsymbol{Q}^{n-1}(t) = \boldsymbol{Q}(s,t) - \varepsilon\,\boldsymbol{Q}'(s,t)$ we arrive at an expression for $\tilde{\boldsymbol{\gamma}}(s,t)$ in terms of the dihedral angles

$$\tilde{\boldsymbol{\gamma}} = -(\overset{\circ}{\boldsymbol{a}}\,\psi' + \overset{\circ}{\boldsymbol{b}}\,\phi'). \tag{1.5}$$

Substituting (1.5) into (1.4) we obtain a system of partial differential equations for the unknowns $\boldsymbol{\omega}$, ψ and ϕ. Projecting onto the orthonormal triad $(\overset{\circ}{\boldsymbol{a}}, \overset{\circ}{\boldsymbol{b}}, \overset{\circ}{\boldsymbol{e}}{}^{\perp})$ and using the notation $\tilde{\omega}_1 = \overset{\circ}{\boldsymbol{a}} \cdot \tilde{\boldsymbol{\omega}}$, $\tilde{\omega}_2 = \overset{\circ}{\boldsymbol{b}} \cdot \tilde{\boldsymbol{\omega}}$, $\tilde{\omega}_3 = \overset{\circ}{\boldsymbol{e}}{}^{\perp} \cdot \tilde{\boldsymbol{\omega}}$, we have

$$\begin{aligned} &\tilde{\omega}_1' + \tilde{\omega}_3\phi' = -\dot{\psi}' \\ &\tilde{\omega}_2' - \tilde{\omega}_3\psi' = -\dot{\phi}' \\ &\omega_3' + \tilde{\omega}_2\psi' - \tilde{\omega}_1\phi' = 0. \end{aligned} \tag{1.6}$$

Equation (1.6) implies a number of scalar relations, namely,

$$\left(\frac{1}{2}|\tilde{\boldsymbol{\gamma}}|^2\right)^{\cdot} = \tilde{\boldsymbol{\omega}}' \cdot \tilde{\boldsymbol{\gamma}}, \quad \left(\frac{1}{2}|\tilde{\boldsymbol{\omega}}|^2\right)' = \tilde{\boldsymbol{\omega}}' \cdot \dot{\tilde{\boldsymbol{\gamma}}}, \quad \left[\frac{1}{2}(\psi'^2 + \phi'^2)\right]^{\cdot} = \psi'\omega_1' + \phi'\omega_2'.$$

The purely kinematic relation (1.4) is a part of the complete system governing the motion of the peptide chain. We note here that (1.4) is satisfied identically if $\tilde{\boldsymbol{\omega}}$ and $\tilde{\boldsymbol{\gamma}}$ (and then ψ and ϕ as well) depend on a single variable, say $\xi(s,t)$, rather than on s and t separately. In particular, when $\xi(s,t) = s - vt$, v constant, we are faced with travelling waves. In this case we may introduce the potential $\boldsymbol{\kappa}(\xi)$,

$$\boldsymbol{\kappa} = -(\overset{\circ}{\boldsymbol{a}}\,\psi + \overset{\circ}{\boldsymbol{b}}\,\phi), \tag{1.7}$$

so that

$$\boldsymbol{\gamma} = \boldsymbol{\kappa}' = \xi'\boldsymbol{\kappa}_{,\xi}, \qquad \boldsymbol{\omega} = \dot{\boldsymbol{\kappa}} = \dot{\xi}\boldsymbol{\kappa}_{,\xi} \tag{1.8}$$

We shall now proceed to the equations of dynamics. We recall the definitions of the inertia tensor of the peptide unit in the triad $(\overset{\circ}{\boldsymbol{a}}, \overset{\circ}{\boldsymbol{b}}, \overset{\circ}{\boldsymbol{e}}{}^{\perp})$ (Figs. 1, 2):

$$\overset{\circ}{\boldsymbol{I}} = \sum_{P^{(n)}} m(P^{(n)})\, \overset{\circ}{\boldsymbol{\rho}}\,(P^{(n)})\, \overset{\circ}{\boldsymbol{\rho}}\,(P^{(n)})$$

$$\overset{\circ}{\boldsymbol{J}} = \boldsymbol{\delta}\ \mathrm{tr}\ \overset{\circ}{\boldsymbol{I}} - \overset{\circ}{\boldsymbol{I}}\ .$$

Since the unit is assumed to be plane and contains vectors $\overset{\circ}{\boldsymbol{a}}$ and $\overset{\circ}{\boldsymbol{b}}$, $\overset{\circ}{\rho}_3 = 0$, $\overset{\circ}{\boldsymbol{I}}_{i3} = 0$ for $i = 1, 2, 3$, and

$$\overset{\circ}{J} = \begin{pmatrix} \overset{\circ}{I}_{22} & -\overset{\circ}{I}_{12} & 0 \\ -\overset{\circ}{I}_{12} & \overset{\circ}{I}_{11} & 0 \\ 0 & 0 & \mathrm{tr}\ \overset{\circ}{I} \end{pmatrix}. \tag{1.9}$$

The inertia tensor $\boldsymbol{J}$ entering the equations of motion is $\boldsymbol{J} = \boldsymbol{Q}\ \overset{\circ}{\boldsymbol{J}}\ \boldsymbol{Q}^{\top}$ and depends on time.

The system of equations we present here refers to an inextensible deformable line with two degrees of freedom $\psi(s, t)$ and $\phi(s, t)$. Since these (dihedral) angles are introduced from the beginning (see (1.5)) the system presented in [1] can be considerably simplified. The angular momentum equation and the inextensibility condition can be written in the form

$$\overset{\circ}{\rho}\ (\boldsymbol{J} \cdot \boldsymbol{\omega})^{\cdot} = \left(\frac{\partial V}{\partial \boldsymbol{\gamma}}\right)' + \boldsymbol{Q} \cdot \boldsymbol{M}^{(e)}, \quad \boldsymbol{r}' = \boldsymbol{Q} \cdot \overset{\circ}{\boldsymbol{r}}{}' . \tag{1.10}$$

Here $V(\boldsymbol{\gamma})$ is the internal energy and $\boldsymbol{M}^{(e)} = \boldsymbol{M}^{\text{ext}} + \overset{\circ}{\boldsymbol{e}}{}^{\perp}\ m$. The external part of $\boldsymbol{M}^{(e)}$ is assumed to be known. The second part of $\boldsymbol{M}^{(e)}$ is the reaction moment required to freeze the third angular degree of freedom while $\overset{\circ}{\boldsymbol{r}}{}'$ is the tangent to the line considered in its reference state.

In view of (1.5) the projections of (1.10) onto $\overset{\circ}{\boldsymbol{a}}$ and $\overset{\circ}{\boldsymbol{b}}$ effected in (1.6) constitute five equations for $\boldsymbol{\omega}$, ψ and ϕ. Then we can in principle determine the rotation matrix from the definitions (1.1) and hence the current configuration from the inextensibility condition (1.10).

The system (1.10) can be considerably simplified. Differentiating the inertia tensor $\boldsymbol{J} = \boldsymbol{Q}\ \overset{\circ}{\boldsymbol{J}}\ \boldsymbol{Q}^{\top}$,

$$\overset{\circ}{\rho}\ (\boldsymbol{J} \cdot \boldsymbol{\omega})^{\cdot} = \overset{\circ}{\rho}\ \boldsymbol{Q} \cdot (\overset{\circ}{\boldsymbol{J}} \cdot \dot{\tilde{\boldsymbol{\omega}}} + (\overset{\circ}{\boldsymbol{J}} \cdot \tilde{\boldsymbol{\omega}}) \times \tilde{\boldsymbol{\omega}}),$$

Furthermore,

$$\left(\frac{\partial V}{\partial \boldsymbol{\gamma}}\right)' = \boldsymbol{Q} \cdot \left(\frac{\partial V}{\partial \tilde{\boldsymbol{\gamma}}}\right)' + \boldsymbol{Q}' \cdot \frac{\partial V}{\partial \tilde{\boldsymbol{\gamma}}} = \boldsymbol{Q} \cdot \left(\left(\frac{\partial V}{\partial \tilde{\boldsymbol{\gamma}}}\right)' + \frac{\partial V}{\partial \tilde{\boldsymbol{\gamma}}} \times \tilde{\boldsymbol{\gamma}}\right).$$

Hence,

$$\overset{\circ}{\rho}\,(\overset{\circ}{\boldsymbol{J}}\cdot\dot{\tilde{\boldsymbol{\omega}}}+(\overset{\circ}{\boldsymbol{J}}\cdot\tilde{\boldsymbol{\omega}})\times\tilde{\boldsymbol{\omega}})=\left(\frac{\partial V}{\partial\tilde{\boldsymbol{\gamma}}}\right)'+\frac{\partial V}{\partial\tilde{\boldsymbol{\gamma}}}\times\tilde{\boldsymbol{\gamma}}+\boldsymbol{M}^{\text{ext}}+\overset{\circ}{\boldsymbol{e}}{}^{\perp}\,m. \tag{1.11}$$

It is reasonable to assume that the internal energy $V(\tilde{\boldsymbol{\gamma}})$ contains no terms referring to direction $\overset{\circ}{\boldsymbol{e}}{}^{\perp}$. For instance, it may depend on the scalar $|\tilde{\boldsymbol{\gamma}}|^2$ or $\overset{\circ}{\boldsymbol{\lambda}}\cdot\tilde{\boldsymbol{\gamma}}$, where $\overset{\circ}{\boldsymbol{\lambda}}=\lambda_1\,\overset{\circ}{\boldsymbol{a}}+\lambda_2\,\overset{\circ}{\boldsymbol{b}}$. Then

$$\overset{\circ}{\boldsymbol{a}}\cdot\left(\frac{\partial V}{\partial\tilde{\boldsymbol{\gamma}}}\times\tilde{\boldsymbol{\gamma}}\right)=\overset{\circ}{\boldsymbol{b}}\cdot\left(\frac{\partial V}{\partial\tilde{\boldsymbol{\gamma}}}\times\tilde{\boldsymbol{\gamma}}\right)=0.$$

Therefore, projecting (1.11) onto $\overset{\circ}{\boldsymbol{a}}$ and $\overset{\circ}{\boldsymbol{b}}$ we finally obtain (denoting $\overset{\circ}{\boldsymbol{a}}\cdot\boldsymbol{J}.$ by $\overset{\circ}{J}_{1\cdot}$, etc.)

$$\begin{aligned}\overset{\circ}{\rho}\,(\overset{\circ}{J}_{1\beta}\,\dot{\tilde{\omega}}_\beta+\overset{\circ}{J}_{2\beta}\,\tilde{\omega}_\beta\tilde{\omega}_3)&=\overset{\circ}{\boldsymbol{a}}\cdot\left(\frac{\partial V}{\partial\tilde{\boldsymbol{\gamma}}}\right)'+\boldsymbol{M}_1^{\text{ext}}\\ &\qquad\qquad\alpha,\ \beta=1,2.\\ \overset{\circ}{\rho}\,(\overset{\circ}{J}_{2\beta}\,\dot{\tilde{\omega}}_\beta-\overset{\circ}{J}_{1\beta}\,\tilde{\omega}_\beta\tilde{\omega}_3)&=\overset{\circ}{\boldsymbol{b}}\cdot\left(\frac{\partial V}{\partial\tilde{\boldsymbol{\gamma}}}\right)'+\boldsymbol{M}_2^{\text{ext}}\end{aligned} \tag{1.12}$$

Equations (1.6) and (1.12) constitute the required system of partial differential equations for ψ, ϕ, $\tilde{\omega}_1$, $\tilde{\omega}_2$ and $\tilde{\omega}_3$.

This system is nonlinear for two reasons. First there is the kinematic nonlinearity expressed by the products $\omega\omega$. Second, we are faced with possible physical nonlinearity due to the form of the constitutive relation for the internal energy $V(\tilde{\gamma})$.

However, the kinematic nonlinearity is removed when all unknowns depend on one variable $\xi(s,t)$ (cf. (1.7)) rather than on s and t. For information on internal and external energies see [2, Chap. 3]. In fact, then (1.6) has the solution

$$\tilde{\boldsymbol{\omega}}=\tilde{\boldsymbol{\omega}}(\xi)=\dot{\xi}\boldsymbol{\kappa}_{,\xi},\quad \tilde{\boldsymbol{\gamma}}=\tilde{\boldsymbol{\gamma}}(\xi)=\xi'\boldsymbol{\kappa}_{,\xi}, \tag{1.13}$$

where

$$\boldsymbol{\kappa}=-(\overset{\circ}{\boldsymbol{a}}\,\psi(\xi)+\overset{\circ}{\boldsymbol{b}}\,\phi(\xi)).$$

Now, (1.12) is a system of two ordinary differential equations for $\psi(\xi)$ and $\phi(\xi)$ which we write in 2D vector form. We confine ourselves to travelling waves, $\xi=s-vt$. Then (1.12) can be written in the form of a two-dimensional system of ordinary differential equations for $\kappa(\xi)$,

$$\overset{\circ}{\rho}\,v^2\,\overset{\circ}{\boldsymbol{J}}\cdot\boldsymbol{\kappa}_{'\xi\xi}-\partial\cdot\left(\frac{\partial V}{\partial\boldsymbol{\kappa}_{'\xi}}\right)_{'\xi}=-\boldsymbol{M}^{\text{ext}}. \tag{1.14}$$

In (1.14) $V=V(\boldsymbol{\kappa}_{'\xi})$ and we assume that $\boldsymbol{M}^{\text{ext}}=\boldsymbol{M}^{\text{ext}}(\boldsymbol{\kappa})$.

There are particular cases in which (1.14) can be further simplified. We present one example: we want the left-hand side of (1.14) to be linear in κ'', and to this end we assume that $V(\kappa') = \frac{1}{2}(\overset{\circ}{g} \cdot \kappa')^2$ (or else $V = \frac{V_0}{2}|\kappa'|^2$), where $\overset{\circ}{g}$ is a constant vector. Moreover, let $\boldsymbol{M}^{\text{ext}}$ be a function of the scalar $\boldsymbol{\kappa} \cdot \overset{\circ}{h}$, where $\overset{\circ}{\boldsymbol{h}}$ is another constant vector. Then, setting $\overset{\circ}{\boldsymbol{J}}{}^{g} = \overset{\circ}{\boldsymbol{J}} - \overset{\circ}{g}\overset{\circ}{g}$, we have

$$\overset{\circ}{\rho}\, v^2\, \overset{\circ}{\boldsymbol{J}}{}^{g} \cdot \boldsymbol{\kappa}'' = -\boldsymbol{M}^{\text{ext}}(\overset{\circ}{\boldsymbol{h}} \cdot \boldsymbol{\kappa}). \tag{1.15}$$

If $\overset{\circ}{\boldsymbol{J}}{}^{g}$ is invertible, inverting (1.15) and performing a scalar multiplication by $\overset{\circ}{\boldsymbol{h}}$ we obtain a scalar differential equation for $x = \overset{\circ}{\boldsymbol{h}} \cdot \boldsymbol{\kappa}$,

$$\overset{\circ}{\rho}\, v^2 x'' = -\overset{\circ}{h} \cdot (J^g)^{-1} \cdot \partial \boldsymbol{M}^{\text{ext}}(x). \tag{1.16}$$

If $x(\xi)$ is known, $\boldsymbol{\kappa}(\xi)$ follows from (1.15) by integration.

Frequently ([2], Ch.3) the external moment is periodic in angles ψ and ϕ. Then we assume that, in the simplest case, $\boldsymbol{M}^{\text{ext}}(x) = \boldsymbol{M}_0 \sin x$ and so (1.16) is the pendulum equation.

2 The internal energy argument: extensible chains

An examination of the interaction between adjacent cells in the discrete model leads to a suggestion concerning the argument of the internal energy, i.e. a deformation parameter depending on $\tilde{\boldsymbol{\gamma}}$ and the gradient of the additional displacement $\boldsymbol{u}'(s,t)$.

Consider two adjacent cells n and $n+1$ (Fig. 2). The vector between the carbons $C_\alpha^{(n)}$ and $C_\alpha^{(n+1)}$ is assumed to have the form

$$\boldsymbol{r}(C_\alpha^{(n)}, C_\alpha^{(n+1)}) = \boldsymbol{Q}^{(n)} \cdot [\overset{\circ}{r}(C_\alpha^{(n)}, C_\alpha^{(n+1)}) + \tilde{\boldsymbol{u}}(C_\alpha^{(n+1)}) - \tilde{\boldsymbol{u}}(C_\alpha^{(n)})]. \tag{2.1}$$

The vectors $\boldsymbol{\rho}(P^{(n)})$ and $\boldsymbol{\rho}(\bar{P}^{(n+1)})$ between the carbons and typical atoms in the cells n and $n+1$ are

$$\boldsymbol{\rho}(P^{(n)}) = \boldsymbol{Q}^{(n)} \cdot \overset{\circ}{\boldsymbol{\rho}}(P^{(n)})$$
$$\boldsymbol{\rho}(\bar{P}^{(n+1)}) = \boldsymbol{Q}^{(n+1)} \cdot \overset{\circ}{\boldsymbol{\rho}}(\bar{P}^{(n+1)}).$$

In (2.1) $\boldsymbol{Q}^{(n)}$ is the rotation matrix of cell n defining its rigid rotation and $\tilde{\boldsymbol{u}}(C_\alpha^{(n)})$ is a displacement of $C_\alpha^{(n)}$ from its reference state, additional to the rigid rotation. Note that the difference between the displacements in (2.1) is multiplied by $\boldsymbol{Q}^{(n)}$.

Now,

$$\boldsymbol{r}(P^{(n)}, \bar{P}^{(n+1)}) = \boldsymbol{r}(C^{(n)}, C^{(n+1)}) + \boldsymbol{Q}^{(n+1)} \cdot \overset{\circ}{\boldsymbol{\rho}}(\bar{P}^{(n+1)}) - \boldsymbol{Q}^{(n)} \cdot \overset{\circ}{\boldsymbol{\rho}}(\bar{P}^{(n)}).$$

Since we are interested in the continuum limit we set (as in Sect. 1) $\boldsymbol{Q}^{(n)} = \boldsymbol{Q}(s)$, $\boldsymbol{Q}^{(n+1)} = \boldsymbol{Q}(s) + \varepsilon \boldsymbol{Q}'(s)$ and we introduce $\boldsymbol{\Gamma}(s) = \boldsymbol{Q}'\boldsymbol{Q}^{\perp}$ and $\tilde{\boldsymbol{\Gamma}}(s) =$

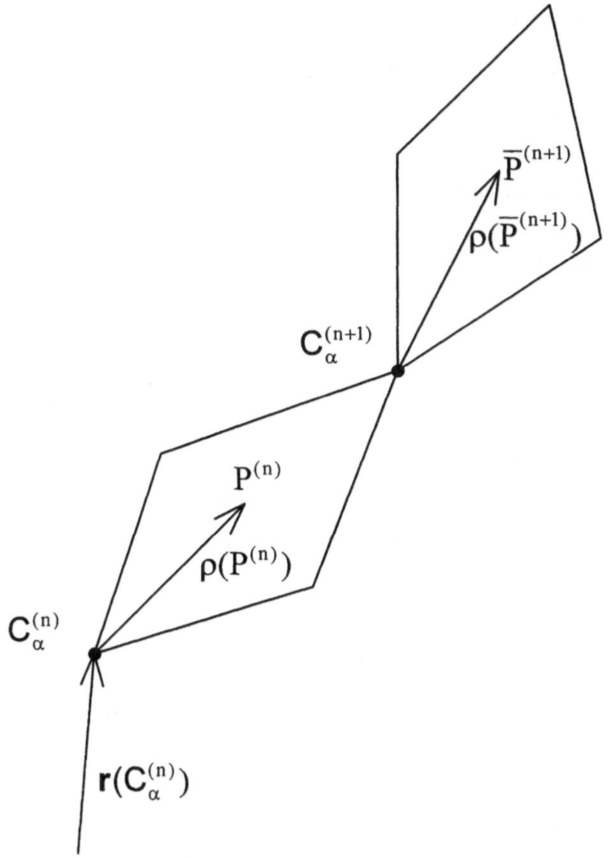

Fig. 2. Adjacent units. $P^{(n)}$ and $P^{(n+1)}$ are typical atoms in the units n and $n+1$, respectively, $\rho(P^{(n)}) = Q^{(n)} \cdot \overset{\circ}{\rho}(P^{(n)})$, $\rho(\bar{P}^{(n+1)}) = Q^{(n+1)} \cdot \overset{\circ}{\rho}(\bar{P}^{(n+1)})$

$Q^\top Q'$. Then, denoting $\overset{\circ}{r}(C_\alpha^{(n)}, C_\alpha^{(n+1)})$ by $\overset{\circ}{d}{}^{(n)}$, we have

$$r(P^{(n)}, \bar{P}^{(n+1)}) = Q \cdot \Big[\overset{\circ}{d}{}^{(n)} + \overset{\circ}{\rho}(\bar{P}^{(n+1)}) - \overset{\circ}{\rho}(\bar{P}^{(n)}) + \varepsilon(\tilde{\Gamma} \cdot \overset{\circ}{\rho}(\bar{P}^{(n+1)}) + \tilde{u}') \Big], \tag{2.2}$$

where $\varepsilon \tilde{u}'(s) \approx \tilde{u}(C_\alpha^{(n+1)}) - \tilde{u}(C_\alpha^{(n)})$ is the first term in the Taylor series on the line.

Equation (2.2) contains two macroscopic deformation parameters $\tilde{\Gamma}(s)$ and $\tilde{u}'(s)$; their combination however still contains the microscopic vector $\overset{\circ}{\rho}(\bar{P}^{(n+1)})$. To eliminate the latter we introduce a weight function identical for each cell, $w(P^{(n)})$, such that $\sum_{P^{(n)}} w(P^{(n)}) = 1$ and we define the average

$$r_w = \sum_{P^{(n)}} r(P^{(n)}, \bar{P}^{(n+1)}) w(P^{(n)})$$

whence, for identical cells n and $n+1$,

$$r_w = r_w(\tilde{\Gamma}, \tilde{u}'; \mathring{l}) = Q \cdot [\mathring{d} + \varepsilon(\tilde{\Gamma} \cdot \mathring{l} + \tilde{u}')], \tag{2.3}$$

where

$$\mathring{l} = \sum_{P^{(n)}} \mathring{\rho}\,(P^{(n)}) w(P^{(n)})$$

is a new macroscopic vector parameter.

We assume that the internal energy, i.e., the energy of interaction, depends on the argument $e = \tilde{\Gamma} \cdot \mathring{l} + \tilde{u}'$. Thus, since (cf. (1.1) and (1.5)) $\tilde{\gamma} = \operatorname{vect} \tilde{\Gamma} = -(\psi' \mathring{a} + \phi' \mathring{b})$,

$$e = \mathring{l} \times \tilde{\gamma} + \tilde{u}' = -(\psi' \mathring{l} \times \mathring{a} + \phi' \mathring{l} \times \mathring{b}) + \tilde{u}'. \tag{2.4}$$

We note that, for the internal energy depending only on the length of the vector $|e|$,

$$\begin{aligned} \frac{\partial}{\partial \tilde{\gamma}} V(|e|) &= \frac{V'(|e|)}{|e|} e \times \mathring{l}, \\ \frac{\partial}{\partial u'} V(|e|) &= \frac{V'(|e|)}{|e|} e, \\ \frac{\partial}{\partial \tilde{\gamma}} V(|e|) &= \frac{\partial}{\partial u'} V(|e|) \times l i_0. \end{aligned} \tag{2.5}$$

Observe finally that instead of averaging the vector $r(P^{(n)}, \bar{P}^{(n+1)})$ we may average the following square:

$$\begin{aligned} |Q^{\top} \cdot r(P^{(n)}, \bar{P}^{(n+1)}) - [\mathring{d} + \mathring{\rho}\,(\bar{P}^{(n+1)}) - \mathring{\rho}\,(\bar{P}^{(n)})]|^2 \\ = \varepsilon^2 |\tilde{\Gamma} \cdot \mathring{\rho}\,(\bar{P}^{(n+1)}) + \tilde{u}'|^2. \end{aligned}$$

Then the average, which we denote by $2\varepsilon^2 \Delta$, is

$$2\varepsilon^2 \Delta = \varepsilon^2 \sum_{P^{(n)}} w(P^{(n)}) |\tilde{\Gamma} \cdot \mathring{\rho}\,(P^{(n)}) + \tilde{u}'|^2. \tag{2.6}$$

Since

$$|\tilde{\Gamma} \cdot \mathring{\rho}\,(P^{(n)})|^2 = \tilde{\gamma} \cdot [\delta |\mathring{\rho}\,(P^{(n)})|^2 - \mathring{\rho}\,(P^{(n)}) \mathring{\rho}\,(P^{(n)})] \cdot \tilde{\gamma},$$

we have

$$\sum_{P^{(n)}} |\tilde{\boldsymbol{\Gamma}} \cdot \overset{\circ}{\boldsymbol{\rho}} (P^{(n)})|^2 w(P^{(n)}) = \tilde{\boldsymbol{\gamma}} \cdot \overset{\circ}{\boldsymbol{J}}_w \cdot \tilde{\boldsymbol{\gamma}}$$

$$\sum_{P^{(n)}} \tilde{\boldsymbol{u}}' \cdot \tilde{\boldsymbol{\Gamma}} \cdot \overset{\circ}{\boldsymbol{\rho}} (P^{(n)}) w(P^{(n)}) = \overset{\circ}{\boldsymbol{\sigma}}_w \cdot (\tilde{\boldsymbol{\gamma}} \times \tilde{\boldsymbol{u}}'),$$

where

$$\begin{aligned} \overset{\circ}{\boldsymbol{J}}_w &= \sum_{P^{(n)}} [\boldsymbol{\delta} |\overset{\circ}{\boldsymbol{\rho}} (P^{(n)})|^2 - \overset{\circ}{\boldsymbol{\rho}} (P^{(n)}) \overset{\circ}{\boldsymbol{\rho}} (P^{(n)})] w(P^{(n)}) \\ \overset{\circ}{\boldsymbol{\sigma}}_w &= \sum_{P^{(n)}} \overset{\circ}{\boldsymbol{\rho}} (P^{(n)}) w(P^{(n)}) \end{aligned} \tag{2.7}$$

and then the argument of the internal energy is

$$\Delta = \Delta(\tilde{\boldsymbol{\gamma}}, \tilde{\boldsymbol{u}}') = \frac{1}{2} \tilde{\boldsymbol{\gamma}} \cdot \overset{\circ}{\boldsymbol{J}}_w \cdot \tilde{\boldsymbol{\gamma}} + \overset{\circ}{\boldsymbol{\sigma}}_w \cdot (\tilde{\boldsymbol{\gamma}} \times \tilde{\boldsymbol{u}}') + \frac{1}{2} |\tilde{\boldsymbol{u}}'|^2. \tag{2.8}$$

Note that (2.8) is obtained from the square of (2.4) by replacing $\overset{\circ}{\boldsymbol{J}}_w$ by $|\overset{\circ}{\boldsymbol{l}}|^2 \boldsymbol{\delta} - \overset{\circ}{\boldsymbol{l}}\overset{\circ}{\boldsymbol{l}}$ and denoting $\overset{\circ}{\boldsymbol{l}}$ by $\overset{\circ}{\boldsymbol{\sigma}}$.

Observe also that, if $w(P^{(n)}) = \frac{m(P^{(n)})}{\sum_{P^{(n)}} m(P^{(n)})}$, where $m(P^{(n)})$ is the mass, then Eqs. (2.7) give the ordinary moment of inertia and the static moment of cell, respectively.

The equation of linear momentum, external forces neglected, is

$$\overset{\circ}{\rho} \, \ddot{\tilde{\boldsymbol{u}}} = \left(\frac{\partial V}{\partial \tilde{\boldsymbol{u}}'} \right)' \tag{2.9}$$

and for $V(|\boldsymbol{e}|)$ it takes the form (cf. (2.5))

$$\overset{\circ}{\rho} \, \ddot{\tilde{\boldsymbol{u}}} = \left(\frac{V'(|\boldsymbol{e}|)}{|\boldsymbol{e}|} \boldsymbol{e} \right)', \tag{2.10}$$

whence

$$\overset{\circ}{\rho} \, \ddot{\tilde{\boldsymbol{u}}} \times \overset{\circ}{\boldsymbol{l}} = \left(\frac{\partial V}{\partial \tilde{\boldsymbol{\gamma}}} \right)', \quad \overset{\circ}{\rho} \, \ddot{\tilde{\boldsymbol{u}}} \cdot \overset{\circ}{\boldsymbol{l}} = \left(\frac{V'(|\boldsymbol{e}|)}{|\boldsymbol{e}|} \tilde{\boldsymbol{u}}' \cdot \overset{\circ}{\boldsymbol{l}} \right)'. \tag{2.11}$$

The above equations, following from (2.5), show that the general system (1.6), (1.12), (2.9) can be considerably simplified, owing to the fact that V depends on $|\boldsymbol{e}|$ only.

Appendix

To make this paper self-contained, following [1] we present here in some detail both the discrete and the continuous models of the peptide chain (Fig. 1). For the geometry of peptides see [3].

A peptide unit (Fig. 1, top) is a system of five atoms: two carbon (C_α and C'), one nitrogen (N), one hydrogen (H) and oxygen (O). Identical peptide units are linked at the carbon atoms C_α to yield a peptide chain. The units are numbered by $n = 0, 1, , 2, \dots , N-1$. An atom in the nth unit will generally be denoted by $P^{(n)}$, e.g., $P^{(n)} = H$ or $P^{(n)} = N$. The upper carbon atom $C_\alpha^{(n+1)}$ belongs to the next, $(n+1)$st, unit. The position vectors are $\boldsymbol{r}(P^{(n)}, t)$ and the time variable will usually be suppressed; we shall also use the notation $\overset{\circ}{\boldsymbol{r}}\ (P^{(n)}) = \boldsymbol{r}(P^{(n)}, 0)$ to represent the position vector in a natural state identical with the state at $t = 0$ in the generating unit.

The peptide chain at time t is generated from a single peptide unit completed by its upper C_α atom in the following way: to determine the position of the nth peptide unit ($n = 0, 1, 2, \dots , N-1$), displace the lower C_α atom to $C_\alpha^{(n)}$ and rotate the unit by means of the rotation tensor $\boldsymbol{Q}^{(n)}$. The upper C_α atom of the nth unit and the lower C_α of the $(n+1)$st unit (Fig. 2) coincide. The following are the two kinematic assumptions on which we base the dynamics.

(1) The peptide unit and the upper C_α atom move as a rigid system of material points. The rigidity and moreover the planarity are due to the double bond character of $C' - N$ which excludes rotation about the $C' - N$ line.
(2) The relative rotation of the adjacent peptide units, say the $(n-1)$st and the nth, linked at the carbon atom $C_\alpha^{(n)}$, is restricted to the following two rotations:
   - around the rigid bond between the atom N in the $(n-1)$st unit and $C_\alpha^{(n)}$, by angle $\phi^{(n-1)}$,
   - around the rigid bond between $C_\alpha^{(n)}$ and C' in the nth unit, by an angle $\psi^{(n)}$.

   The angles $\phi^{(n-1)}$ and $\psi^{(n)}$ are called dihedral angles. The two above axes of rotation are fixed in the $(n-1)$st and nth units, respectively. They are therefore material axes and if the two unit vectors defining them are denoted by $\boldsymbol{b}^{(n-1)}$ and $\boldsymbol{a}^{(n)}$ we have $b^{(n-1)} = \boldsymbol{Q}^{(n-1)} \cdot \overset{\circ}{\boldsymbol{b}}$ and $a^{(n)} = \boldsymbol{Q}^{(n)} \cdot \overset{\circ}{\boldsymbol{a}}$, where $\overset{\circ}{\boldsymbol{a}}$ and $\overset{\circ}{\boldsymbol{b}}$ are constant vectors determined by the geometry of the generating unit. Also $\boldsymbol{Q}^{(n)}\,\boldsymbol{Q}^{(n)\top} = \delta$. No side chains are considered here. It is assumed that they are sufficiently small, e.g., glycine, to be regarded as rigidly attached to the C_α atom, just increasing the mass of the backbone chain.

The (locally rigid) continuum model is as follows (Fig. 1, bottom). A triad ($\overset{\circ}{\boldsymbol{a}}, \overset{\circ}{\boldsymbol{b}}, \overset{\circ}{\boldsymbol{e}}{}^{\perp} = \overset{\circ}{\boldsymbol{a}} \times \overset{\circ}{\boldsymbol{b}}$) is associated with points s on the line. For convenience we assume that ($\overset{\circ}{\boldsymbol{a}}, \overset{\circ}{\boldsymbol{b}}$) is the orthogonal and normalised system ($\overrightarrow{C_\alpha^{(n)}C'}, \overrightarrow{NC_\alpha^{(n)}}$). The dihedral angles ψ and ϕ are the rotation angles around $\overset{\circ}{\boldsymbol{a}}$ and $\overset{\circ}{\boldsymbol{b}}$, respectively; their spatial gradients therefore are linear combinations of the original ones around ($\overrightarrow{C_\alpha^{(n)}C'}$) and ($\overrightarrow{NC_\alpha^{(n)}}$). Using dashes to denote the original quantities we have

$$\overset{\circ}{\boldsymbol{a}} = \frac{1}{2\cos\beta/2}(\bar{\boldsymbol{a}} + \bar{\boldsymbol{b}}), \quad \overset{\circ}{\boldsymbol{b}} = \frac{1}{2\sin\beta/2}(-\bar{\boldsymbol{a}} + \bar{\boldsymbol{b}}), \quad \overset{\circ}{\boldsymbol{a}} \times \overset{\circ}{\boldsymbol{b}} = \frac{1}{\sin\beta}(\bar{\boldsymbol{a}} \times \bar{\boldsymbol{b}}),$$

$$\psi' = \overset{\circ}{\boldsymbol{a}} \cdot (\bar{\boldsymbol{a}}\bar{\psi}' + \bar{\boldsymbol{b}}\bar{\phi}'), \quad \phi' = \overset{\circ}{\boldsymbol{b}} \cdot (\bar{\boldsymbol{a}}\bar{\psi}' + \bar{\boldsymbol{b}}\bar{\phi}'),$$

where β is the angle between $\bar{\boldsymbol{a}}$ and $\bar{\boldsymbol{b}}$.

There is no rotation around $\overset{\circ}{\boldsymbol{e}}^{\perp}$; the problem therefore is two-dimensional. The peptide unit is rigid and planar and contains the vectors $\overset{\circ}{\boldsymbol{a}}$ and $\overset{\circ}{\boldsymbol{b}}$.

Acknowledgements. The author is grateful to the Polish Committee for Scientific Research for financial support, grant no. 7T 07A 01913.

Reference

[1] Zorski, H., Infeld, E. (1997): Continuum dynamics of a peptide chain. Internat. J. Non-Linear Mech. **32**, 769–801

[2] van Holde, K.E., Johnson, W.C., Ho, P.S. (1998): Principles of physical biochemistry. Prentice Hall, Upper Saddle River, NJ

[3] Branden C., Tooze J. (1991): Introduction to protein structure. Garland, New York